BIOCHEMISTRY OF THERMOPHILY

ACADEMIC PRESS RAPID MANUSCRIPT REPRODUCTION

BIOCHEMISTRY OF THERMOPHILY

edited by

S. MARVIN FRIEDMAN

Department of Biological Sciences
Hunter College of the City University of New York
New York, New York

ACADEMIC PRESS New York San Francisco London 1978
A Subsidiary of Harcourt Brace Jovanovich, Publishers

ACADEMIC PRESS, INC.
111 Fifth Avenue, New York, New York 10003

United Kingdom Edition published by
ACADEMIC PRESS, INC. (LONDON) LTD.
24/28 Oval Road, London NW1 7DX

Library of Congress Cataloging in Publication Data

Main entry under title:

Biochemistry of thermophily.

Papers presented at a seminar held under the auspices of the U.S.-Japan Cooperative Science Program at the East-West Center in Honolulu, June 22–24, 1977.
1. Microbiological chemistry—Congresses.
2. Micro-organisms, Thermophilic—Congresses.
I. Friedman, Selwyn Marvin, Date
II. United States–Japan Cooperative Science Program.
QR84.B455 76'.11'9162 78-23214
ISBN 0-12-268250-5

PRINTED IN THE UNITED STATES OF AMERICA

CONTENTS

ECOLOGICAL AND GENETIC ASPECTS

PROTEIN SYNTHESIS

PROTEINS

LIST OF PARTICIPANTS

Alfred F. Esser, Scripps Clinic and Research Foundation, La Jolla, California.

S. Marvin Friedman, Department of Biological Sciences, Hunter College of the City University of New York, New York, New York.

Kazutomo Imahori, Department of Biochemistry, Faculty of Medicine, University of Tokyo, Hongo, Bunkyoku, Tokyo.

Yasuo Kagawa, Department of Biochemistry, Jichi Medical School, Minamikawachi, Tochigi.

Yoshito Kaziro, Institute of Medical Science, University of Tokyo, Takanawa, Minatoku, Tokyo.

Thomas A. Langworthy, Department of Microbiology, University of South Dakota School of Medicine, Vermillion, South Dakota.

Lars G. Ljungdahl, Department of Biochemistry, University of Georgia, Athens, Georgia.

Yoshiaki Nosoh, Laboratory of Chemistry of Natural Products, Tokyo Institute of Technology, Meguro, Tokyo.

Mieko Oshima, Department of Biochemistry, Kitasato University School of Medicine, Sagamichara, Kanagawa.

Tairo Oshima, Mitsubishi-Kasei Institute of Life Sciences, Machida, Tokyo.

Robert F. Ramaley, Department of Biochemistry, University of Nebraska Medical Center, Omaha, Nebraska.

Takashi Saiki, Department of Agricultural Chemistry, Faculty of Agriculture, University of Tokyo, Yayoi, Bunkyoku, Tokyo.

Earle Stellwagen, Department of Biochemistry, University of Iowa, Iowa City, Iowa.

Masamichi Tsuboi, Faculty of Pharmaceutical Sciences, University of Tokyo, Hongo, Bunkyoku, Tokyo.

Frederick C. Wedler, Chemistry Department, Cogswell Laboratory, Rensselaer Polytechnic Institute, Troy, New York.†

†Present address: Department of Biochemistry and Biophysics, Pennsylvania State University, University Park, Pennsylvania

Neil E. Welker, Department of Biochemistry and Molecular Biology, Northwestern University, Evanston, Illinois.

Makoto Yaguchi, Division of Biological Sciences, National Research Council, Ottawa, Canada

Katsuhide Yutani, Institute for Protein Research, Faculty of Science, Osaka University, Suita, Osaka.

Herbert Zuber, Institute für Molekularbiologie und Biophysik, Eidgenössische Technische Hochschule, Zürich, Switzerland.

PREFACE

Thermophilic microorganisms have attracted the attention of biologists for several important reasons. Thermophiles are capable of proliferating at elevated temperatures, and the vast majority of macromolecules that they synthesize are heat-stable. Because of these characteristics, they are often found as contaminants of canned food and dairy products. Since thermophiles can survive under extreme conditions, they are considered as possible candidates for extraterrestrial life. In addition, thermophiles are interesting from an evolutionary standpoint since they appear to represent primordial growth forms. Finally, thermophiles pose a growing ecological threat as thermal pollutants of our aquatic environments.

Research on the molecular basis for thermophily can provide valuable insights into other areas of biology. The elucidation of mechanisms underlying the thermostability of cell constituents will help us to understand the interrelationships among chemical structure, conformation, and biological function of biopolymers. Furthermore, information about those structural alterations in cellular components which allow for survival at high temperatures enables one to identify the biochemical strategies that cells employ in adapting to thermal stress. These strategies are also relevant to the process of biological evolution since thermophiles were probably among the earliest inhabitants of this planet. Solving the puzzle of thermophily can also yield basic knowledge for the application of thermophilic microorganisms to industrial processes.

This volume contains the papers presented at a seminar held under the auspices of the U.S.–Japan Cooperative Science Program at the East–West Center in Honolulu, Hawaii from June 22 through June 24, 1977. It is especially appropriate that a seminar devoted to the biochemistry of thermophily should be sponsored by the two countries that have the most extensive research programs in this area. As the U.S. coordinator, I felt a compelling responsibility to publicize the proceedings of this meeting since the work reported clearly represents a new level of achievement in the effort to solve an intriguing but difficult problem.

The planning and execution of a scientific conference entails many elements of cooperation. I would like to express my deep appreciation to the U.S.–Japan

Cooperative Science Program for making this seminar possible and to all the participants for contributing their unique talents. I am particularly indebted to Dr. K. Imahori, who not only conceived this meeting but was also instrumental in its successful realization, to Dr. Y. Kaziro, who served as the Japanese coordinator, and to Dr. T. Oshima for his unrelenting efforts and dedication. I am obliged to Dr. Allen Holt of the National Science Foundation for many helpful suggestions, including the delightful location for the meeting. I am also grateful to Mr. James P. McMahon and his gracious staff at the East–West center for providing us with extremely well-managed facilities. It is a pleasure to thank the staff of Academic Press for their care in the production of this volume, and to acknowledge the financial support for typing expenses that was provided by an NIH Biomedical Support Research Grant awarded to the Department of Biological Sciences of Hunter College. Finally, I wish to express my heartfelt gratitude to my wife Rivka for her invaluable assistance.

From left to right: T. Saiki, F. Wedler, Y. Kagawa, E. Stellwagen, M. Yaguchi, R. Ramaley, Y. Nosoh, K. Yutani, A. Esser, K. Imahori, H. Zuber, M. Friedman, M. Oshima, L. Ljungdahl, T. Oshima, Y. Kaziro, M. Tsuboi, N. Welker, T. Langworthy

STRUCTURE AND FUNCTION OF MEMBRANE LIPIDS IN THERMOPHILIC BACTERIA

Mieko Oshima

Department of Biochemistry
Kitasato University
School of Medicine
Asamizodai, Sagamihara

This article describes comparative analyses of fatty acids in several strains of moderately and extremely thermophilic bacteria and the effect of growth temperature upon the fatty acid composition and complex lipid content of the extreme thermophile, *Thermus thermophilus* HB-8. Physicochemical studies of liposomes prepared from membrane lipids of thermophiles are also discussed.

FATTY ACID COMPOSITION OF THERMOPHILIC BACTERIA

The fatty acids of various microorganisms are known to be affected by the temperature at which the cells are grown. Bacterial cells which contain unsaturated fatty acids have higher proportions of unsaturated acids when grown below their optimum temperature (Marr and Ingraham, 1962; Kates and Baxter, 1962; Bishop and Still, 1963; Kates and Hagen, 1964). The cis-vaccenic acid content in *Escherichia coli* phospholipids was found to increase with a concomitant decrease of palmitic acid when cultures were transferred from 40°C to 20°C (Haest *et al.*, 1969; Aibara *et al.*, 1972). Similar phenomena were observed with *Bacillus lichenformis* (Chang and Fulco, 1973) and *Serratia marcescens* (Kates and Hagen, 1964). In *B. megaterium* cells, which contain only saturated fatty acids at 37°C, induction of the enzyme causing desaturation at the C-5 position of the fatty acid chain was observed at 20°C (Fulco, 1970, 1972).

ISBN 0-12-268250-5

On the other hand, when cells are grown at high temperature, there is a shift in fatty acid synthesis to the direction of more saturated fatty acids (Marr and Ingraham, 1962). The replacement of linolenic acid by linoleic acid was observed in the thermophilic alga, *Cyanidium caldarium*, when grown at 40°C instead of 20°C (Adams *et al.*, 1971).

Both in moderately and extremely thermophilic bacteria, unsaturated fatty acids or hydroxy fatty acids were not detected. Cyclopropane and 10-methyl branched fatty acids, which are biosynthesized from monounsaturated fatty acids, could not be detected in any of the thermophilic bacterial strains. The percentage composition of fatty acids in thermophilic bacteria are listed in Table 1. Fatty acids in moderate and extreme thermophiles were qualitatively similar, consisting mainly of C_{14}-C_{17} iso and anteiso branched acids. Iso and anteiso branched acids are present in many bacterial strains, such as *Bacillus* (Kaneda, 1967), *Micrococcus* and *Staphylococcus* (Ishizuka *et al.*, 1966) and *Streptomyces* (Ballio *et al.*, 1965). These fatty acids in most bacteria so far examined are mainly C_{15} acids (Ueta *et al.*, 1970). Iso $C_{17:0}$ acids were the most abundant in both moderate thermophiles (10-33%) and extreme thermophiles (50-61%). The predominance of iso $C_{17:0}$ acids in thermophiles may be attributed to the hot environment of these bacteria. The presence of branched chain C_{17} acids in thermophilic organisms was also reported by Bauman and Simmonds (1969) and Heinen *et al.* (1970).

The relative abundance of branched chain fatty acids is known to reflect the availability of amino acid precursors and the age of the culture (Lennarz, 1966; Law *et al.*, 1963). Therefore, in this work, all strains were grown in a medium of the same nutrient composition and harvested at the same stage of growth (late log phase).

EFFECT OF TEMPERATURE ON THE FATTY ACID COMPOSITION OF *T. THERMOPHILUS* HB-8

The effect of growth temperature upon the fatty acid composition of the extreme thermophile, *T. thermophilus* HB-8 is shown in Table 2. The amount of longer chain acids increased with elevating temperatures. The percentage of iso $C_{17:0}$ increased at high growth temperatures, while those of iso $C_{15:0}$ and iso $C_{16:0}$ decreased. At the optimum growth temperature of 75-80°C, two-thirds of the total acids were iso $C_{17:0}$. The iso C_{19} acid was also detected at temperatures over 75°C. The ratios of iso C_{17}/iso C_{15} and anteiso C_{17}/anteiso C_{15} increased greatly at high growth temperatures,

TABLE 1
Fatty Acid Content in Thermophilic Bacteria Cultured at Their Optimum Growth Temperature

Fatty acid/Temp.	*B.st* *64°*	*V-2* *60°*	*HB-8* *75°*	*AT-62* *70°*	*YT-1* *71°*	*TA-6* *65°*
14:n	--	1.4	--	--	--	--
15:iso	31.0	24.0	22.0	26.0	33.0	--
15:anteiso	11.5	11.0	4.8	3.3	3.7	--
16:iso	2.5	6.9	5.2	1.0	3.2	--
16:n	7.0	14.0	--	--	4.2	1.4
17:iso	33.1	10.9	56.8	61.0	50.1	5.5
17:anteiso	14.3	23.3	9.2	7.1	4.8	16.0
18:n	--	1.5	--	--	--	11cy 35.5
19:anteiso	--	2.8	--	--	--	
19:iso	--	--	1.2	trace	--	13cy 38.8

B. st: *B. stearothermophilus* IAM 1035, V-2: *B. stearothermophilus* isolated from hot springs, HB-8: *T. thermophilus* HB-8, AT-62: *T. flavus* AT-62, YT-1: *T. aquaticus* YT-1, TA-6: *B. acidocaldarius* TA-6. 11cy and 13cy:ω-cyclohexyl fatty acids.

indicating chain elongation of fatty acids at high temperature (Fig. 1).

The pair of fatty acids, iso $C_{15:0}$ and iso $C_{17:0}$, constitute 76-87% of the total fatty acids in extreme thermophiles and 34-64% in moderate thermophiles. Anteiso $C_{15:0}$, anteiso $C_{17:0}$, iso $C_{15:0}$ and iso $C_{17:0}$ have melting points of 25.8°, 38.0°, 52.2° and 60.5°C, respectively. Extreme thermophiles contained appreciably more of the high melting point iso branched acids than did moderate thermophiles. A difference was also apparent in the amount of anteiso acids. Moderate thermophiles contained twice as much of the fatty

TABLE 2
Fatty Acid Composition of the Extreme Thermophile, T. thermophilus HB-8, Cultured at Different Growth Temperatures

Fatty Acid	Growth Temperature					
	49°	53°	65°	69°	75°	82°
14:iso	1.5	2.1	1.1	--	--	--
15:iso	31.1	34.1	38.9	25.9	21.8	12.2
15:ante-I	13.1	7.9	8.7	11.8	4.8	2.4
16:iso	6.4	9.1	7.7	6.0	5.2	1.8
16:normal	1.7	T	1.3	T	T	1.1
17:iso	31.3	35.1	38.8	40.7	56.8	64.2
17:ante-I	11.2	6.8	7.9	9.8	9.2	11.8
18:normal	1.0	2.0	T	T	T	T
19:iso	--	--	--	T	1.2	5.0
19:ante-I	--	--	--	T	T	1.3

Content of less than 1% is designated as T.

acid pair, anteiso $C_{15:0}$ and anteiso $C_{17:0}$ (25-34%), than did extreme thermophiles (8-15%). Anteiso acids have much lower melting points than iso acids. In thermophiles, long chain iso branched acids apparently are synthesized for maintaining the physical properties of the membrane lipids at high temperature.

STRUCTURE OF THE NOVEL GLYCOLIPID AND THE EFFECT OF GROWTH TEMPERATURE ON ITS CONTENT

All of the extreme thermophiles contain unusual complex lipids. The major complex lipid of *T. thermophilus* and *T. flavus* was found to be a novel glycolipid. This glycolipid constituted 70% of the total lipid and its structure was

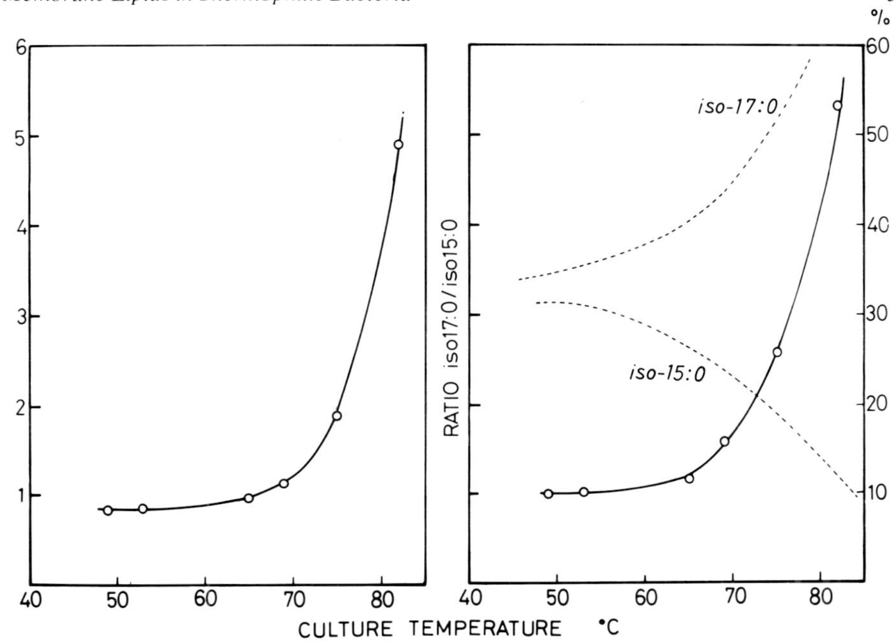

FIGURE 1. Effect of temperature on fatty acid content. Ratio of the content of C_{17}iso/C_{15}iso fatty acid (right) and ratio of C_{17}anteiso/C_{15}anteiso acid (left). Dotted lines show increase in the percentage of iso C_{17} and decrease in that of iso C_{15} fatty acid. Taken from Oshima and Miyagawa (1974) and reproduced with permission from the American Oil Chemists Society.

determined to be: Gal(f)(1$\xrightarrow{\beta}$2)Gal(p)(1$\xrightarrow{\alpha}$6)GlcN(15-methylhexadecanoyl)(1$\xrightarrow{\beta}$2)Glc(p)$\xrightarrow{\alpha}$2,3-diglyceride, as shown in Fig. 2 (Oshima and Yamakawa, 1972; Oshima and Miyagawa, 1974; Oshima and Ariga, 1976). The content of this glycolipid increases with increasing culture temperatures (Fig. 3). This indicates that temperature may affect polar lipid content as well as chain elongation of fatty acids.

The content of the novel glycolipid is very high compared to those of other glycolipids found in bacteria. Mono-, di- and triglycosyl diglycerides are widely distributed in microorganisms and the content of diglycosyl diglycerides (usually the predominant glycolipid in bacteria) varies from few percent of the total lipid in *Lactobacillus* and *Staphylococcus* (Brundish *et al.*, 1966, 1967) to 50% in *Streptococcus hemolyticus* (Ishizuka and Yamakawa, 1968).

Brundish *et al.* (1967) studied the molecular shape of diglycosyl diglyceride and proposed its structural significance. They suggested that the glycolipid could adopt a conformation in which all of the hydroxyl groups are oriented on

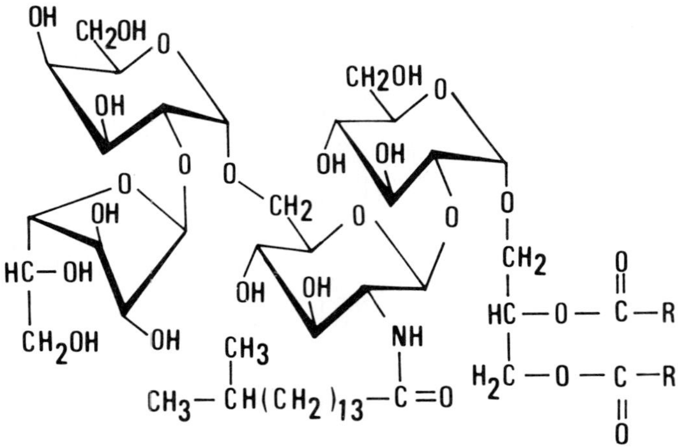

FIGURE 2. Structure of the novel glycolipid isolated from T. thermophilus *HB-8. Taken from Oshima and Ariga (1976) and reproduced with permission from the Federation of European Biochemical Societies.*

one side of the molecule and the lipophilic components (i.e., fatty acids, ring oxygens of sugars and glycosidic oxygen) are oriented on the other side. The hydrophilic regions of several molecules could come together to form pores in the membrane through which small molecules might pass (Show, 1970). These pores might be within the membrane or even on the surface, where some involvement in binding or anchoring of intracellular components could be possible. The tetraglycosyl diglyceride of the extreme thermophile may have some function in regulating the size of these pores.

PHYSICO-CHEMICAL STUDIES OF MEMBRANE LIPIDS

The lipid bilayer is known to undergo a reversible thermotropic gel to liquid crystalline phase transition. This rigid-to-fluid phase transition arises mainly from a melting of hydrocarbon chains in the interior of the bilayer.

In the biological membrane, there are many kinds of complex lipid components with different phase transition temperatures as well as protein and enzyme components. Hence the process of rigid-to-fluid phase transition of biomembranes is very complex. These complex phase changes in

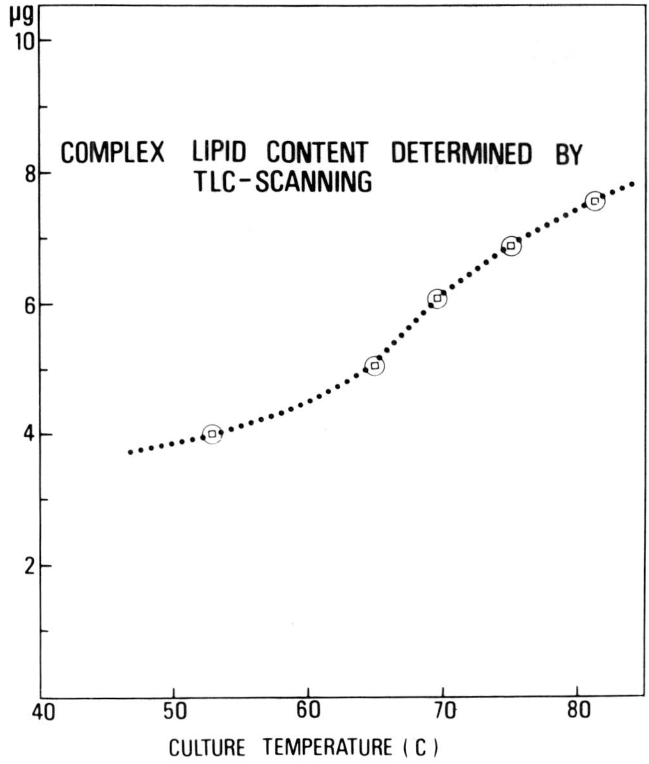

FIGURE 3. Glycolipid content of T. thermophilus *HB-8 cultured at different temperatures as determined by thin layer densitometry.*

biomembranes cause the phase separation at the temperature of cell growth. The phase separation of membrane components and the constant fluidity are known to be most important for membrane functions.

The rigid-to-fluid transition of the lipid of thermophiles was studied using liposomes (layered latticed liquid crystals) in dilute electrolyte solution. The penetration of glycerol into the liposomes (initial swelling rate) is detected by turbidity changes at 450 nm. Fig. 4 indicates the swelling of liposomes prepared from the total lipid mixture of several thermophile strains. Liposomes prepared from the lipid of thermophiles with higher optimum growth temperatures showed the swelling at higher temperatures. The liposome membranes prepared from the total lipid of *T. thermophilus* HB-8 grown at 53°C were permeable at relatively low

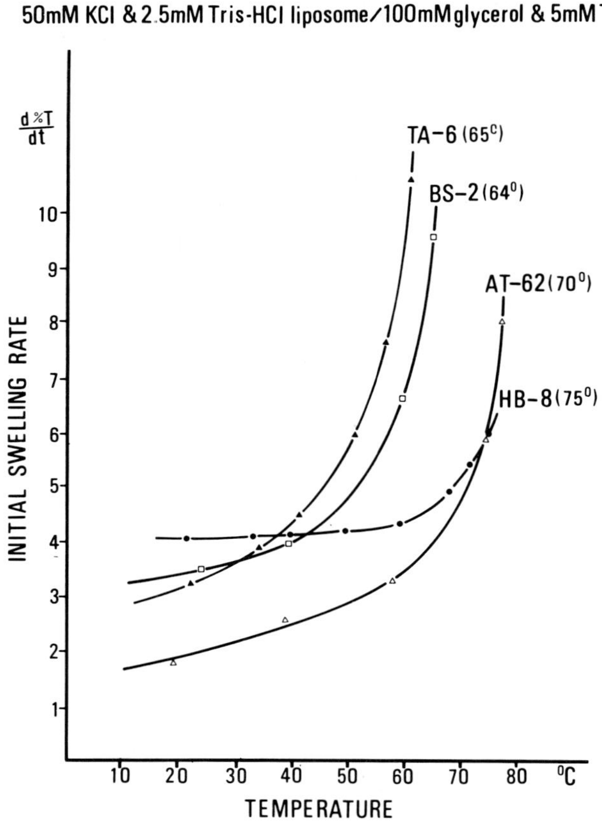

FIGURE 4. Initial swelling rates of liposomes prepared from the total lipid mixtures of various strains of thermophiles. TA-6: B. acidocaldarius *TA-6; BS-2:* B. stearothermophilus *IAM 1035; AT-62:* T. flavus *AT-62; HB-8:* T. thermophilus *HB-8. Each strain was cultured at the temperature shown in parentheses.*

temperatures, whereas the liposome membranes prepared from the lipid of cells grown at 82°C showed swelling at relatively high temperatures (Fig. 5).

These results apparently indicate that the phase transition temperature of the liposome is correlated with the temperature at which cells are grown. However, the temperature at which liposomes showed the dramatic swelling is much lower than that of the culture temperature of the cells. Wakayama and Oshima (1975) studied the phase transition of spheroplast membranes of *T. thermophilus* HB-8 by the method

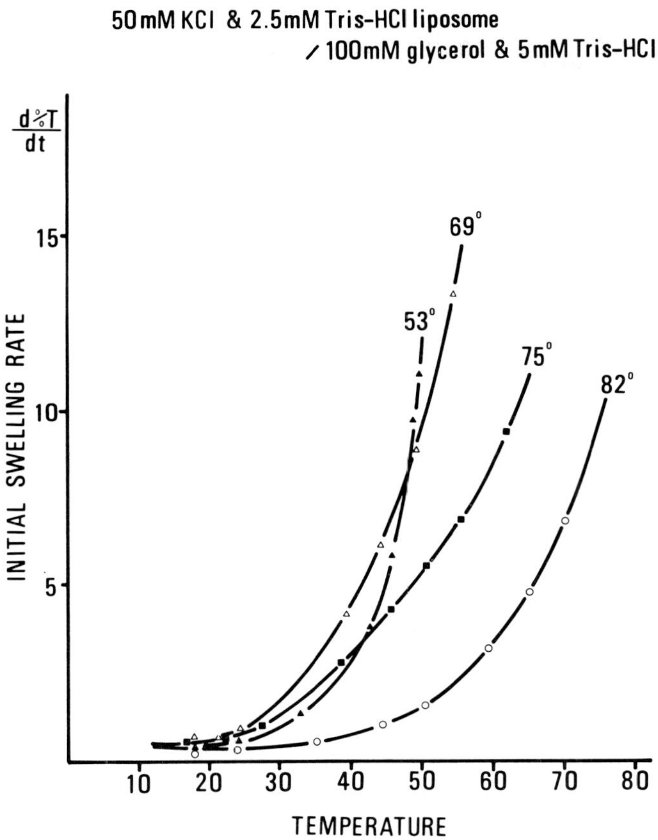

FIGURE 5. Initial swelling rates of liposomes prepared from the lipids of T. thermophilus *HB-8 cultured at different temperatures.*

of fluorescent viscosity. They observed two phase transition temperatures. The upper phase transition temperature was a few degrees higher than the culture temperature of the cells. This upper phase transition temperature may correlate with the phase separation (mixing and demixing) temperature suggested by Esser and Souza (1974).

The fluidity of bacterial biomembranes is always maintained constant in growing cells. This phenomenon is termed "homeoviscous adaptation" by Sinensky (1974). The structural changes in fatty acid components and the variations in complex lipid content reported in this work are advanced as mechanisms for achieving homeoviscous adaptation in bacteria.

REFERENCES

Adams, B. L., McMahon, V. and Seckback, J. (1971). Biochem. Biophys. Res. Comm. *42*, 359.

Aibara, S., Kato, M., Ishinaga, M. and Kito, M. (1972). Biochim. Biophys. Acta *270*, 301.

Ballio, A. Barcellona, A. and Boniforti, L. (1965). Biochem. J. *94*, 11c.

Bauman, A. J. and Simmonds, P. G. (1969). J. Bacteriol. *98*, 528.

Bishop, D. G. and Still, J. L. (1963). J. Lipid Res. *4*, 87.

Brundish, D. E., Show, N. and Baddiley, J. (1967). Biochem. J. *105*, 885.

Chang, N. C. and Fulco, A. J. (1973). Biochim. Biophys. Acta *296*, 287.

Esser, A. F. and Souza, K. A. (1974). Proc. Natl. Acad. Sci. USA. *71*, 4111.

Fulco, A. J. (1970). J. Biol. Chem. *218*, 558.

Fulco, A. J. (1972a). J. Biol. Chem. *247*, 3503.

Fulco, A. J. (1972b). J. Biol. Chem. *247*, 3511.

Haest, C. W. M., De Gier, J. and Van Deenen, L. L. M. (1969). Chem. Phys. Lipid *3*, 413.

Heinen, W., Klein, H. P. and Volkman, C. M. (1970). Ark. Mikrobiol. *72*, 199.

Ishizuka, I., Ueta, N. and Yamakawa, T. (1966). Japan. J. Exp. Med. *36*, 73.

Kaneda, J. (1967). J. Bacteriol. *93*, 894.

Kates, M. and Baxter, R. H. (1962). Can. J. Biochem. *40*, 1213.

Kates, M. and Hagen, P. O. (1964). Can. J. Biochem., *42*, 48.

Law, J. H., Zalkin, H. and Kaneshiro, T. (1963). Biochim. Biophys. Acta *70*, 143.

Lennarz, W. J. (1966). In *Lipid Metabolism in the Bacteria*, Adv. in Lipid Res. *4*, 175, Academic Press, New York.

Marr, A. G. and Ingraham, I. L. (1962). J. Bacteriol. *84*, 1260.

Oshima, M. and Yamakawa, T. (1972). Biochem. Biophys. Res. Comm. *49*, 185.

Oshima, M. and Yamakawa, T. (1964). Biochemistry *13*, 1140.

Oshima, M. and Miyagawa, A. (1974). Lipids *9*, 476.

Oshima, M. and Ariga, T. (1976). FEBS Letters *64*, 440.

Show, N. (1970). Bacteriol Rev. *34*, 465.

Sinensky, M. (1974). Proc. Natl. Acad. Sci. USA. *71*, 522.

Ueta, N., Ishizuka, I. and Yamakawa, T. (1970). *Proceedings 1st International Conference of Culture Collections*, p. 371, Univ. of Tokyo Press, Tokyo.

Wakayama, N. and Oshima, T. (1975). In The Special Issue for the Temperature Adaptation of Microorganisms: *Proteins, Nucleic Acids and Enzymes* (in Japanese) *20*, 169.

MEMBRANES AND LIPIDS OF EXTREMELY THERMOACIDOPHILIC MICROORGANISMS

Thomas A. Langworthy

Department of Microbiology
School of Medicine
University of South Dakota
Vermillion, South Dakota

The thermoacidophilic bacteria have been relatively recent discoveries typified by three morphologically distinct types: *Thermoplasma acidophilum*, a cell wall-less mycoplasma (Darland et al., 1970); *Sulfolobus acidocaldarius*, an atypically walled facultative autotroph (Brock et al., 1972) and *Bacillus acidocaldarius*, a true cell wall-containing, spore-forming bacillus (Darland and Brock 1971). These organisms are not only resistant to the combination of high temperature and low pH but require both environmental extremes for growth and reproduction. Because intracellular pH values (near neutrality) are not in equilibrium with the external pH, interest to date has focused on the membranes and surface structures of these organisms in attempts to help explain thermoacidophily.

I would like to begin this presentation with a general introduction to the thermoacidophiles followed by a discussion of the known features and possible roles in thermoacidophily of the membranes and lipid structures of each organism on an individual basis. In conclusion, I would like to assess the emerging comparative aspects of the membrane and lipid components.

GENERAL FEATURES

Thermoplasma, a member of the *Mycoplasmatales*, is of special interest because its cytoplasm is surrounded by only a

ISBN 0-12-268250-5

trilaminar membrane. It grows optimally at pH 2 and 59°C and within the limits of 59°C and within the limits of 45°C to 62°C and pH 0.5 to 4.5. Growth is exceedingly slow at the extremes. So far, self-heating coal refuse piles appear to be its only known natural habitat (Belly et al., 1973). *Thermoplasma* is an aerobe possessing only c- and O-type cytochromes and the naphthoquinone K_2-7. It is a heterotroph but specifically requires a partially characterized basic polypeptide from yeast extract for growth (Smith et al., 1975). Because of the large quantities of this component required for growth, the "growth factor" may not be a nutritional requirement at all, but may actually function as some type of ion scavenger or even be associated with some type of H^+ ion protection capacity on the cellular surface. The thermoplasmal genome approximates 1×10^9 daltons in size with a quanine plus cytosine content (G+C) of about 46% (Christiansen et al., 1975). Interestingly, Searcy (1975) has shown the association of a basic histone-like protein with isolated DNA, although its association has not been demonstrated intracellularly. The intracellular pH of *Thermoplasma* appears to approach neutrality. Hsung and Haug (1975) reported an internal pH value of 6.4 to 6.8 using [^{14}C]-5,5-dimethyl-2,4-oxazolidinedione (DMO). Isolated cytoplasmic malate dehydrogenase also had a broad optimum at pH 8. These values were reportedly maintained even in the presence of metabolic inhibitors or cells boiled for 1 h. Searcy (1976), on the other hand, has estimated a lower internal pH value of 5.4 by titration. In addition, the internal pH of boiled cells was in equilibrium with the external pH. In either case, a pH gradient appears to exist across the membrane of viable cells which approximates 3.5 to 4.5 pH units.

Sulfolobus grows within a temperature range of about 55°C to 85°C and displays the highest temperature optimum among the thermoacidophiles, namely 75°C. Optimal pH for growth is 3 with a pH range of about 0.9 to 5.0. In contrast to the restricted habitat of *Thermoplasma*, *Sulfolobus* has been found around the world associated with volcanic areas, acid hot springs and sulfatara soils (Brock et al., 1972; de Rosa et al, 1975). *Sulfolobus* is a facultative autotroph capable of growing on either yeast extract, sulfur or iron. Brierley and Brierley (1973) have also isolated strictly autotrophic strains which are similar to *Sulfolobus*. The organism is characterized morphologically by its atypical cell wall. Peptidoglycan is absent. The cytoplasm instead is overlaid by a polyhexagonal subunit ultrastructural element (Weiss, 1974). Depending upon strains, the G+C content is 46% to 80%. The internal pH also appears to be near neutral by direct measurement of broken cell contents (de Rosa et al., 1975).

Bacillus acidocaldarius has been isolated from the same habitats as *Solfolobus* (Darland and Brock, 1971; de Rosa et al., 1971a; Oshima et al., 1977). Its growth is optimal at 60°C to 65°C and pH 3. The general limits to growth are between 40°C to 70°C and pH 2 to 6. *B. acidocaldarius* is morphologically distinguished from the other thermoacidophiles by virtue of its peptidoglycan cell wall (or at least the presence of muramic acid) and its sensitivity to vancomycin, ristocetin and cycloserine. The organism is a heterotrophic, aerobic, spore-forming rod resembling in all respects a typical *Bacillus*. Its G+C content is about 60%. An internal pH of 5.5 to 6.0 has been reported using [^{14}C]-DMO (Oshima et al., 1977) and by measurement of the intracellular excitation spectra of flourescene diacetate (Thomas et al., 1976). Oshima et al. (1977) have also isolated glyceraldehyde-3-phosphate dehydrogenase (optimum pH 8.0) and shown the maintenance of the near neutral internal pH in the presence of metabolic inhibitors. The general features of the three types of thermoacidophiles are summarized in Table 1.

In each of the three organisms the internal environment appears to be one of a hot neutral cytoplasm while a H^+ ion gradient exists across the membrane or envelopes of about 3 to 4 units. Apparently the internal pH is maintained near neutrality even in the presence of metabolic inhibitors (it seems sodium azide, potassium cyanide and iodoacetate are still effective in hot acid). This might suggest the view that H^+ ions may be inhibited from passage across the membranes or envelopes by structural features rather than energy dependent H^+ exclusion mechanisms. With this in mind, we will next examine features of the membranes and lipid structures which, at this stage of the art, seem to be as unusual as the existence of these organisms might imply.

THERMOPLASMA

Electron micrographs of thin-sectioned cells reveal nuclear material dispersed throughout the cytoplasm. The cytoplasm is surrounded by only a trilaminar membrane about 4 nm to 10 nm thick (Darland et al., 1976). Cells subjected to freeze-fracture, on the other hand, are generally observed in cross-fracture revealing a cross section of cytoplasm and membrane. Inner and outer layers of the membrane surface are apparently not observed (P. Ververgaert, personal communication). This phenomenon suggests a rigid membrane structure. The rigid nature of the membrane is supported by EPR studies (Smith, G.G. et al., 1974) in which the mobility of spin-labels in the lipid domain of the membrane appears to be even

TABLE 1
General Features of the Three Morphologically Different Thermoacidophiles

	Thermoplasma acidophilium[a]	*Sulfolobus acidocaldarius*[b]	*Bacillus acidocaldarius*[c]
Temperature optimum	59°C	75°C	65°C
pH optimum	2	3	3
Cell wall	No	Atypical	Yes
Heterotroph	Yes	Facultative; S, Fe	Yes

[a]Darland et al. (1970).

[b]Brock et al. (1972).

[c]Darland and Brock (1971).

less than that reported for the extreme halophile *Halobacterium cutirubrum* (Esser and Lanyi, 1973). Indeed, the stability of *Thermoplasma* is further exemplified by its extreme resistance to mechanical breakage and its osmotic stability (conditions under which mesophilic mycoplasmas and protoplasts lyse). Even cells heated at 100°C for 5 h maintain structural integrity, although cells do lose viability. In addition, *Thermoplasma* is resistant to non-ionic detergents, ethylenediaminetetraacetic acid, pronase, trypsin, digitonin and lysozyme (Belly and Brock, 1972; Smith et al., 1973a).

Despite its extreme resistance to a variety of treatments, *Thermoplasma* is lysed by neutrality (Belly and Brock, 1972; Smith et al., 1973a). Cells or prepared membranes are stable between pH values of 1 to 5 but cellular lysis ensues at pH values greater than 6 (Fig. 1). The requirement for H^+ ions is specific. Other monovalent or divalent cations cannot substitute. In this respect *Thermoplasma* resembles the extreme halophiles which require salt for cellular integrity (Larsen, 1967). At high pH values the membranes apparently dissaggregate as Smith et al. (1973a) could not recover any membranous

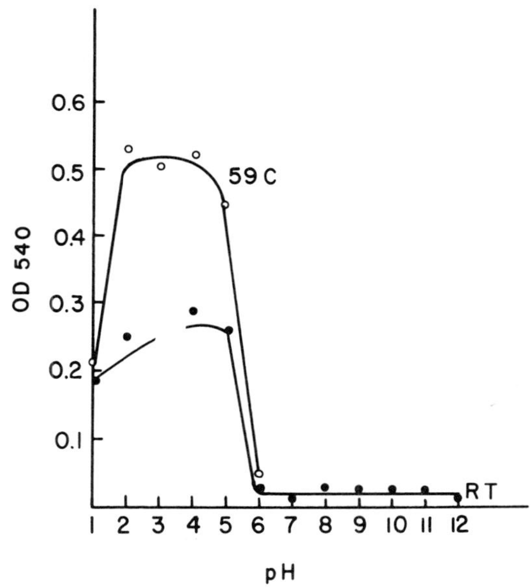

Fig. 1. Effect of pH on the stability of Thermoplasma. *Buffers at 0.1 M concentration: pH 1 to 3, KCl-HCl buffers, pH 4 to 5, acetate buffers; pH 6 to 8, phosphate buffers; pH 9 to 12, borate buffers. Stability determined at optimal temperature, 59°C and at room temperature, RT.*

material after centrifugation at 100,000 x g after 4 h. Interestingly, membrane-like vesicles could be recovered by dialysis of high pH lysates against deionized water. Ruwart and Haug (1975) have later reported that membranes can be recovered from high pH lysates by carefully controlling cellular protein concentrations and if one "works rapidly" before membranes completely dissaggregate. Because of the extreme stability of cells to mechanical breakage at low pH and dissagregation at high pH, native membranes are probably best prepared by sonication of cell suspensions at moderate ionic strength (0.05) and pH 5, a value which borders cellular instability (Smith et al., 1973a).

The nature of the membrane surface charges appears related to the obligatory H^+ ion requirement for cellular stability (Smith et al., 1973a). The ratio of free -COOH to free $-NH_2$ groups on the membrane surface (4:1) is identical to non-thermoacidophilic mycoplasmal membranes, although the total quantity of these groups per milligram of protein is less than half in *Thermoplasma*. Whereas membrane disruption normally occurs above pH 6, alteration of the surface charges by blocking free -COOH groups with an apolar radical such as glycine methyl ester, which removes a potential negative charge, renders cells or membranes stable over the entire pH range (pH 2 to 10). Further, when reacted with ethylene diamine, which not only removes a potential negative charge, but adds an additional positive charged radical ($-NH_2$), normal stability is reversed, e.g., cells or membranes become stable at high pH but lyse at low pH. Taken together, these results suggest first that the surface is very hydrophobic. Secondly, the obligatory H^+ ion requirement for stability may in part be explained by the necessity for protonation of -COOH groups on the surface. At high pH, repulsion of ionized $-COO^-$ groups likely causes membrane disruption.

The cytoplasmic membrane of *Thermoplasma* is comprised of about 60% protein, 25% lipid and 10% carbohydrate (Smith et al., 1973a). Membrane proteins appear as a heterogeneous mixture by polyacrylamide gel electrophoresis and possess an amino acid composition similar to non-thermoacidophilic mycoplasmal membrane proteins. As yet, no outstanding features of the membrane proteins are readily apparent.

In contrast, the complex lipids of *Thermoplasma* are unusual. Although not entirely unexpected, the lipids are characterized by the presence of ether-linkages rather than fatty acid ester residues. The assembly of the ether residues as diglycerol tetraethers, however, is unique (Langworthy, 1977a). The diglycerol tetraethers are structurally composed of two glycerol molecules bridged through through ether-linkages by two identical pairs of fully saturated, isopranoid branched, C_{40}-terminal alkyl diols (Fig. 2). The alkyl chains may be either $C_{40}H_{82}$ (acyclic), $C_{40}H_{80}$ (monocyclic) or $C_{40}H_{78}$ (bicyclic). De Rosa et al. (1977) have shown that cyclization within the hydrocarbons is in the form of pentacyclic rings. In *Thermoplasma*, the acyclic (65%) and monocyclic (33%) hydrocarbons predominate giving rise to $C_{86}H_{172}O_6$, M.W. 1300 and $C_{86}H_{168}O_6$, M.W. 1296 as the predominant diglycerol tetraether species. The tetraethers themselves account for about 15% of the membrane by weight.

Elucidation of the tetraether assembly has not been without some controversy. Initially, based primarily upon their apparent high moelcular weights, the ethers were thought to be long-chain glycerol diethers similar to *sn*-2,3-di-O-phytanylglycerol as found in the extreme halophiles (Kates et al., 1965) but containing two C_{40} rather than C_{20} alkyl chains (Langworthy et al., 1972). Further work by de Rosa et al. (1974a, 1976a), however, established the presence of equimolar proportions of glycerol and C_{40}-terminal alkyl diols in the ether which lead to the suggestion of a cyclic glycerol diether in which a single C_{40} hydrocarbon was combined through ether-linkages with *sn*-2,3-glycerol forming a forty carbon macrocyclic loop. The structure was now thought to resemble di-O-phytanylglycerol in which the terminal ends of the C_{20} chains had been joined by covalent linkage forming the C_{40} macrocyclic loop. But, because the molecular weight (approx. 1300) was much higher than that required by a cyclic diether (646 to 650) and because chromatographic properties suggested more than one -OH group, further work was undertaken establishing the diglycerol tetraether structure (Langworthy,

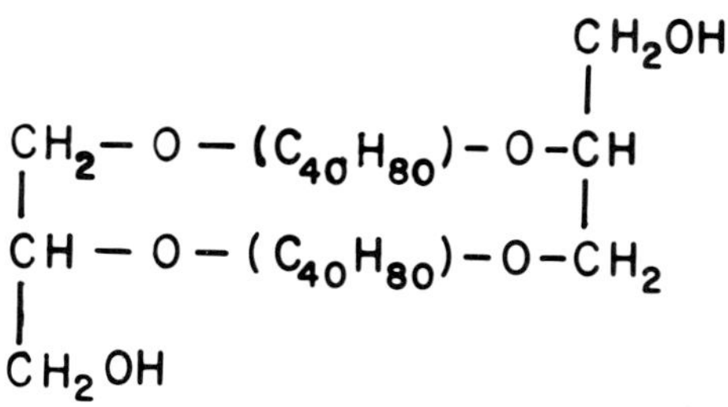

*Fig. 2. Diglycerol tetraether structure (Langworthy, 1977a). Configuration (*cis *or* trans*) of the hydroxyl groups has not been established. The identical pairs of fully saturated, isopranoid branched, hydrocarbon chains may be either $C_{40}H_{82}$ (acyclic), as shown, $C_{40}H_{80}$ (monocyclic) or $C_{40}H_{78}$ (bicyclic). Cyclization within the alkyl chains is in the form of pentacyclic rings (de Rosa et al., 1977).*

1977a). Without describing all the details here, the primary data establishing the tetraether assembly are: (i) molecular weight 1290 to 1308 by vapor phase osmometry and gel permeation chromatography, (ii) equimolar proportions of glycerol and C_{40} alkyl diols, (iii) demonstration of two -OH groups by isolation of both mono- and diacetates and mono- and dimethoxy derivatives from partial reaction mixtures, (iv) analysis of the monoacetate derivative giving proportions of 2 glycerol:2 alkyl:1 acetate, (v) NMR analysis of the monoacetate and monomethoxy derivatives which unequivocally established the presence of six oxygens, four in ether-linkages, two as hydroxyl groups and the presence of two C_{40} alkyl chains. The tetraether structure is also in critical accord with all previously established data (Langworthy et al., 1972, 1974; de Rosa et al., 1974a, 1976a). De Rosa et al. (1977) have now confirmed the diglycerol tetraether assembly. Biosynthesis of the tetraethers has not been fully elucidated at this time although alkyl chains incorporated ^{14}C-mevalonate, indicating that the mevalonate pathway is operative.

The presence of diglycerol tetraethers implies several functional features. Their occurrence suggests a membrane assembly which might be considered to be formed as a lipid monolayer rather than a typical lipid bilayer. A monolayer would appear to be formed by virtue of the covalent extension of the C_{40} alkyl chains across the membrane rather than by intercalation of separate and opposite diglyceride residues. The tetraethers do in fact resemble two molecules of di-O-phytanylglycerol which have been joined by head to head linkage of the C_{20} phytanyl chains. The physical effect of the tetraethers in the membrane might correlate with several of the aforementioned membrane properties. Although physical features of the membrane are certainly influenced by the nature of the proteins and carbohydrate, the presence of the tetraethers correlates well with the extreme membrane rigidity suggested by the EPR data as well as the characteristic cross-fracturing of freeze-fracture cells. The diglycerol tetraethers might be considered as providing three functions. In effect, the inner and outer faces of the lipid domain of the membrane are held together by virtue of the covalent linkage of hydrocarbon chains across the membrane, thereby providing stability. The presence of acid resistant ether-linkages likely relates to acidophily while the long hydrocarbon chains would appear to provide appropriate membrane fluidity at the high temperature required for growth. Diglycerol tetraethers would therefore seem to be well suited for a membrane directly exposed to hot acid.

The extractable complex lipids of *Thermoplasma* (Table 2) are comprised of neutral lipids (17.5%), glycolipids (25.1%) and acidic lipids (56.6%) (Langworthy et al., 1972). Only

TABLE 2
Partially Characterized Lipids of Thermoplasma[a]

Lipid Class	*% Total Lipids*	*Components*	*% Lipid Class*
Neutral lipids	17.5	vitamin K_2-7	-
		diglycerol tetraethers	-
Glycolipids	26.5	monoglycosyl diglycerol tetraether	-
		diglycosyl diglycerol tetraether	-
Acidic lipids	56.9	glycerolphosphoryl glycosyl diglycerol tetraether	~ 80.0
		amine-containing phosphoglycolipids	~ 20.0

[a]*From Langworthy et al. (1972). Total extractable lipid represents about 3% of cell dry weight.*

the naphthoquinone K_2-7 and free diglycerol tetraethers have been identified in the neutral lipid fraction. The glycolipids and acidic lipids all contain diglycerol tetraether residues. The glycolipids consist of at least six components none of which have been identified although two may be mono- and diglycosyl diglycerol tetraethers. Typical phosphatides are absent in *Thermoplasma*. All of the lipid phosphorous is accounted for by glycerolphosphate-substituted glycolipids in the form of phosphoglycolipids. Acid resistant phosphonate (C-P) bonds are not present. Approximately 80% of the phospholipids and almost half of the total lipid in the organism is accounted for by a partially characterized phosphoglycolipid: glycerolphosphoryl glycosyl diglycerol tetraether. The structure of the carbohydrate residue has not been established. Four minor phosphoglycolipids are also present. These appear to contain 2 to 4 carbohydrates including possible amino sugars as well as glycerolphosphate. Other than the initial compositional analyses (Langworthy et al., 1972), the exact structures and nature of the unusual sugars in these tetraether derived complex lipids have not been established.

The existence of diglycerol tetraethers, containing two primary hydroxyl groups, allows several possibilities for the assembly of the complex lipid structures. Polar constituents such as carbohydrates in the glycolipids might be substituted asymmetrically to only one -OH group (Fig. 2) or symmetrically to both -OH groups on each side of the tetraether. The phosphoglycolipids might contain a glycerolphosphoryl → glycoside linked to one or both -OH groups, or possibly a glycerolphosphate and carbohydrate attached to opposite sides of the molecule. Since the symmetrically assembled tetraethers extend across the membrane providing an -OH group on each side, membrane asymmetry will be represented in the individual glycolipid and phosphoglycolipid structures. Their structural elucidation awaits further studies.

Carbohydrate content of the membrane arises from the glycolipids, phosphoglycolipids, as well as a new type of lipopolysaccharide (LPS). Its structure has been tentatively determined by physical, chemical and enzymatic techniques to be [mannose-(1 $\overset{\alpha}{\to}$ 2)-mannose-(1 $\overset{\alpha}{\to}$ 2)-mannose-(1 $\overset{\alpha}{\to}$ 3)]$_8$-glucose diglycerol tetraether (Mayberry-Carson et al., 1974). Again, the exact linkage of sugars to the tetraether has not been established. The LPS represents about 3% of the cell dry weight and it is stable at pH 2 and 60°C for at least several weeks. Although this LPS might be considered as a glycolipid containing 25 sugar residues, transmission electron microscopy reveals a trilaminar structure about 5 nm in width similar to gram negative LPS (Mayberry-Carson et al., 1975). It behaves as a specific hapten in complement fixation tests using anti-membrane sera (Sugiyama et al., 1974). In addition, concanavalin A binds to cells, suggesting its location on the cell surface (K.J. Mayberry-Carson, personal communication).

SULFOLOBUS

Sulfolobus can be considered to be a type of wall defective organism. It contains no peptidoglycan but is enveloped by a cytoplasmic membrane which is further overlaid by an ultrastructural element consisting of a regular polyhexagonal array of protein subunits 13 to 15 nm in diameter (Weiss, 1974). Pili may also be present which are extremely acid and heat stable. Less is known about *Sulfolobus*. In contrast to *Thermoplasma*, it is not lysed by neutrality. It is, however, quite resistant to mechanical breakage. Weiss (1974) has successfully separated the subunit "cell wall" from the cell membrane and shown it to be comprised of a heterogeneous amino acid mixture with about a 10% excess of acidic over basic amino acids, similar to organisms living in neutral habitats. Hexosamines and neutral sugars are also present. The nature of surface polymers is unknown. Unfortunately, "protoplasts" or membranes have not been prepared. Typical interior wall layers or inner hydrophobic cytoplasmic membrane are not exposed by freeze-fracturing. Instead, cells corss-fracture exposing a cross section of surface-membrane and cytoplasm as in *Thermoplasma*.

Complex lipids from organisms grown heterotrophically at pH 3 and 70°C account for about 3% of the cell dry weight. The extractable lipids (Table 3) are composed of about 10.5% neutral lipids, 67.6% glycolipids and 21.7% phosphoglycolipids (Langworthy et al., 1974). As in *Thermoplasma*, all of the glycolipids and acidic lipids contain diglycerol tetraether residues. Whereas the acyclic and monocyclic hydrocarbon chains predominate in *Thermoplasma* tetraethers the monocyclic and bicyclic chains predominate in *Sulfolobus* (de Rosa et al., 1976a). Only free diglycerol tetraethers and a tri-O-phytanylglycerol (de Rosa et al., 1976b) have been identified in the neutral lipid fraction. Glycolipids are composed of two components: glucosyl galactosyl diglycerol tetraether and a glucosyl glycerolpolyoltetraether. The latter glycolipid contains an unidentified octitol or nonitol attached through ether-linkages to the tetraether. This polyol is very likely substituted for one of the glycerol residues in the tetraether structure. Additionally, this ether residue contains primarily tricyclic and tetracyclic hydrocarbon chains (de Rosa et al., 1977). Identification of the

TABLE 3
Partially Characterized Lipids of Sulfolobus[a]

Lipid Class	*% Total Lipid*	*Components*	*% Lipid Class*
Neutral	10.5		
		diglycerol tetraethers	-
		tri-O-phytanylglycerol[b]	-
Glycolipids	67.6		
		glucosyl galactosyl diglycerol tetraether	43.6
		glucosyl glycerolpolyoltetraether	56.3
Acidic lipids	21.7		
		inositolphosphoryl diglycerol tetraether	38.0
		inositolphosphoryl glucosyl galactosyl diglycerol tetraether	27.0
		inositolphosphoryl glucosyl glycerolpolyoltetraether	28.9
		glucosyl glycerolpolyoltetraether monosulfate	6.0

[a] *From Langworthy et al. (1974). Total extractable lipid represents about 3% of cell dry weight.*

[b] *Found by de Rosa et al. (1976b).*

glycerolpolyoltetraether moiety should reveal a new type of tetraether assembly. The acidic lipids are phosphoglycolipids comprised primarily of the inositolphosphoryl derivatives of the two glycolipids, namely inositolphosphoryl glucosyl galactosyl diglycerol tetraether and inositolphosphoryl glucosyl glycerolpolyoltetraether. Also present are lesser amounts of a monosulfate derivative of glucosylpolyoltetraether and the diglycerol tetraether derivative of phosphatidylinositol. Phosphonate or sulfonate bonds are absent. As with *Thermoplasma*, it is not known how the polar residues in any of the complex lipids are linked to the apolar tetraether moieties.

Initial examination of the complex lipids from *Sulfolobus* grown autotrophically on elemental sulfur (Langworthy, 1977b) indicate the same structures are present as in heterotrophically grown cells although differing in distribution. The glucosyl glycerolpolyoltetraether and its inositolphosphoryl derivative account for nearly 80% of their respective glycolipid and acidic lipid fractions. A new acidic lipid does appear in autotrophically grown cells. It accounts for about 13% of the acidic lipids and appears to be a derivative of glucosyl glycerolpolyoltetraether which contains sulfate and/or acid resistant phosphonate (C-P) residues. The occurrence only in autotrophically grown cells might suggest a role in high temperature sulfur oxidation.

BACILLUS ACIDOCALDARIUS

Bacillus acidocaldarius is so far the sole example of a typical cell wall-containing bacterium requiring a hot acid environment for growth. As might be expected, the organism is stable at neutral pH. Unfortunately, detailed information relating to its cell wall composition and structure, which is in direct contact with the hot acid, has not been reported. Neither have "protoplasts" or membranes been prepared. The nature of the extractable lipids, however, has been detailed (Table 4). Total lipids account for about 8.1% of the cell dry weight and are comprised of approximately 15.7% neutral lipids, 64.0% glycolipids and 20.3% acidic lipids (Langworthy et al., 1976).

In direct contrast to *Thermoplasma* and *Sulfolobus*, the lipids of *B. caldocaldarius* contain no ether bonds, only ester- or amide-linked fatty acids. The fatty acid composition does differ, however, from non-thermoacidophilic organisms by the presence of unusual ω-cyclohexyl C_{17} and C_{19} fatty acids, namely, 11-cyclohexylundecanoic and 13-

TABLE 4
Lipids of Bacillus acidocaldarius[a]

Lipid Class	*% Total Lipid*	*Components*	*% Lipid Class*
Neutral lipids[b]	15.7		
		menaquinone-9, C_{50} and C_{55} polyprenols, squalene, pentacyclic triterpene hydrocarbons; hopene-b [hop-22(29)-ene].	
Glycolipids	64.0		
		tetrahydroxybacteriohopane	1.5
		N-acylGlcN(1 $\overset{\beta}{\rightarrow}$ 1)tetrahydroxybacteriohopane	26.0
		GlcN(1 $\overset{\beta}{\rightarrow}$ 1)tetrahydroxybacteriohopane	4.8
		Glc(1 $\overset{\beta}{\rightarrow}$ 4)N-acylGlcN(1 $\overset{\beta}{\rightarrow}$?)diacylglycerol	24.9
		Glc(1 $\overset{\beta}{\rightarrow}$ 4)N-acylGlcN(1 $\overset{\beta}{\rightarrow}$?)monoacylglycerol	41.3
		Glc(1 $\overset{\beta}{\rightarrow}$ 4)N-acylGlcN(1 $\overset{\beta}{\rightarrow}$?)glycerol	1.5
Acidic lipids	20.3		
		diphosphatidylglycerol	32.3
		lysodiphosphatidylglycerol	5.3
		phosphatidic acid	5.8
		phosphatidylglycerol	13.4
		sulfonolipid	43.2

[a] *From Langworthy et al. (1976). Total extractable lipids represent about 8.1% cell dry weight.*

[b] *Neutral lipid components found by de Rosa et al. (1971b; 1973).*

cyclohexyltridecanoic acid (de Rosa et al., 1972; Oshima and Ariga, 1975). These fatty acids have been shown to be biosynthesized via the shikimate pathway and the cyclohexyl rings originating from glucose (de Rosa et al., 1972, 1974b; Oshima and Ariga, 1975).

Several lipids of the neutral class have been identified including C_{50} and C_{55} polyprenols, menaquinone-9, squalene, and pentacyclic triterpene hydrocarbons belonging to the hopane class, the major one being hop-22(29)-ene (de Rosa et al., 1971b, 1973). The occurrence of pentacyclic triterpenes and their hydroxylated bacteriohopane derivatives, as noted later, is unusual as they have only been found in a few other procaryotes, e.g., *Acetobacter xylinium*, *Methylococcus capsulatus* and *Acetomonas oxydans* (Bird et al., 1971; Förster et al., 1973; Rohmer and Ourisson, 1976).

The glycolipids are characterized by the presence of glucosamine residues. Analysis of the glycolipids has identified the existence of two major types as shown in Fig. 3 (Langworthy et al., 1976; Langworthy and Mayberry, 1976). The first has been established as a 1-(O-β-N-acylglycosaminyl)-2,3,4-tetrahydroxypentane-29-hopane. This component contains N-acylglucosamine which is β-linked to the primary hydroxyl of a fully saturated 1,2,3,4-tetrahydroxypentane-substituted pentacyclic triterpene nucleus ($C_{35}H_{62}O_4$, M.W. 546). This new type of hydrophobic aglycone has been termed

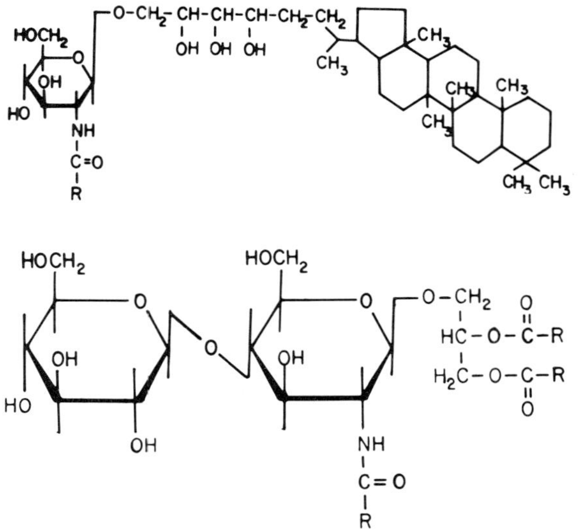

Fig. 3. The two basic glycolipid type structures in B. acidocaldarius *(Langworthy and Mayberry, 1976; Langworthy et al., 1976). Top, 1-(O-β-N-acylglucosaminyl)-2,3,4-tetrahydroxypentane-29-hopane. Bottom, O-β-D-glucopyranosyl-(1 → 4)-O-2-acylamido-2-deoxy-β-D-glucopyranosyldiacylglycerol.*

tetrahydroxybacteriohopane (Förster et al., 1973). A small amount of free tetrahydroxybacteriohopane is also present in the glycolipid fraction. Glycolipids containing tetrahydroxybacteriophopane residues account for nearly 20% of the total extractable lipids. The second type of glycolipid structure has been identified as glucosyl(1 $\overset{\beta}{\rightarrow}$ 4)N-acylglucosaminyl (1 $\overset{\beta}{\rightarrow}$?)diacylglycerol. Several lyso forms of the two glycolipids also exist (Table 4). Nearly all of the $-NH_2$ groups in the glucosamidyl glycolipids are N-acylated with the amide-linked fatty acids having a distribution similar to the esterified fatty acids in the glyceride portion of the glycolipids, e.g., principally ω-cyclohexyl C_{19} (19%) and C_{17} (42%) and C_{17}-branched (33%) fatty acids. In contrast, the glucosamine-containing pentaglycosyl glycolipid, gal*f*-(1 → 2) -gal*p*-(1 → 6)-glcN(15 methylhexadecanoyl)-(1 → 2)-glc*p*-diacylglycerol found in *Thermus thermophilus* (Oshima and Yamakawa, 1974) specifically contains the C_{17} branched fatty acid 15-methylhexadecanoic acid in the amide linkage to glucosamine.

The phospholipids (Table 4) are comprised of the common phosphatides, mainly diphosphatidyl glycerol and phosphatidylglycerol with lesser amounts of lysodiphosphatidylglycerol and phosphatidic acid. Phosphoglycolipids are absent in *B. acidocaldarius*. Of particular interest, however, is the presence of a sulfonolipid containing the acid resistant C-S which accounts for close to half of the total acidic lipids. Structurally it appears to be very similar to, if not the same as, the plant sulfonolipid 6-sulfoquinovosyldiacylglycerol. Apparently its occurrence is the first in a non-photosynthetic procaryote. Such a component has been found only in the photosynthetic bacteria, algae, plants and several marine forms (Haines, 1971). In these organisms it is thought to possibly play some role in photosynthesis or respiration. Sulfonolipids are also the most acidic of the polar lipids being ionized in aqueous solution at all pH values. At this point its function in *B. acidocaldarius* is not known. Perhaps it represents a precursor to some very acidic wall polymer or, alternatively, possibly function in stimulation of respiration at the high growth temperature and/or function in some H^+ ion exclusion capacity.

The overall picture of *B. acidocaldarius* lipid structures appears to be one of intermediate conventionality between the diglycerol tetraether derived lipids of *Thermoplasma* and *Sulfolobus* and the typical diglyceride-containing complex lipids of non-thermoacidophilic organisms. The existance in and requirement for a hot acid environment must certainly include features of the cell wall structure which is in direct contact with the environment. Perhaps the presence of the unusual lipid structures suggests a likely requirement for unusual wall polymers as well.

CONCLUSION

An overall consideration of the membrane structures of the three thermoacidophiles is of course speculative at best in view of the variable amounts of information presently available. Most information is available on *Thermoplasma*. So far, studies suggest that the underlying basis for *Thermoplasma's* obligate thermoaciophily may in part be accounted for by its hydrophobic surface, specificity of H^+ ions for protonation of surface -COOH groups required for the maintenance of cellular integrity and by the presence of long chain diglycerol tetraethers providing increased membrane stability, acid resistance and appropriate membrane fluidity. In comparison, information on the composition and properties of the walls, prepared protoplasts or isolated membranes from *Sulfolobus* and *B. acidocaldarius* is presently lacking. In terms of lipid composition, however, the membranes of *Thermoplasma* and *Sulfolobus* appear to be related by their synthesis of diglycerol tetraethers. These two organisms are also characterized by the presence of phosphoglycolipids rather than typical phosphatides. In *Thermoplasma* these are glycerolphosphoryl derived glycolipids and in *Sulfolobus* inositolphosphoryl derived glycolipids. In effect, all but the neutral lipids in each contain carbohydrate residues. The role of the phosphoglycolipids is not clear. Their occurrence may relate to the absence of a true cell wall in these two organisms since certain mycoplasmas and L-forms do accumulate phosphoglycolipids (Smith et al., 1973b). In contrast, *B. acidocaldarius* does not contain diglycerol tetraethers or phosphoglycolipids. It possesses glucosamidly glycolipids, a sulfonic acid derived glycolipid and typical phosphatides. Rather than ether-linkages, saturated ω-cyclohexyl and branched chain fatty acids and tetrahydroxybacteriohopane residues are present. Perhaps ethers are not required due to the presence of a true cell wall.

A possible common denominator of the three morphologically different types of thermoacidophiles is the occurrence of cyclization and methyl branching within the saturated lipid hydrocarbon residues. Cyclization occurs within the isopranoid branched, C_{40} alkyl chains of *Thermoplasma* and *Sulfolobus* and in the form of ω-cyclohexyl fatty acids and pentacyclic triterpene derivatives in *B. acidocaldarius*. The isopentenyl pyrophosphate pathway is operative in all three organisms. Perhaps a function for the prevalent cyclization might be that of reducing rotational freedom in the hydrocarbon chains thereby promoting tighter molecular packing in these membranes providing an even more rigid and likely less permeable membrane than found in non-acidophilic thermophiles. It is of interest to note that the mesophilic extreme acidophiles are marked by an increased cyclopropane fatty acid content (Levin, 1971) and thermophiles by an increased proportion of saturated and branched chain fatty acids (McElhaney, 1976). The hydrophobic residues of the thermoacidophiles display features of both cyclization and branching.

The occurrence of high amounts of carbohydrate containing lipids in each of the different thermoacidophiles might also reflect a common feature. Even *B. acidocaldarius* possesses about 75% of its complex lipids as carbohydrate derivatives. Similarly, the glycolipids in *Thermus thermophilum* represent almost 70% of the total cell lipid content (Oshima and Yamakawa, 1974). Ray et al. (1971) have also shown that the glycolipids in *Thermus aquaticus* (mono- and diglucosyl diacylglycerols) increased about four fold with a temperature increase from 50°C to 75°C which suggested the glycolipids may aid in imparting increased membrane stability. The preponderance of carbohydrate derived lipids in the thermoacidophiles may also be a necessary response to high temperature and function by raising membrane rigidity and stability.

It is probably safe to conclude at this point that the types of lipids demanded by these organisms appear to be influenced by the presence or absence of some type of cell wall or surface layer. Undoubtedly, the presence of diglycerol tetraethers or cyclohexyl fatty acids and triterpene derivatives in the complex lipids together with the nature of the membrane carbohydrate and proteins play a significant role in thermoacidophily. So far, studies indicate that the thermoacidophiles synthesize new types of membrane lipids and they appear to possess a new type of membrane assembly in the case of *Thermoplasma* and *Sulfolobus*. Very likely *Sulfolobus* and *B. acidocaldarius* synthesize new or modified wall or surface structures as well. Further work on the membrane and surface lipid, cabohydrate and protein structure should provide the necessary background for future studies on their relationship to envelope assembly, metabolism and function. This information would expand our understanding of the mechanisms of thermoacidophily.

ACKNOWLEDGMENT

The author wishes to acknowledge P.F. Smith, W.R. Mayberry, K.J. Mayberry-Carson, M.R. Smith and J. Ratzlaff who have been involved in several aspects of the work presented here.

REFERENCES

Belly, R. T. and Brock, T. D. (1972). J. Gen. Microbiol. *73*, 465.

Belly, R. T., Bohlool, B. B. and Brock, T. D. (1973). Ann. N. Y. Acad. Sci. *225*, 94.

Bird, C. W., Lynch, J. M., Pirt, S. J. and Reid, W. W. (1971). Tetrahedron Letters *34*, 3189.

Brierley, C. L. and Brierley, J. A. (1973). Can. J. Microbiol. *19*, 183.

Brock, T. D., Brock, K. M., Belly, R. T. and Weiss, R. L. (1972). Arch, Mikrobiol. *84*, 54.

Christiansen, C., Freundt, E. A. and Black, F. T. (1975). Int. J. Syst. Bacteriol. *25*, 99.

Darland, G. and Brock, T. D. (1971). J. Gen. Microbiol. *67*, 9-15.

Darland, G., Brock, T. D., Samsonoff, W. and Conti, S. F. (1970). Science *170*, 1416.

De Rosa, M., Gambacorta, A., Minale, L. and Bu'lock, J.D. (1971a). J.C.S. Chem. Comm. *1971*, 1334.

De Rosa, M., Bambacorta, A., Minale, L. and Bu'lock, J.D. (1971b). J.C.S. Chem. Comm. *1971*, 619.

De Rosa, M., Gambacorta, A., Minale, L. and Bu'lock, J.D. (1972). Biochem. J. *128*, 751.

De Rosa, M., Gambacorta, A., Minale, L. and Bu'lock, J.D. (1973). Phytochemistry *12*, 1117.

De Rosa, M., Gambacorta, A., Minale, L. and Bu'lock, J.D. (1974a). J.C.S. Chem. Comm. *1974a*, 543.

De Rosa, M., Gambacorta, A. and Bu'lock, J.D. (1974b). Phytochemistry *13*, 905.

De Rosa, M., Gambacorta, A. and Bu'lock, J.D. (1975). J. Gen. Microbiol. *86*, 156.

De Rosa, M., Gambacorta, A. and Bu'lock, J.D. (1976a). Phytochemistry *15*, 143.

De Rosa, M., De Rosa, S., Gambacorta, A. and Bu'lock, J.D. (1976b). Phytochemistry *15*, 1995.

De Rosa, M., De Rosa, S., Gambacorta, A. and Bu'lock, J.D. (1977). J.C.S. Chem. Comm. *1977*, in press.

Esser, A. F. and Lanyi, J. K. (1973). Biochemistry *12*, 1933.

Förster, H. J., Biemann, K., Haigh, W. G., Tattric, N. J. and Colvin, J. R. (1973). Biochem. J. *135*, 133.

Haines, T. H. (1971). In *Progress in the Chemistry of Fats and Other Lipids* (R. T. Holman, ed.) *11*, 297.

Hsung, J. C. and Haug, A. (1975). Biochim. Biophys. Acta *389*, 477.

Kates, M., Yengoyan, L. S. and Sastry, P. S. (1965). Biochim. Biophys. Acta *98*, 252.

Langworthy, T. A. (1977a). Biochim. Biophys. Acta *487*, 37.

Langworthy, T. A. (1977b). J. Bacteriol. *130*, 1326.

Langworthy, T. A. and Mayberry, W. R. (1976). Biochim. Biophys. Acta *431*, 570.

Langworthy, T. A., Smith, P. F. and Mayberry, W. R. (1972). J. Bacteriol. *112*, 1193.

Langworthy, T. A., Mayberry, W. R. and Smith, P. F. (1974). J. Bacteriol. *119*, 106.

Langworthy, T. A., Mayberry, W. R. and Smith, P. F. (1976). Biochim. Biophys. Acta *431*, 550.

Larsen, H. (1967). Adv. Microbiol. Physiol. *1*, 97.

Levin, R. A. (1971). J. Bacteriol. *108*, 992.

Mayberry-Carson, K. J., Langworthy, T. A., Mayberry, W. R. and Smith, P. F. (1974). Biochim. Biophys. Acta *360*, 217.

Mayberry-Carson, K. J., Roth, I. L. and Smith, P. F. (1975). J. Bacteriol. *121*, 700.

McElhaney, R. W. (1976). In *Extreme Environments: Mechanisms of Microbial Adaption* (M. R. Heinrich, ed.), p. 255, Academic Press, New York.

Oshima, M. and Yamakawa, T. (1974). Biochemistry *13*, 1140.

Oshima, M. and Ariga, T. (1975). J. Biol. Chem. *250*, 6963.

Oshima, T., Arakawa, H. and Baba, M. (1977). J. Biochem. *81*, 1107.

Ray, P. H., White, D. C. and Brock, T. D. (1971). J. Bacteriol. *108*, 227.

Rohmer, J. and Ourisson, G. (1976). Tetrahedron Letters *40*, 3637.

Ruwart, M. J. and Haug, A. (1975). Biochemistry *14*, 860.

Searcy, D. G. (1975). Biochim. Biophys. Acta *395*, 535.

Searcy, D. G. (1976). Biochim. Biophys, Acta *451*, 278.

Smith, P. F., Langworthy, T. A., Mayberry, W. R. and Hougland, A. E. (1973a). J. Bacteriol. *116*, 1019.

Smith, P. F., Langworthy, T. A. and Mayberry, W. R. (1973b). Ann. N. Y. Acad. Sci. *225*, 22.

Smith, P. F., Langworthy, T. A. and Smith, M. R. (1975). J. Bacteriol. *124*, 884.

Smith, G. G., Ruwart, M. J. and Haug, A. (1974). FEBS Letters *45*, 96.

Sugiyama, T., Smith, P. F., Langworthy, T. A. and Mayberry, W. R. (1974). Infect. Immun. *10*, 1273.

Thomas, J. A., Cole, R. E. and Langworthy, T. A. (1976). Fed. Proc. *35*, 1455.

Weiss, R. L. (1974). J. Bacteriol. *118*, 275.

ω-CYCLOHEXYL FATTY ACIDS IN ACIDO-THERMOPHILIC BACTERIAL MEMBRANES AND PHAGE CAPSIDS

Mieko Oshima

Department of Biochemistry
Kitasato University
School of Medicine
Asamizodai, Sagamihara

Yoshiyuki Sakaki

Department of Biochemistry
Kyushu University
Fukuoka

Tairo Oshima

Mitsubishi-Kasei Institute
of Life Sciences
Machida, Tokyo

The bacterium, *Bacillus acidocaldarius*, and phage φNS11 (infectious for strain TA-6) both possess a lipid that contains unusual ω-cyclohexyl fatty acids. The identification, biosynthesis and possible function of these novel fatty acids will be described in this paper.

BACILLUS ACIDOCALDARIUS AND ITS PHAGE

Acidophilic-thermophilic bacteria are characterized by their tolerance to high temperature and acidity (55-90°C, pH 1-5). So far, three morphologically distinct types of acido-thermophiles have been recently discovered in acidic hot springs (Uchino and Doi, 1967; Darland *et al.*, 1970;

ISBN 0-12-268250-5

Brock *et al.*, 1972; Furuya *et al.*, 1977).

One of the three types of acido-thermophiles, *B. acidocaldarius*, is a true peptidoglycan cell wall containing, spore forming, rod shaped bacterium which can grow at pH 2-5 and from 55-70°C. Several strains of *B. acidocaldarius* been isolated from the acidic hot springs of Yellowstone and Hawaii Volcanic National Park, U.S.A. (Darland and Brock, 1971), Agnano, Italy (De Rosa *et al.*, 1972), Beppu and Tamagawa, Japan (Uchino and Doi, 1967; Oshima *et al.* 1977).

A novel bacteriophage infectious for *B. acidocaldarius* strain TA-6 was first isolated from Beppu hot springs, Japan by Sakaki and Oshima (1976). The phage, designated ϕNS11, showed optimum growth at 60°C and pH 3.5. It can grow in a medium over a pH range from 2-5, but it is unstable above pH 6. Phage ϕNS11 has a polyhedral shape with a spike-like structure and its diameter is 60-70 nm (Fig. 1).

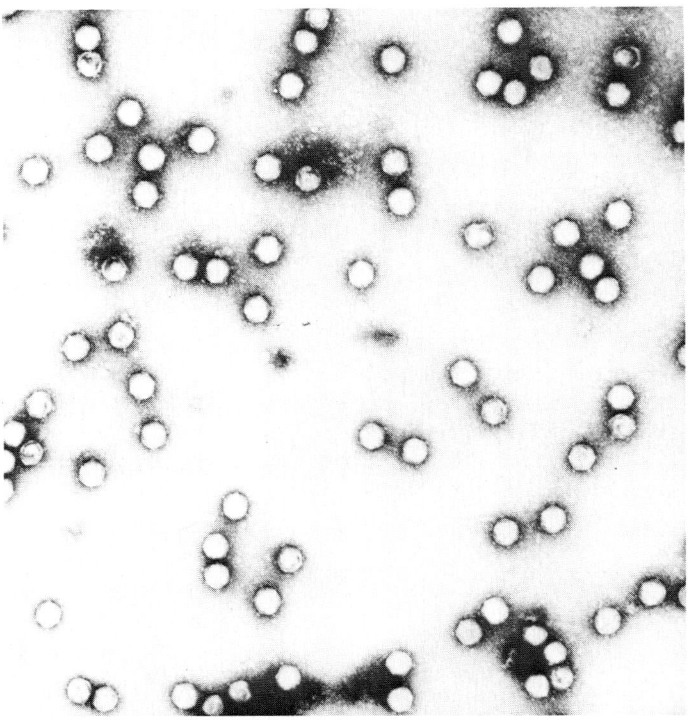

FIGURE 1. Electron micrograph of purified ϕNS11. The purified phage was negatively stained with uranyl acetate.

Based on sensitivity to nucleases, the nucleic acid of ϕNS11 was shown to be double stranded DNA. Pancreatic DNase degraded more than 90% of the phage DNA, while RNases or nucleases specific for single stranded nucleic acid degraded less than 1% of the phage DNA.

The buoyant density of ϕNS11 in CsCl is about 1.27g/cm^3, which is similar to that of lipid-containing phages PM2 (Espejo and Canelo, 1968), PR4 (Bradley and Rutherford, 1975) and ϕ6 (Vidaver *et al.*, 1973). Phage ϕNS11 is highly sensitive to organic solvents. The phage lipid extracted by chloroform-methanol(2:1, v/v) was estimated to be about 12.5% by weight of the dried phage particle (Sakaki *et al.*, 1977).

Polyacrylamide gel electrophoresis of the purified phage showed five protein components. The relative amount of these five proteins did not change after further purification of phage particles by sucrose or CsCl density gradients. The molecular weight of the major protein component (ca. 77-78% of total proteins) is 44,000 as determined by SDS polyacrylamide gel electrophoresis.

IDENTIFICATION OF ω-CYCLOHEXYL FATTY ACIDS

The fatty acid composition of 10 strains of *B. acidocaldarius* isolated from different Japanese hot springs and geyser basins was examined. All strains contained 74-93% of fatty acid methyl esters of unknown structure (Table 1, Fig. 2). About 20 mg each of these fatty acid methyl esters were isolated by preparative gas chromatography from the mixture of fatty acid methyl esters prepared from *B. acidocaldarius* strain TA-6.

The infra-red spectra of these unknown fatty acids show the $-CH_2-$ scissoring vibration of the cyclohexane ring at 1455 cm^{-1}. The absorption bands at 892-840 cm^{-1} reveal the presence of the cycloalkane ring.

The proton nuclear magnetic resonance spectra show the methylene protons of a cycloalkane ring (broad signal at 1.64 ppm). There are no terminal methyl protons at around 0.95 ppm. These facts provide evidence for the cycloalkane ring at the ω-end of the fatty acid.

The fragment ions at m/e 83, 69, 55 and 41 shown in the GC-mass spectra (Figs. 3 and 4) of the two unknown fatty acid methyl esters provide evidence for fragmentation of the cyclohexane ring. The intensity of these fragment ions was the same as in the mass spectrum of authentic cyclohexyl carboxylate, once again indicating cyclohexane ring fragmentation. The molecular weights of these two fatty acid methyl esters were 282 and 310, respectively. Taken

TABLE 1

Fatty Acid Composition of 10 Strains of B. acidocaldarius *Isolated from Different Japanese Hot Springs*

Strain	*Fatty acid content (%)*								
	15:I	*15:A*	*16:I*	*16:N*	*17:I*	*17:A*	*11:CY*	*13:CY*	*Cyclic*
TA-6	T	T	T	1.4	5.8	15.8	35.5	38.8	74.3
BA-152	-	0.2	0.6	2.2	0.7	2.7	37.1	55.8	92.9
FA-1	3.0	3.5	7.8	0.7	2.3	6.5	49.1	25.6	74.7
BA-18	4.2	3.8	1.1	0.4	3.9	6.6	54.7	25.1	79.8
BA-151	0.2	0.2	0.9	0.8	1.6	3.0	43.0	50.0	93.0
BA-11	1.1	0.3	1.8	1.0	6.2	4.9	39.3	44.3	83.6
BA-5	4.3	3.0	2.0	0.9	4.0	7.5	48.1	30.0	78.1
T-7	0.6	1.1	0.3	0.5	3.4	14.5	43.7	34.3	78.0
T-17	0.5	0.8	0.5	0.6	2.4	10.3	46.3	37.7	84.0
TM-24	0.7	0.9	1.0	1.0	4.2	12.0	31.9	47.0	78.9

Percentages were calculated as % areas by gas-liquid chromatographic analysis. T: less than 0.4% of the total fatty acids. The abbreviations used are: 11-cy:11-cyclohexyl undecanoate, 13-cy:13-cyclohexyl tridecanoate, 15:I, C_{15} iso acid. 17:A, C_{17} anteiso acid. Taken from Oshima and Ariga (1975) and reproduced with permission from The American Society of Biological Chemists.

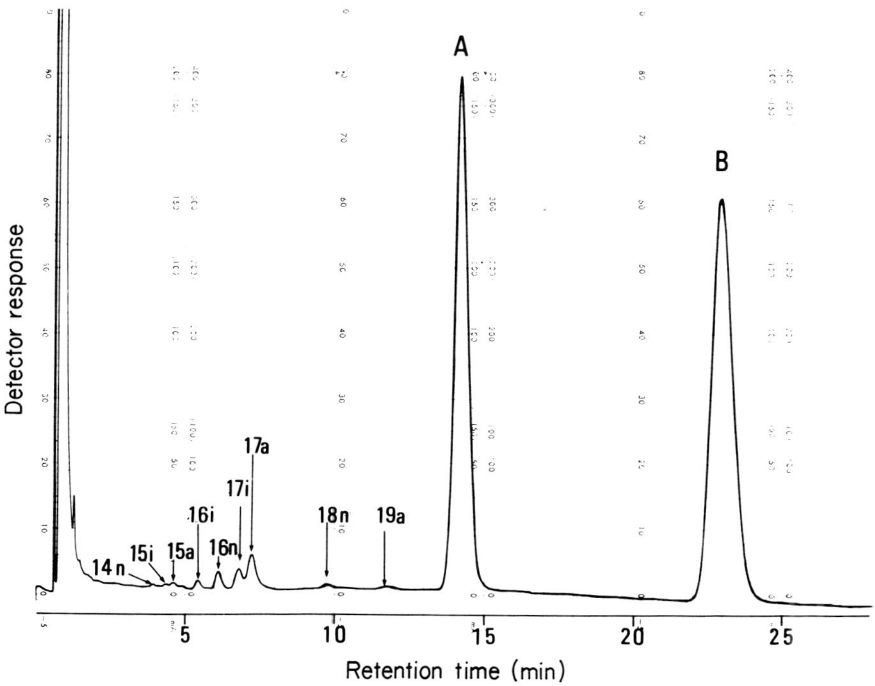

FIGURE 2. Gas-liquid chromatography of 11-cyclohexyl undecanoate (A) and 13-cyclohexyl tridecanoate (B). The fatty acid methyl esters were analyzed on a 25% polyethylene glycol succinate column (3.5 m) at 170°C.

together, these findings indicate that the unusual fatty acids are 11-cyclohexyl undecanoic acid and 13-cyclohexyl tridecanoic acid.

The fatty acid methyl esters prepared by the same manner from the dried phage particles were analysed. The novel fatty acids, 11-cyclohexyl undecanoic acid and 13-cyclohexyl tridecanoic acid, were also detected in the phage (Fig. 5). The complex lipids of the phage resembles that of B. acidocaldarius TA-6 (Fig. 6).

BIOSYNTHESIS OF ω-CYCLOHEXYL FATTY ACIDS

Glucose had a marked effect on the biosynthesis of ω-cyclohexyl fatty acids. The amount of ω-cyclohexyl fatty acids synthesized increases when increasing amounts of

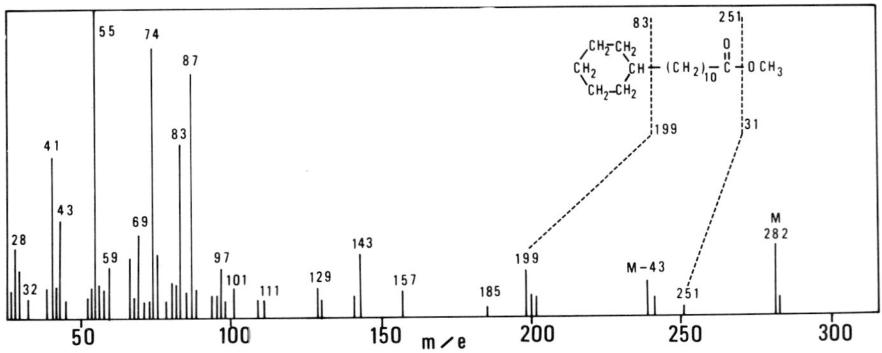

FIGURE 3. GC-MS spectrum of methyl 11-cyclohexyl undecanoate recorded using a Shimadzu-LKB model 9000 spectrometer with a column of 4% XE-60 on Chromosorb W. Taken from Oshima and Ariga (1975) and reproduced with permission from the American Society of Biological Chemists.

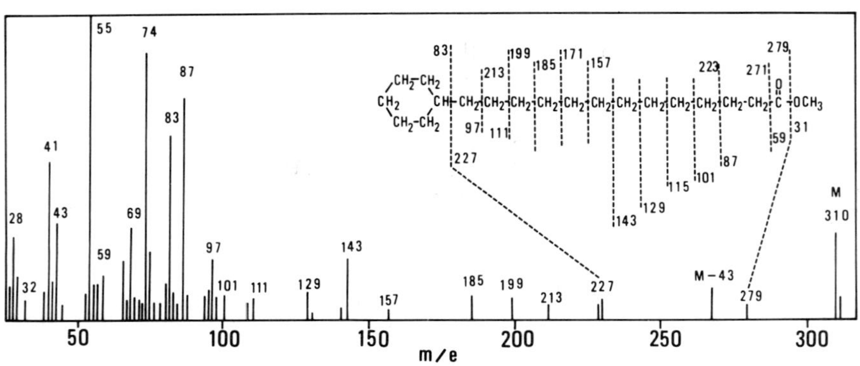

FIGURE 4. GC-MS spectrum of methyl 13-cyclohexyl tridecanoate. Taken from Oshima and Ariga (1975) and reproduced with permission from The American Society of Biological Chemists.

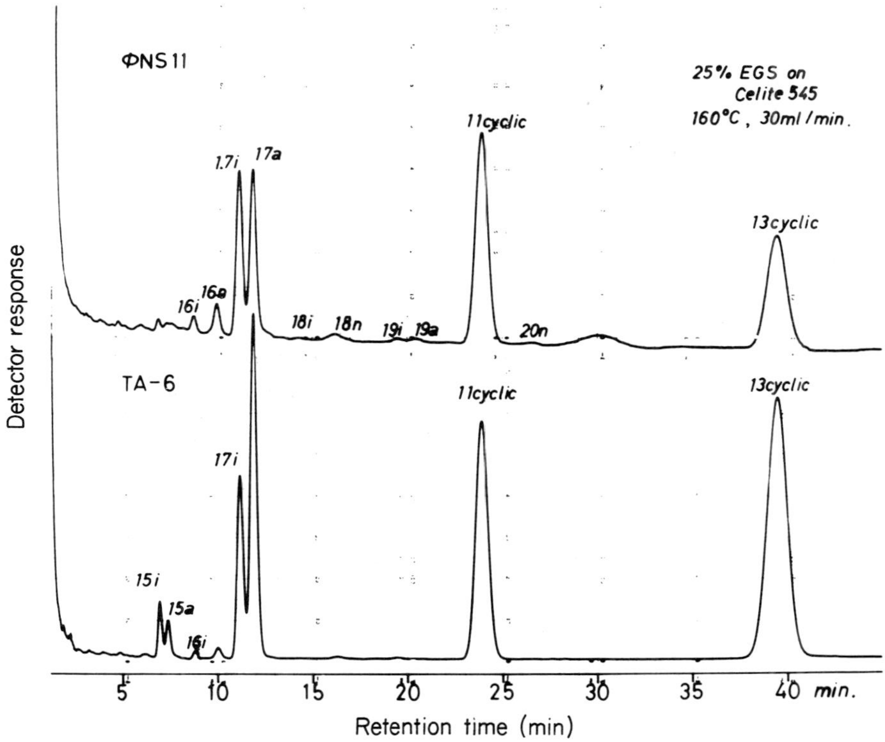

FIGURE 5. Fatty acid composition of phage ϕNS11 and B. acidocaldarius *strain TA-6. 11cyclic: 11-cyclohexyl undecanoate. 13cyclic:13-cyclohexyl tridecanoate.*

glucose are added to the medium as shown in Fig. 7.

The stable isotope incorporations from possible precursors, such as D-[2-D]glucose, methyl-α-D-[-6-D_2]glucopyranoside and [2-D]shikimic acid were studied by GC-mass spectrometry. The mass fragment analyses indicated the deuterium was incorporated into the specific circled position of ω-cyclohexyl fatty acid as shown in Fig. 8.

Table 2 shows the incorporation of ^{14}C from labeled precursors. Incorporation of radioactivity into ω-cyclohexyl fatty acids was high using [U-^{14}C]shikimic acid and [1-^{14}C]cyclohexyl carboxylic acid. These results suggest that ω-cyclohexyl fatty acids were synthesized by reduction of shikimate and elongation of the alkyl chain of cyclohexyl carboxylate. It appears that simple chain elongation on cyclohexyl carboxylate occurred rather than via a starter unit of cyclohexyl propionate formed by decarboxylation of prephenate, an intermediate of phenylalanine biosynthesis. De Rosa *et al.* (1972) also reported ^{14}C

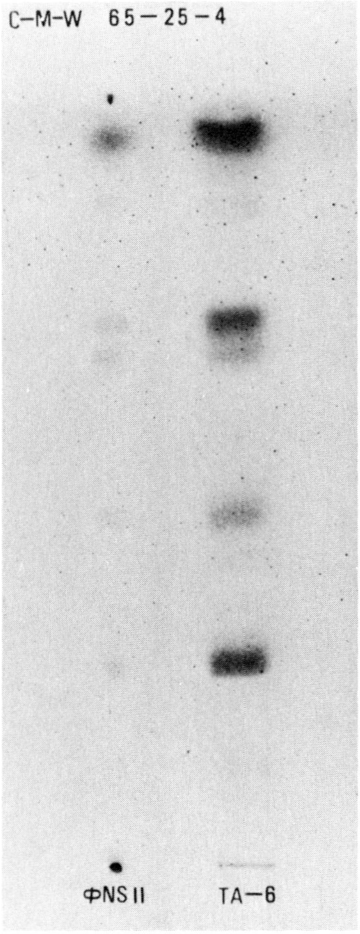

FIGURE 6. Complex lipids of phage ϕ*NS11 and* B. acidocaldarius *TA-6 analyzed by thin layer chromatography.*

incorporation from shikimate into the cyclohexyl fatty acids found in the Italian strain of *B. acidocaldarius*. Thus, studies with isotope labeled precursors showed that ω-cyclohexyl fatty acids are synthesized from glucose through shikimate and cylcohexyl carboxylate as shown in Fig. 8.

Many bacterial fatty acids, such as iso or anteiso-branched acids, cyclopropane fatty acids, 10-methyl-branched acids, etc., are known to be derived from the corresponding amino acids, i.e., valine, isoleucine, leucine and methionine. Thus, the fatty acid composition of bacteria is linked to their amino acid metabolism (Lennarz, 1966). However, the metabolism of ω-cyclohexyl fatty acids is related to the

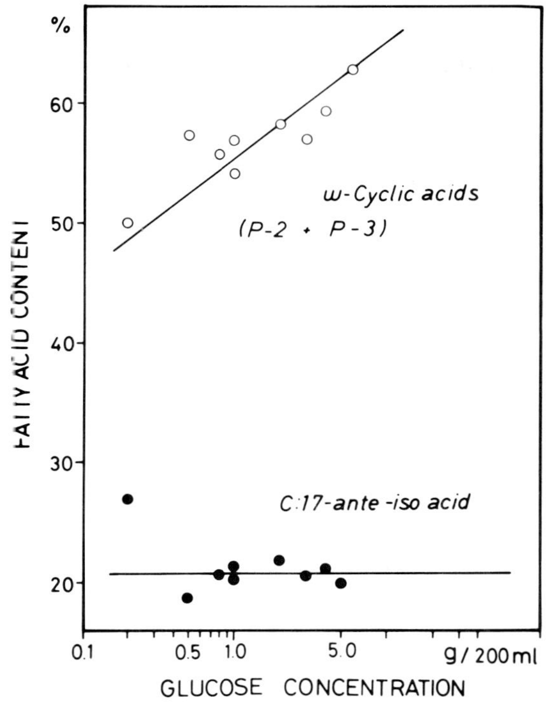

FIGURE 7. Effect of glucose concentration on the biosynthesis of fatty acids. ω-cyclic acids: % area of 11-cyclohexyl undecanoate plus 13-cyclohexyl tridecanoate. C:17 anteiso acid: 14-methyl hexadecanoate. Taken from Oshima and Ariga (1975) and reproduced with permission from The American Society of Biological Chemists.

metabolism of glucose, not to amino acids. This may represent an adaptation of acidothermophilic bacteria to life in hot springs, where the amino acid nutrient content is low.

INTRACELLULAR pH

The intracellular pH of *B. acidocaldarius* TA-6 cells was estimated using [^{14}C]-5,5-dimethyl-2,4-oxazolidinedione (DMO). Extracellular pellet water was measured using [^{3}H]-inulin. The estimated pH was about 6 when cells were suspended in 10 mM glycine-HCl buffer of pH 2.6 (Oshima *et al.*, 1977). The intracellular pH was not markedly changed by the

TABLE 2

Incorporations of Radioactivity into Acyclic and Cyclic Fatty Acids

Precursor	*Specific activity (dpm/mol) x 10^{-9}*			
	15:anteiso	*17:anteiso*	*11:cyclic*	*13:cyclic*
[1-^{14}C]Acetate	1880	1620	561	298
[U-^{14}C]Glucose	773	769	658	447
[U-^{14}C]Shikimic acid			1870	1250
[1-^{14}C]Cyclohexyl carboxylic acid			4750	2510

Dpm per mol of fatty acids determined with stearic acid as an internal standard. Taken from Oshima and Ariga (1975) and reproduced with permission from The American Society of Biological Chemists.

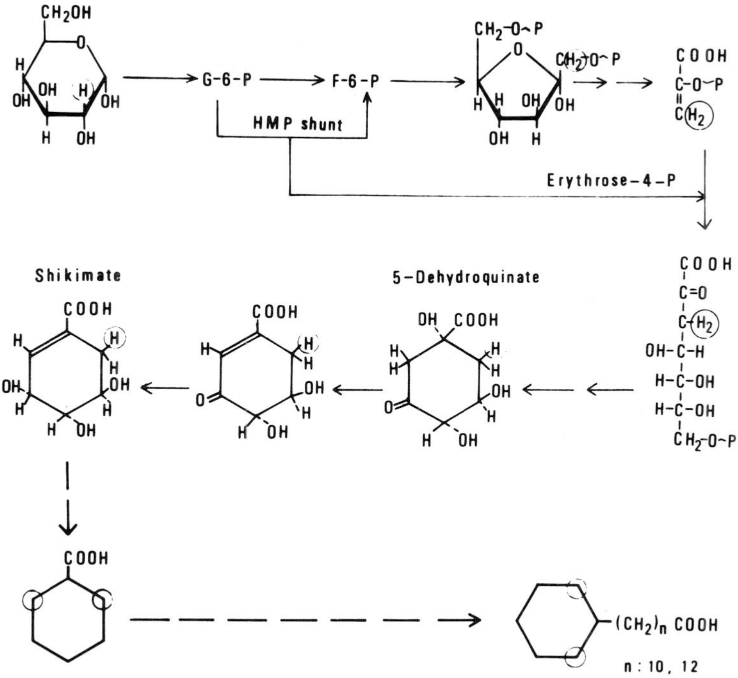

FIGURE 8. Scheme of biosynthesis of ω-cyclohexyl fatty acids. Circled protons indicate positions labeled by the deuterium of [2-D]glucose or [2-D]shikimic acid. Taken from Oshima and Ariga (1975) and reproduced with permission from The American Society of Biological Chemists.

addition of uncouplers, such as CCCP or FCCP (Table 3). The addition of other inhibitors of energy metabolism, such as gramicidin A, sodium azide or potassium cyanide had no effect on the intracellular pH. The pH gradient is maintained even if respiration or oxidative phosphorylation is inhibited. Experiments using DMO in the presence of inhibitors suggest that the mechanism of the pH gradient across the membrane is a static and not a dynamic one, such as a hydrogen ion pump coupled with energy metabolism.

The intracellular pH of the phage particle ϕNS11 must also be neutral since the double stranded DNA molecules in the phage can easily be depurinated at pH 3, 60°C, conditions which are optimal for reproduction of the phage (Sakaki *et al.*, 1977).

TABLE 3
Intracellular pH of B. acidocaldarius *TA-6 cells suspended in 10 mM Glycine-HCl Buffer, pH 2.6, in the Absence and Presence of Uncouplers*

Uncoupler Added	*Experiment*	
	I	*II*
None	6.3	6.2
CCCP (200 μg/ml)	6.1	
FCCP (2 μg/ml)		6.0

POSSIBLE FUNCTION OF THE MEMBRANE LIPIDS

An enzyme (phosphatase) in the periplasm of *B. aciocaldarius* cells was active at acidic pH and was resistant to the hot and acidic environment. These facts suggest that the cytoplasmic membrane rather than the cell wall of the bacillus may play a special role in maintaining the hydrogen ion gradient. Since the intracellular pH is kept neutral in *B. acidocaldarius* and phage ϕNS11, biomembranes containing large amounts of ω-cyclohexyl fatty acids may have important properties which maintain the internal pH at neutrality.

Liposomes prepared from the lipid fraction of *B. acidocaldarius* TA-6 showed dramatic swelling above pH 5 or shrinking below pH 2 in assays carried out at the same salt concentration (Fig. 9). This phenomenon indicates that liposomes are stable within the pH range from 2 to 5, corresponding to the pH range at which TA-6 cells can grow. Further studies are needed in order to clarify the function of the lipid in acido-thermophilic biomembranes.

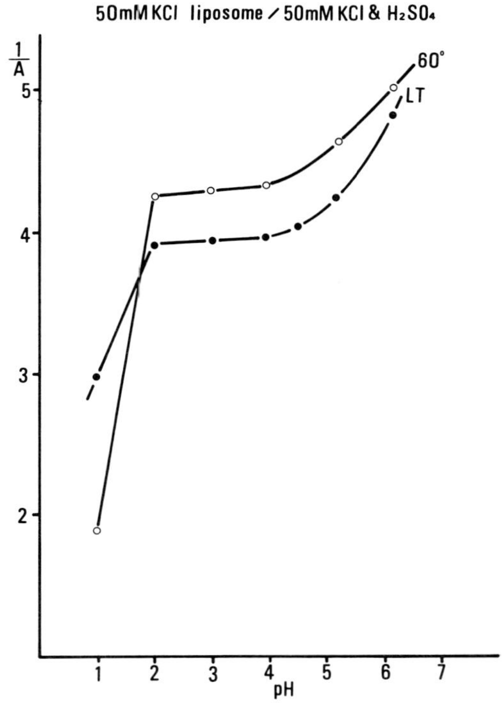

FIGURE 9. Relationship between the reciprocal of the absorbance at 450 nm of liposome systems and the pH of the medium. LT: recorded at room temperature.

REFERENCES

Bradley, D. E. and Rutherford, E. L. (1975). Can. J. Microbiol. *21*, 152.

Brock, T. D., Brock, K. M., Belly, R. T. and Weiss, R. L. (1972). Arch. Mikrobiol. *84*, 54.

Darland, G., Brock, T. D., Samaonoff, W. and Conti, S. F. (1970). Science *170*, 1416.

Darland, G. and Brock, T. D. (1971). J. Gen. Microbiol. *67*, 9.

De Rosa, M., Gambacorta, A., Minale, L. and Bu'Lock, J.D. (1971). Chem. Commun. *1971*, 1334.

De Rosa, M., Gambacorta, A., Minale, L. and Bu'Lock, J. D. (1972). Biochem. J. *128*, 751.

De Rosa, M., Gambacorta, A. and Bu'Lock, J. D. (1974). J. Bacteriol. *117*, 212.

Espejo, R. T. and Canelo, E. S. (1968). Virology *34*, 738.

Furuya, T., Nagumo, T., Itoh, T. and Kaneko, H. (1977). Agric. Biol. Chem. (Japan) *41*, 1607.

Lennarz, W. J. (1966). In *Lipid Metabolism in the Bacteria*, Adv. in Lipid Res. *4*, 175, Academic Press, New York.

Oshima, T., Arakawa, H. and Baba, M. (1977). J. Biochem. *81*, 1107.

Oshima, M. and Ariga, T. (1975). J. Biol. Chem. *250*, 6963.

Sakaki, Y. and Oshima, T. (1976). Virology *75*, 256.

Sakaki, Y., Yamada, K., Oshima, M. and Oshima, T. (1977). J. Biochem. *82*, 1451.

Sakaki, Y., Oshima, M., Yamada, K. and Oshima, T. (1977). J. Biochem. *82*, 1457.

Uchino, F. and Doi, S. (1967). Agr. Biol. Chem. *31*, 817.

Vidaver, A. K., Koshi, R. K. and Van Etten, J. L. (1973). J. Virol. *11*, 799.

THE INFLUENCE OF GROWTH TEMPERATURE AND LIPID STATE ON THE PLANAR DISTRIBUTION OF LIPIDS AND PROTEINS IN *BACILLUS STEAROTHERMOPHILUS* MEMBRANES*

Alfred F. Esser

Research Institute of Scripps Clinic
La Jolla, California

INTRODUCTION

Most of the lipids in cell membranes are present in the form of a bilayer and are known to undergo phase transitions and phase separations.[1] The temperatures at which these changes occur depend upon (a) the chemical composition of

**This is publication no. 1454 from the Research Institute of Scripps Clinic.*

[1]Throughout this text I will use the term "phase transition" to indicate temperature dependent changes in the state of matter, i.e., fluid-solid transitions, and "phase separations" to indicate formation of domains having either different compositions or a different state of matter, i.e., mixing and demixing. It is clear that both phenomena are interdependent.

Abbreviations

ESR, electron spin resonance; DSC, differential scanning calorimetry; DTA, differential thermal analysis; doxyl, 4'4'-dimethyloxazolidine-N-oxyl; PBS, 150 mM NaCl, 5.0 mM Na_2HPO_4, 0.5 mM $MgCl_2$, 1.8 mM $CaCl_2$, pH 7.2; NBD, 4-nitrobenzo-2-oxa-1,3-diazole. PC, phosphatidylcholine; PE, phosphatidyl ethanolamine; PG, phosphatidylglycerol; CL, cardiolipin.

ISBN 0-12-268250-5

the phospholipids, e.g., the length of the fatty acid chains, the degree of unsaturation, the presence of methyl- or cyclopropyl sidechains, the charge and size of the head group, and (b) the presence of neutral lipids such as cholesterol, and other substances such as ions and proteins (1, 2). Because so many different compounds are present in even the simplest biological membrane, it is obviously impossible to construct an exact phase diagram that encompasses all possible phases and transitions. For this reason, the importance of phase transitions and phase separations to the function of biological membranes remains to be established. Nevertheless, recent experiments strongly suggest that phase changes might serve as a potent way to amplify small signals into a large effect on membranes (3) and play a determining factor in the outcome of immunologic reactions occurring at the cell surface (4).

Insight into the significance of thermotropic phase changes can be gained by investigating the response of poikilothermic organisms to changes in environmental temperature. This approach has been used to establish a close correlation between membrane function and phospholipid composition in microorganisms. Either by varying the growth temperature and subsequently determining lipid changes or by growing cells on defined fatty acid mixtures and then determining minimal and maximal growth temperatures, several investigators (reviewed in 5, 6) have shown the direct dependence of physiological functions on lipid melting characteristics. The general consensus at present seems to be that a minimum amount of lipids in the liquid-crystalline state is required for growth (7) of most microorganisms (the exception perhaps being *Halobacteria* (8), thereby setting a limit for the minimal growth temperature. Whether the maximal growth temperature is also defined by lipid melting points, as was suggested half a century ago by Heilbrunn (9) and Bělehrádek (10), remains in dispute (11). Obviously, the melting points of pure lipids are not important here but rather the temperatures at which fully hydrated lipid layers undergo a cooperative transition from a gel ("solid") to a liquid-crystalline ("fluid") state are of significance (12). Using thermal analysis (DTA) to measure this transition, McElhaney (7) found that *Acheloplasma laidlauwii* cells grow within as well as above the temperature of phase transition. Using the same technique Baldassare *et al.* (13) obtained similar results for *Escherichia coli*, as did Schechter *et al.* (14) using X-ray diffraction. Both DTA and X-ray diffraction measure hydrocarbon chain transitions that result from rotations around C-C bonds and, thereby, provide a direct, quantitative evaluation of the fraction of the paraffin chain in the ordered (all *trans*) and disordered (*gauche*) conformation.

In addition, a variety of spectroscopic probe techniques can be used to gain similar information. Overath and coworkers (15, 16) made extensive use of the fluorescence label technique, and Fox and associates (17, 18) used the ESR spin label technique to detect phase transitions in *E. coli* membranes. The results of both groups generally agree with McElhaney's conclusions (7) that cells can grow above the phase transition temperature, although unsupplemented wild type cells usually grow within. Results with the spin label technique in particular, focused attention on the point that, in addition to phase transitions, one must include phase separations in any discussion on structure-function relationships in membranes. DTA and X-ray diffraction frequently give no information on the number of phases that are present simultaneously at a given temperature and on whether these phases are miscible or cocrystallize. While the solid ↔ fluid transition does not necessarily determine the upper temperature limit at which a given organism grows it is nevertheless conceivable that phase separations might. When microorganisms adjust to different temperatures, three changes in lipid structure (unsaturation, branching, head groups) all of which strongly influence phase separations, are easily detectable. For this reason we have recently postulated that microphase-separations within membranes constitute a prime mechanism for high (or low) temperature adaptation of thermophilic bacteria (19).

Our hypothesis is based mainly on studying membranes from wild type cells of *Bacillus stearothermophilus* and from a temperature sensitive mutant. Using the spin label technique and employing a probe (5-doxyl stearate) that resides within the hydrocarbon phase but close to the polar head group region, we obtained evidence for a strong homeoviscous adaptation mechanism in this organism. In addition, Arrhenius plots indicated the beginning of lateral lipid phase separations at or near the growth temperature. We had suggested initially (20) that the simultaneous presence of fluid and solid lipid phases is necessary for growth and that the formation of only fluid lipids interferes with growth. This proposal was modified somewhat (19) when preliminary DSC measurements indicated that the mutant could grow above the solid-fluid phase transition temperature. The spin label data were still interpreted to indicate that lateral lipid phase separations had occurred, but we concluded that this state could result from either solid-fluid or fluid-fluid phase separation. A subsequent detailed DTA analysis by McElhaney and Souza (21) confirmed that the mutant was capable of growing above the phase transition temperature but that, within the error limits of the method, the growth temperature of the wild type correlated very well with the beginning of the phase transition.

Additional evidence for the presence of lipid phase separations in *B. stearothermophilus* membranes at the growth temperature has been obtained recently by freeze-fracture electron microscopy (Halverson, Esser, and Souza, submitted for publication) and by fluorescence label spectroscopy (Esser, Mack, and Souza, to be published). I would now like to describe these experiments in some detail.

MATERIALS AND METHODS

Bacteria

The source of the thermophilic bacillus (YTG-2), identified as a strain of *B. stearothermophilus* and designated as WT, and the selection of its temperature-sensitive mutant, designated TS-13, have been described (22). Growth conditions and methods for the preparation of spheroplasts are given elsewhere (20). For crosslinking experiments spheroplasts were suspended in PBS, equilibrated at the desired temperature and mixed with an equal volume of 5% glutaraldehyde at the same temperature. After 30 min incubation time, the fixed spheroplasts were washed three times with 150 mM ammonium acetate and used directly for electron microscopy work.

Freeze-fracture electron microscopy

Unfixed spheroplasts, in 150 mM ammonium acetate, were heated to the quench temperature and equilibrated at that temperature for 15 min. Small droplets (approx. 2 μl) were pipetted into long, hollow stem gold planchets (Balzers) resting on a copper block maintained at the same temperature. The temperature of a droplet in a reference planchet was monitored with a fine wire thermocouple. The test planchets were dropped rapidly into partially solidified Freon 22 (du Pont), then transferred to liquid nitrogen, in which they were stored until fractured in a Balzers BA360M freeze-etching device. Fracturing was carried out at -110°C and etching at -100°C. Replicas were floated off and cleaned by standard methods. After mounting on 400 mesh grids the replicas were then examined and photographed in an Hitachi 11A electron microscope at a direct magnification of 30,000 X. Glutaraldehyde-fixed spheroplasts were also suspended in 20% glycerol as a cryoprotectant and frozen from the desired temperatures. The nomenclature of Branton *et al.* (23) is used to identify the different planes.

Fluorescence Label Spectroscopy

The following fluorescent NBD-lipid derivatives (obtained from Avanti, Birmingham, Ala. or Molecular Probes, Roseville, Minn.), were used to label spheroplasts.

I

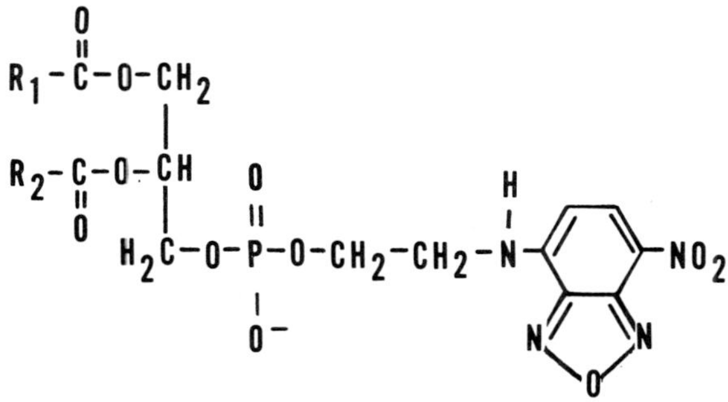

II

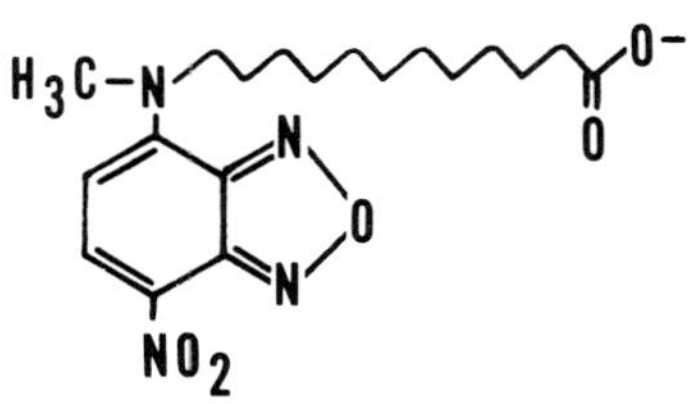

Steady state fluorescence was measured with an Aminco-Bowman spectrofluorometer with a ratio attachment and thermostatable cell holder. The temperature of the sample was measured continuously within the cuvette by using a T type thermocouple that drove the X-axis of an X-Y recorder (Houston). Fluorescence intensity was monitored on the Y-axis; for further details see Overath and Träuble (15).

RESULTS AND DISCUSSION

Table 1 lists the various phase transition temperatures of wild type membranes and of the TS-13 mutant recorded by spin labels and by DTA. Although both techniques recorded similar values for high temperature transitions (T_h) that were very close to or identical with the cells' growth temperature, considerable differences are apparent for the low temperature (T_ℓ) transition.

In order to understand the reason for the observed

TABLE 1

Summary of Lipid Transition Temperatures in Membranes of B. stearothermophilus *WT and TS-13 Grown at Different Temperatures as Detected by ESR Spin Label Experiments*[a] *and by DTA*[b]

Cell Type	Growth Temperature (°C)	High temperature transition (T_h) ESR	DTA	Low temperature transition (T_ℓ) ESR ($T_{\ell-1}$)	DTA $T_{\ell-2}$)
WT	42	42	40	36	15
	52	52	52	40	20
	58	58	58	50	25
	65	65	65	57	24
TS-13	42	47	41	42	20
	52	58	40	52	20
	58	58	40	52	20

[a]*Taken from Ref. 20.*

[b]*Taken from Ref. 21 (reproduced with permission from Elsevier/North-Holland Biomedical Press).*

differences in T_ℓ, we have performed freeze-fracture studies on membranes quick-frozen or crosslinked at these various temperatures. Replicas of wild type spheroplasts quenched from several degrees above the growth temperature show fracture faces that are covered with densely packed particles (Fig. 1a). If quenching occurs at or near the growth temperature (±1.5°C), a slight but distinct aggregation of intramembranous particles becomes apparent (Fig. 1b). This aggregation is more pronounced about 7-10°C below the growth temperature (Fig. 1c), the temperature at which, according to the results of spin label tests, the lateral phase separation is completed (T_ℓ-1). After further decrease in freezing temperature to the midpoint temperature (T_m), as measured by DTA, fracture faces appear that display smooth domains separated by very densely packed network-like regions (Fig. 1d). The total areas covered by smooth regions and particulate regions are approximately equal. When freezing is performed at a temperature at which all lipids are in the solid state (below T_ℓ-2), large smooth areas are visible (Fig. 1e) within the membrane. Although most of the particles are squeezed into one hemisphere, the smooth areas still contain a significant number. Particle segregation can also be observed on replicas of freeze-fractured whole cells (Fig. 1g, h), which proves that this phenomenon is not simply an artifact of spheroplast preparation. Furthermore, quick-freezing at the desired temperature, or crosslinking at the temperature followed by freezing, yields similar pictures. Although the photographs shown here were taken on replicas of spheroplasts prepared from WT cells grown at 65°C, similar results were obtained with WT cells grown at 58°C and 52°C.

Several investigators have suggested that intercalated particles aggregate upon solidification or separation of lipids (14, 24, 25). As lipids solidify, proteins are squeezed out since they would interfere with close hexagonal packing of ordered hydrocarbon chains (i.e., crystal formation). When lipids separate, proteins concentrate in a lipid region for which they show preferential affinity, thereby depleting other lipid domains. The behavior of ATPase from sarcoplasmic reticulum in liposomes exemplifies the solidification events (24), whereas the human myelin basic protein with its preference for acidic phospholipids is an example of the separation sequence (26). In both cases the total energy change for the final state must outweigh the entropy that would be gained by random mixing. The pattern shown in Fig. 1e suggests that the crystal packing energy of the lipids in *B. stearothermophilus* is sufficient to drive the majority of proteins into a separate phase, however some remain scattered.

An interesting pattern of banding is observed on

spheroplast samples that have been stored for prolonged periods at temperatures below $T_{\ell-2}$ prior to freezing and fracturing (Fig. 1f). The repeat period in these banded structures is in the range of 300-500 Å. Repeat distances of 120-500 Å, associated with the $P_{\beta'}$ lipid phase (nomenclature of Tardieu *et al.* (27)) were observed earlier in freeze-fracture electron micrographs of saturated lecithins (25, 28). The transformation of $P_{\beta'}$ to L_{α} correlated with the so-called "pretransition" observed by using DTA (29). The pretransition is not observed in pure phosphatidylethanolamines (PE) or in mixtures of PE and phosphatidylcholine (PC) with more than 10% PE (30), leading Janiak *et al.* (31) to hypothesize that the $P_{\beta'}$ structure is associated only with a tilted configuration of saturated lecithins. *B. stearothermophilus* does not synthesize choline-containing lipids but has about 20% PE, the remainder being cardiolipin (CL) and phosphatidylglycerol (PG)(32). To make the aforementioned hypothesis compatible with the result shown in Fig. 1f, one has to assume that in its banded phase, the membrane is free of PE and that CL and/or PG are also capable of forming $P_{\beta'}$-like structures. Indeed, Papahadjopoulos and coworkers (33, 34) have reported the occurrence of a "pretransition" and banded freeze-fracture faces in pure dipalmitoyl-PG liposomes.

When spheroplasts from WT cells grown at 42°C or from TS-13 cells grown at any temperature were quenched from various temperatures and examined, the large segregations such as shown in Fig. 1 could not be demonstrated. Particularly in the mutant membrane little segregation occurred. Verkleij and Ververgaert (25) have listed three possible explanations for the agsence of particle aggregation: (i) protein-protein interactions are strong enough to prevent concentration by lipids, (ii) protein-lipid interactions prevent the formation of large phases, and (iii) some lipids do not fuse into crystals because they remain loosely packed below the transition point. These authors postulated that branched lipids belong to the latter class of lipids, since incorporation of about 60% of branched fatty acids into *Acheloplasma laidlawii* membranes prevented aggregation of

FIGURE 1. Freeze-fracture faces of membranes from B. stearothermophilus *WT grown at 65°C. Spheroplast samples were quick-frozen from (a) 70°C, (b) 65°C, (c) 58°C, (d) 54°C, (e) 20°C, (f) 20°C after holding the spheroplasts for several days at 4°C, (g) and (h): whole cells frozen at 20°C. Bars in (a) and (e) apply to pictures a through f and in (g) to pictures g and h and correspond to 500 mμ. Arrow indicates direction of shadowing: CA = cell wall.*

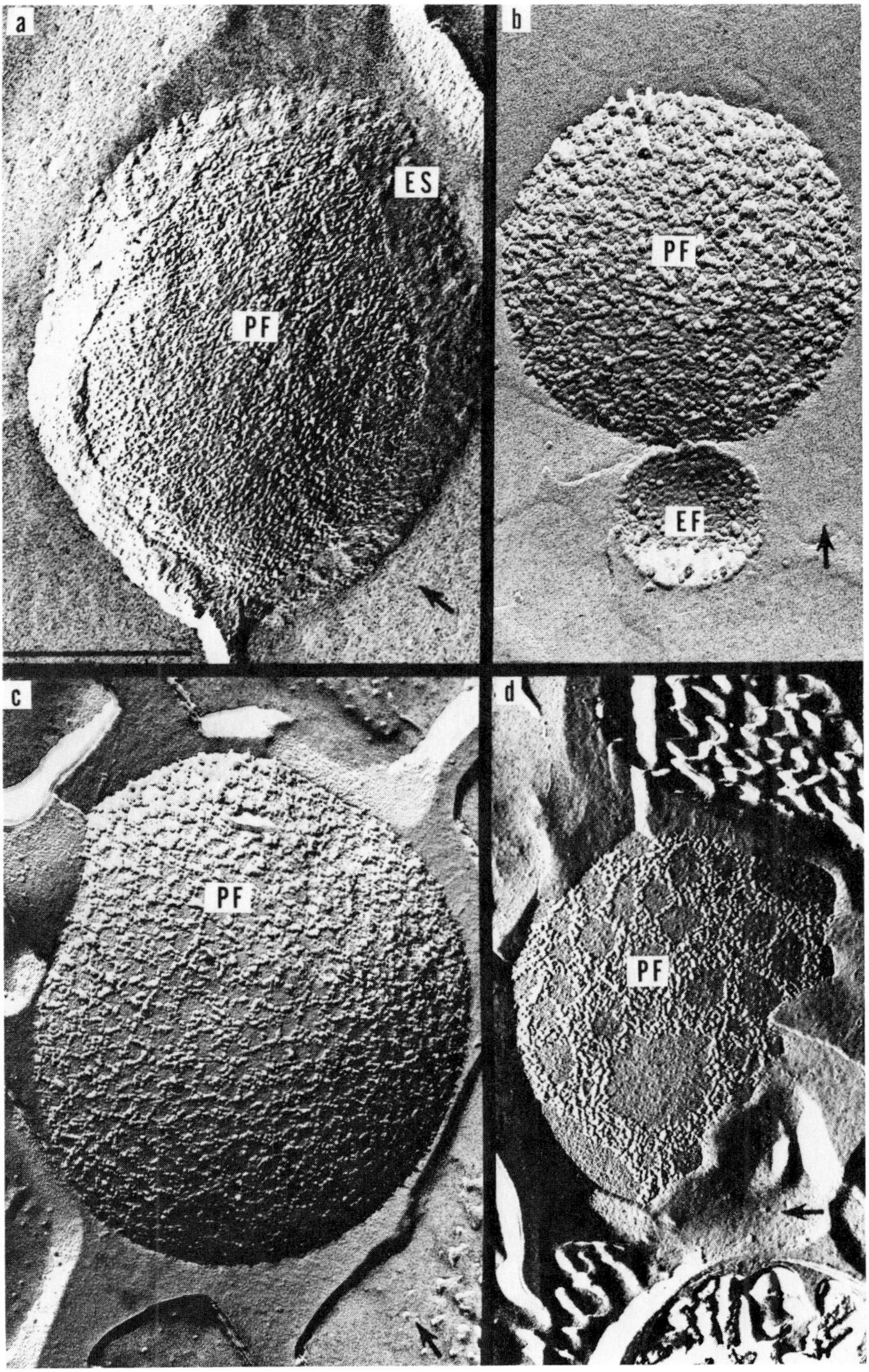
a
ES
PF
b
PF
EF
c
PF
d
PF

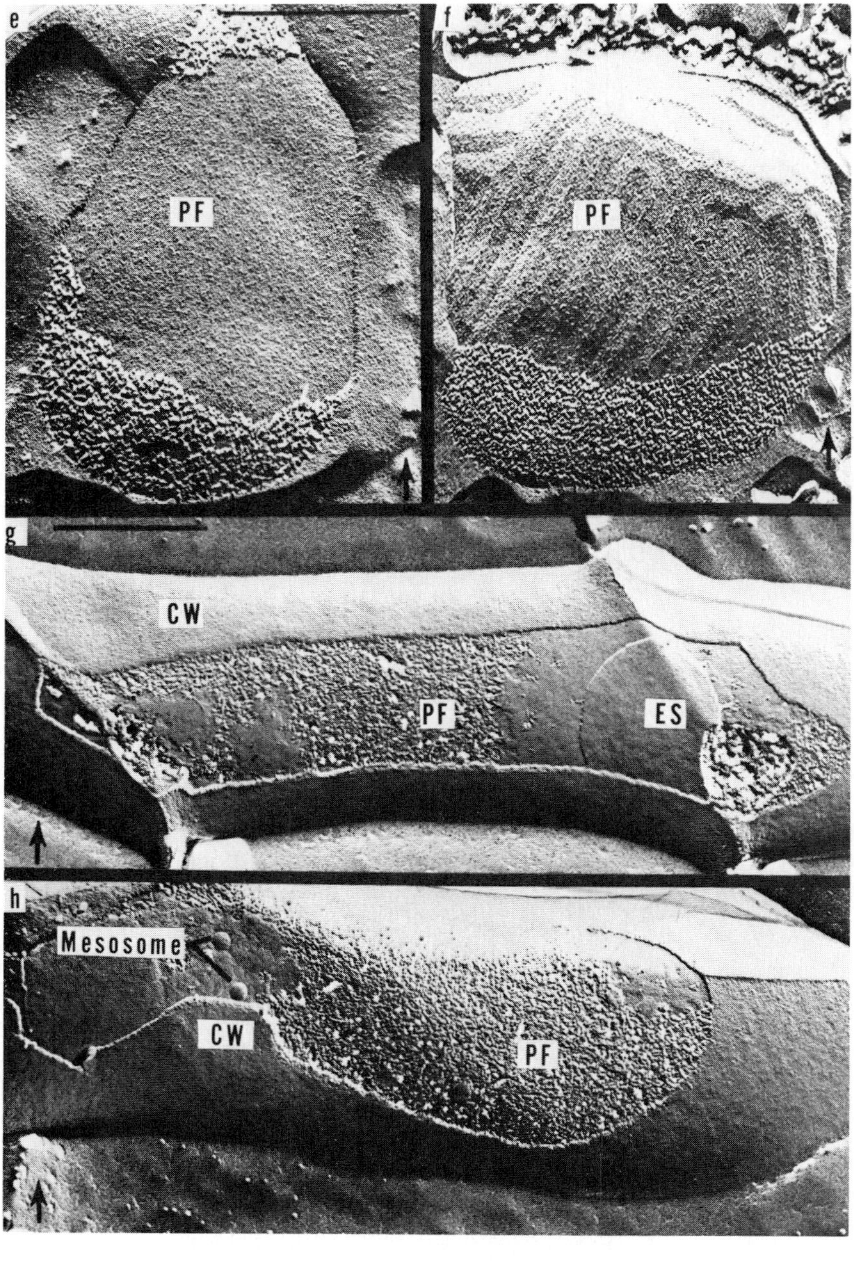

FIGURE 1 (continued). e, f, g and h.

particles. Our results are in excellent agreement with this proposal. As shown in Table 2, the percentage of branched fatty acids in wild type cells increases with decreasing growth temperature, reaching 63% at 42°C, while in the mutant the trend is reversed, but the total amount never falls below 58%. For this reason we could not use freeze-fracture electron microscopy to assess the extent and temperature dependence of lipid phase separation in the mutant.

A comparison of the freeze-fracture patterns shown in Fig. 1 and the lipid transitions tabulated in Table 1 reveals some interesting correlations. When spheroplast membranes from cells grown at 65°C are lowered in temperature to 58°C, only about 10% of their lipids are in the solid state as determined by DTA (see also Fig. 3 in ref. 21). Yet the fracture faces of membranes quenched or crosslinked at that temperature show smooth areas comprising far more than 10% of the total membrane area. What are the forces that keep the particles segregated even when lipids are 90% fluid? A comparatively small temperature increase of about 10°C results in complete randomization, making protein-protein interactions an unlikely explanation. Consideration of the fact that spin label results agree well with DTA data in determining transitions in *isolated, protein-free* lipids (21), but not in membranes, suggests that lipid-lipid interactions should also be dismissed as the impetus for particle separation, leaving lipid-protein interactions as the most likely cause. Apparently then, proteins cause some form of lipid clustering even when almost all lipids are fluid. These clusters are quite large, strongly suggesting that the interactions must be of long range order. The spin label 5-doxyl stearate has the potential to sense such clustering (1) which, when associated with little or no change in enthalpy, could remain unnoticed by DTA.

In addition to spin labels, fluorescence labels are also capable of recording such clustering if the chromophore is located in or near the polar-nonpolar lipid interface (Fig. 2). While the chromophore of label I, located deep in the hydrocarbon phase, records transitions that are in agreement with the DTA data, label II exhibits small changes in fluorescence intensity at the characteristic temperatures recorded by 5-doxyl stearate.

Summarizing these results, we can say that *B. stearothermophilus* WT and especially the mutant TS-13 are capable of growing at a temperature at which most or all of the lipid hydrocarbon chains are in a fluid state. This raises the question: if all the lipids are fluid, can "excess fluidity" hamper growth and function? It should be remembered that even in fluid lipids, as defined by DTA results, interactive forces can prevent membrane constituents from

TABLE 2

Comparison of Branched, Straight Chain, and Unsaturated Fatty Acids of B. stearothermophilus *WT and TS-13*[a]

	TS-13			*WT*			
Growth Temperature:	*42°C*	*52°C*	*58°C*	*42°C*	*52°C*	*58°C*	*65°C*
Fatty Acid Group							
Branched	59	59	65	63	53	46	40
Straight	26	33	33	25	40	46	57
Unsaturated	15	8	2	12	7	8	3

[a]*Values are given as percentages of total fatty acids (taken from Ref. 22 and reproduced with permission from Springer-Verlag).*

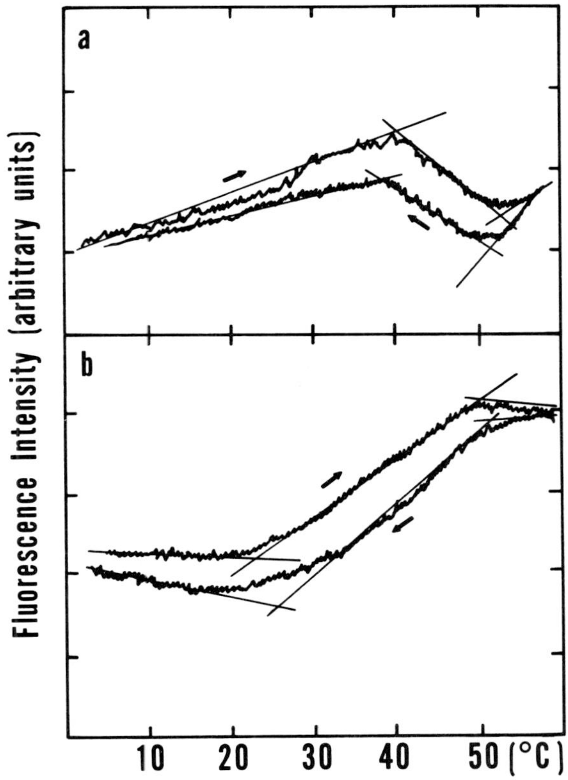

FIGURE 2. Fluorescence intensity of NBD-derivatives incorporated in spheroplasts prepared from WT cells grown at 52°C. (a) Label I (b) Label II.

moving in a completely unrestricted fashion. The nature of the forces that keep some lipids in clusters or domains have been described (1, 33). Further increase in temperature above the lipid melting temperature weakens these forces to a point at which randomization sets in, resulting in loss of function. For this reason, we proposed in answer to the last question, that microorganisms can adapt their membranes to higher temperatures as long as they can synthesize a lipid mixture that does not form an unrestrictive, randomizing two-dimensional fluid in the presence of all other membrane constituents such as water, proteins and ions. Obviously, there are several means by which such restrictions can be created, as Chapman *et al*. (33) have discussed, and such processes can operate in a fluid or partially solid membrane.

The main question that has not been addressed thus far is: what regulatory mechanisms that operate in the thermophilic bacilli during membrane synthesis allow these cells to adjust to extreme variations in temperature? I am not aware of any publications that have dealt directly with this question. Although the literature is flooded with reports describing the changes in lipid composition of thermophiles as growth temperature varies, none identifies the enzyme(s) responsible for these changes. This question has been studied in far more detail in the eukaryotic organism *Tetrahymena*. Thompson and Nozawa (34) recently summarized their extensive studies on temperature adaptation in this organism. Their elegant studies provide convincing evidence that it is the fatty acid desaturase, a membrane bound enzyme, which functions as the "thermometer" in these cells.

However, this enzyme is probably not responsible for adaptative mechanisms in *E. coli*. Okuyama *et al*. (35) made this point after extending earlier studies of Sinensky (36) and Cronan (37) and finding that the fatty acid synthetase complex in particular, and to a lesser degree the acyl transferase system, is most responsive to temperature changes. But the latter experiments cannot be interpreted as clearly as those with *Tetrahymena*. This caveat is necessary since the fatty acid synthetase that showed a strong dependence of activity on growth temperature interacts very little with membranes, whereas no changes in activity could be detected between membrane bound acyl transferases isolated from cells grown at different temperatures.

With respect to the thermophilic bacilli, it will be even more difficult to gain insight into putative control mechanisms. Kaneda (38) recently reviewed aspects of fatty acid synthesis for numerous species of the genus *Bacillus*. Not only do bacilli produce straight chain saturated and unsaturated fatty acids, they also prefer to synthesize branched chain fatty acids. As indicated in Table 1, the relative concentration of all three types of fatty acids varies considerably with temperature adaptation. The question of how the various enzyme systems involved in the synthesis and degradation of these three series of fatty acids respond to temperature variation will provide a stimulating and challenging problem for future research.

ACKNOWLEDGEMENTS

Work in the author's laboratory was supported by NASA's Ames Research Center (NCA2-OR253-601) and the Research Corporation. The author is the recipient of an Established Investigatorship from the American Heart Association (No. 76-225). I am grateful to my collaborators Craig Halverson, Mel Mack and Kenneth Souza whose efforts made this report possible, and to Dennis Chapman for inspiration, advice and helpful comments, and to Phyllis Minick whose knowledge of literary style far exceeds mine.

REFERENCES

1. Lee, A. G. (1977). Biochim. Biophys. Acta *472*, 237.
2. Wu, S. H. and McConnell, H. M. (1975). Biochemistry *14*, 847.
3. Träuble, H. and Eible, H. (1975). In *Functional Linkage in Biomolecular Systems* (F.O. Schmitt, D. M. Schneider and D. M. Crothers, eds.), p. 59, Raven Press, New York.
4. Brûlet, P. and McConnell, H. M. (1977). Biochemistry *16*, 1209.
5. Machtiger, N. A. and Fox, C. F. (1973). Annu. Rev. Biochem. *42*, 575.
6. Cronan, J. E. Jr. and Gelman, E. P. (1975). Bacteriol. Rev. *39*, 232.
7. McElhaney, R. N. (1974). J. Mol. Biol. *84*, 145.
8. Esser, A. F. and Lanyi, J. K. (1973). Biochemistry *12*, 1933.
9. Heilbrunn, L. V. (1924). Am. J. Physiol. *69*, 190.
10. Bělehrádek, J. (1931). Protoplasma *12*, 406.
11. McElhaney, R. N. (1976). In *Extreme Environments: Mechanisms of Microbial Adaptation* (M.R. Heinrich, ed.), p. 255, Academic Press, New York.
12. Chapman, D. (1967). In *Thermobiology* (A. H. Rose, ed.), p. 123, Academic Press, New York.
13. Baldassare, J. J., Rhinehart, K. B., and Silbert, D. F. (1976). Biochemistry *15*, 2986.
14. Schechter, E., Letellier, L., and Gulik-Krzywicki, T. (1974). Eur. J. Biochem. *49*, 61.
15. Overath, P. and Träuble, H. (1973). Biochemistry *12*, 2625.
16. Thilo, L. and Overath, P. (1976). Biochemistry *15*, 328.
17. Linden, C. D., Keith, A. D., and Fox, C. F. (1973). J. Supramol. Struc. *1*, 523.

18. Linden, C. D., Wright, K. L., McConnell, H. M., and Fox, C. F. (1973). Proc. Natl. Acad. Sci. USA *70*, 2271.
19. Esser, A. F. and Souza, K. A. (1976) In *Extreme Environments: Mechanisms of Microbial Adaptation* (M. R. Heinrich, ed.), p. 283, Academic Press, New York.
20. Esser, A. F. and Souza, K. A. (1974). Proc. Natl. Acad. Sci. USA *71*, 4111.
21. McElhaney, R. N. and Souza, K. A. (1976). Biochim. Biophys. Acta *443*, 348.
22. Souza, K. A., Kostiw, L. L., and Tyson, B. J. (1974). Arch. Microbiol. *97*, 89.
23. Branton, D. *et al.* (1975). Science *190*, 54.
24. Kleemann, W. and McConnell, H. M. (1976). Biochim. Biophys. Acta *419*, 206.
25. Verkleij, A. J. and Ververgaert, Th., P.H.T. (1975). Ann. Rev. Phys. Chem. *26*, 101
26. Boggs, J. M., Moscarello, M. A., and Papahadjopoulos, D. (1977). Biochemistry *16*, 5420.
27. Tardieu, A., Luzzati, V., and Reman, F. C. (1973). J. Mol. Biol. *75*, 711.
28. Luna, E. J. and McConnell, H. M. (1977). Biochim. Biophys. Acta *470*, 303.
29. Rand, R. P., Chapman, D., and Larsson, K. (1975). Biophys. J. *15*, 1117.
30. Chapman, D., Urbina, T., and Keough, K. M. (1974). J. Biol. Chem. *249*, 2512.
31. Janiak, M. J., Small, D. M., and Shipley, G. G. (1976). Biochemistry *15*, 4575.
32. Kostiw, L. L. and Souza, K. A. (1976). Arch. Microbiol. *107*, 49.
33. Chapman, D., Cornell, B. A., and Quinn, P. J. (1977). In *Biochemistry of Membrane Transport*, FEBS Symposium No. 42 (G. Semenza and E. Carafoli, eds.), p. 72, Springer Verlag, Berlin.
34. Jacobson, K. and Papahadjopoulos, D. (1975). Biochemistry *14*, 152.
35. Papahadjopoulos, D., Vail, W. J., Pangborn, W. A., and Poste, G. (1976). Biochim. Biophys. Acta *245*, 265.
36. Thompson, G. A. and Nozawa, Y. (1977). Biochim. Biophys. Acta *472*, 55.
37. Okujama, H., Yamada, K., Kameyama, Y., Ikezawa, H., Akamatsu, Y., and Nojima, S. (1977). Biochemistry *16*, 2668.
38. Sinensky, M. (1971). J. Bacteriol. *106*, 449.
39. Cronan, J. E. Jr. (1975) J. Biol. Chem. *250*, 7074.
40. Kaneda, T. (1977). Bacteriol. Rev. *41*, 391.

THE MEMBRANE OF THERMOPHILIC BACTERIUM PS3
1. PURIFIED H^+-ATPASE AND AMINO ACID CARRIER
2. MOLECULAR SPECIES OF PHOSPHOLIPIDS

Yasuo Kagawa
Nobuhito Sone
Hajime Hirata
Masasuke Yoshida

Department of Biochemistry
Jichi Medical School
Minamikawachi, Tochigi

Membrane systems for energy transformation and transport of solutes are highly unstable and purification of membrane proteins responsible for these functions is extremely difficult in the case of mesophiles. The use of thermophiles enabled us to perform the following experiments which had been previously unsuccessful in mesophilic systems (Kagawa, 1977).

1. Reconstitution of pure H^+-ATPase into vesicles capable of H^+ translocation (Sone *et al.*, 1975, 1976) and net ATP synthesis driven by an artificially imposed H^+ gradient (Sone *et al.*, 1977).
2. Crystallization of F_1 (catalytic portion of H^+ ATPase) (Kagawa *et al.*, 1976) and computerized image reconstruction of F_1 (Wakabayashi *et al.*, 1977).
3. Complete reconstruction of F_1 from its five ($\alpha\beta\gamma\delta\varepsilon$) purified subunits (Yoshida *et al.*, 1977a).
4. Identification of channel function of F_o (membrane portion of H^+-ATPase) (Okamoto *et al.*, 1977) and gate function in $\gamma\delta\varepsilon$-complex of F_1 subunits (Yoshida *et al.*, 1977b).
5. Purification of H^+ driven amino acid carrier and its reconstitution into functioning vesicles (Hirata *et al.*, 1976a, b and 1977).

ISBN 0-12-268250-5

6. Determination of molecular species of phospholipids of thermophilic bacterium PS3. Demonstration of energy transformation and transport in the complete absence of unsaturated lipids (Kagawa and Ariga, 1977).
7. Physical studies, such as hydrogen-deuterium exchange, on isolated proteins during energy transfer reactions (Ohta *et al.*, 1978).

SELECTION OF THERMOPHILIC BACTERIUM PS3

In order to study oxidative phosphorylation which is stable, several thermophilic aerobic bacteria were tested (Table 1). All the bacterial strains were isolated from Japanese hot springs and kindly supplied by Dr. T. Oshima of the Mitsubishi-Kasei Institute of Life Sciences, Machida, Tokyo. The ATPase activities of their washed membrane fraction, expressed in units per mg protein at 80°C and 60°C, were compared. Strain PS3 (Fig. 1) was selected as the starting material because of its high membrane ATPase activity, high sensitivity to DCCD (an energy transfer inhibitor) and rapid growth. At 65°C, PS3 doubled every 35 min as determined by absorption at 600 nm of the culture suspension, which contained 0.8% polypeptone, 0.4% yeast extract and 0.3% NaCl, pH 7.0 and was vigorously aerated. The difference spectrum of washed plasma membranes showed characteristic cytochromes and flavins in the electron transport system (Fig. 2). It is noted that the amount of the terminal oxidase (absorption max. at 605 nm) is only 0.6 nmoles/mg protein and is much less than cytochromes *b* (4.0 nmoles/mg) and $c + c_1$ (2.6 nmoles/mg) and flavin (3.1 nmoles/mg). This may be explained by the rapid diffusion of O_2, despite the low O_2 concentration in hot water.

For purification of membrane components, a large scale culture of 6 kg of cell per 1 ton of medium was kindly performed by Kyowa Hakko Co. The cells (500 g, wet weight) were suspended in 4.5 liters of 50 mM Tris-sulfate, pH 8.0, at 36°C and 500 mg of lysozyme (EC 3.2.1.17) were added. The mixture was stirred for 30 min, and then 25 ml of 1 M $MgCl_2$ and 10 mg of DNase 1 (EC 3.1.4.5) were added. After 2 hours, the resulting lysate was centrifuged at 17,000 x g for 20 min. The precipitate was homogenized and washed three times with 1 liter volumes of 50 mM Tris-sulfate, pH 8.0, since thorough washing was essential in order to remove soluble proteins. The washed membrane fraction thus obtained was the starting material for the isolation of the following proteins or lipids.

TABLE 1

ATPase and Other Properties of Thermophilic Aerobic Bacteria Isolated from Japanese Hot Springs

Strain	*PS3*	*HB8**	*T8*	*B822*	*V2*	*BA152***	*TA6***	E. coli *K12*
Hot Spring	*Mine*	*Mine*	*Tama-gawa*	*Beppu*	*Mine*	*Beppu*	*Gosho-gake*	-------
Prefecture	*Shizu-oka*	*Shizu-oka*	*Akita*	*Ohita*	*Shizu-oka*	*Ohita*	*Akita*	-------
Growth Temperature								
Maximum	80°C	85°C	85°C	74°C	70°C	70°C	70°C	48°C
Optimum	65°C	69°C	70°C	65°C	60°C	65°C	65°C	37°C
Spore Formation	(+)	(-)	(+)	(+)	(-)	(+)	(+)	(-)
Gram Staining	(+)	(-)	(+)	(+)	(-)	(+)	(+)	(-)
ATPase: at 80°C	2.25	1.45	1.94	1.70	0.55	----	----	0.0
at 60°C	4.52	6.19	4.48	4.75	2.03	----	----	0.15
(Pi μmoles/mg)								
DCCD sensitivity								
at 80°C	85	3	82	63	52	----	----	0
at 60°C	87	0	84	69	61	----	----	76

*Thermus thermophilus *HB8, ATCC 27634.*

**Bacillus acidocardarius. *From Kagawa, 1976.*

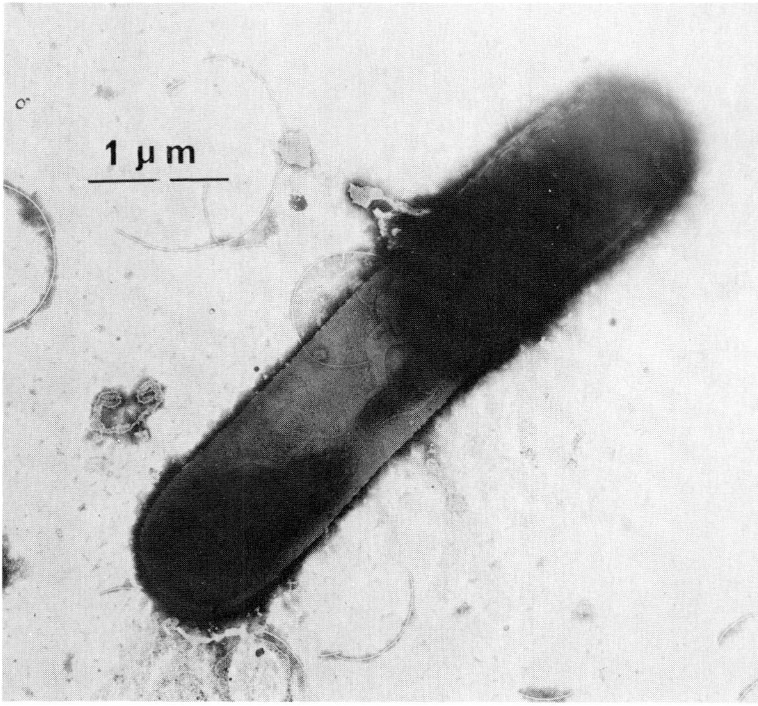

FIGURE 1. Electron micrograph of thermophilic bacterium PS3. Taken from Kagawa (1976) and reproduced with permission from the Wistar Institute Press.

PURIFICATION AND STABILITY OF F_1

The washed membranes (8.7 g) were homogenized in 2 liters of 0.2 mM EDTA, pH 7.4 at 25°C (Yoshida *et al.*, 1975). In other cases, the membranes were extracted with a mixture of chloroform and water and the water layer was collected (Kagawa and Yoshida, 1978). The extract was centrifuged at 105,000 x g for 1 hour and the supernatant fraction was passed through a column of DEAE-cellulose previously equilibrated with Buffer A (50 mM Tris-sulfate, pH 7.8, 0.1 mM EDTA). The column was washed successively with 2.5 liters of Buffer A and 1 liter of 0.1 M NaCl in Buffer A, and then F_1 was eluted with 4 liters of a linear gradient of 0.15 M to 0.35 M NaCl in Buffer A. Fractions containing F_1 were dialyzed against solid ammonium sulfate and the precipitate was dissolved in 22 ml of Buffer A. The solution was then applied to a column (5 x 68 cm) of Sepharose 6B, which had been

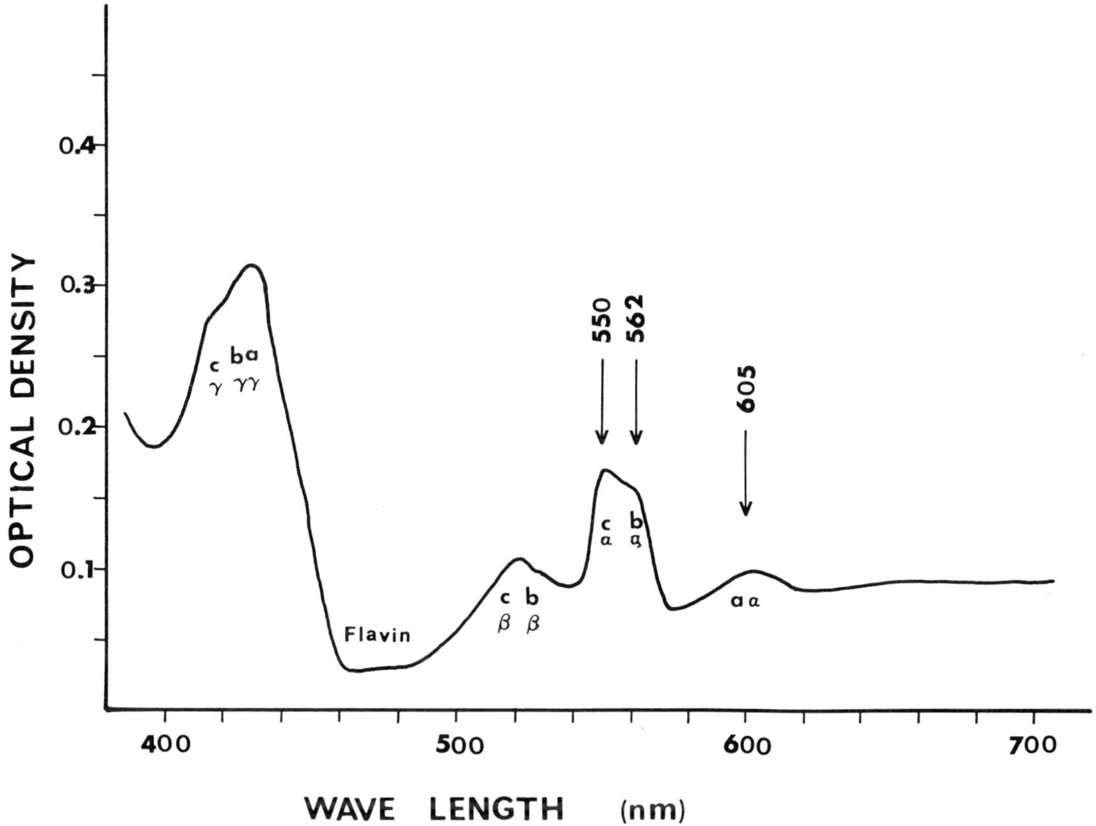

FIGURE 2. Difference spectrum of the washed plasma membrane of thermophilic bacteriam PS3. Sample side is reduced with hydrosulfite and the reference side is oxidized with air. Protein concentration is 8.19 mg/ml of 0.1M phosphate buffer (K salt), pH 7.5, at 25°C.

equilibrated with Buffer A. Fractions containing F_1 were suspended in 55% ammonium sulfate.

A comparison of thermophilic and mesophilic membrane proteins is usually difficult because the mesophilic counterpart is usually not found in a pure state. However, in the case of F_1, similar preparations have been purified from beef heart, chloroplasts and several bacterial membranes. In order to differentiate F_1 from beef heart and thermophilic bacterium PS3, the former is called F_1 and the latter is called TF_1. TF_1 is more resistant to heat than F_1, and TF_1 is not cold labile (Fig. 2). The secondary structure of TF_1 is also resistant to heat as shown in Fig. 3. The CD spectrum of TF_1 was hardly affected by increasing the temperature from 2°C to 75°C, but changes were noted at 81°C, and the extent of the trough decreased markedly at 96°C.

The stabilities of F_1 and TF_1 were compared in the presence of various concentrations of urea (Fig. 4a), chaotropic anions (Fig. 4b), detergents (Fig. 4c) and organic solvents (Fig. 4d). These properties make it convenient to purify and reconstitute F_1. The reason for the stability of F_1 is unknown, but amino acid analysis revealed a higher content of salt bridge-forming amino acid residues (Arg 6.02 in TF_1, 5.49 in F_1; Glu 13.29 in TF_1, 12.54 in F_1) and a very low content of SH (Cys 0.08 in TF_1, 0.35 in F_1). Values in parentheses are expressed as mole% (Yoshida *et al.*, 1975). Helix content at 25°C was 31.5% for TF_1 and 34.5% for F_1. The effect of temperature on the secondary structure of TF_1 is shown in Fig. 5.

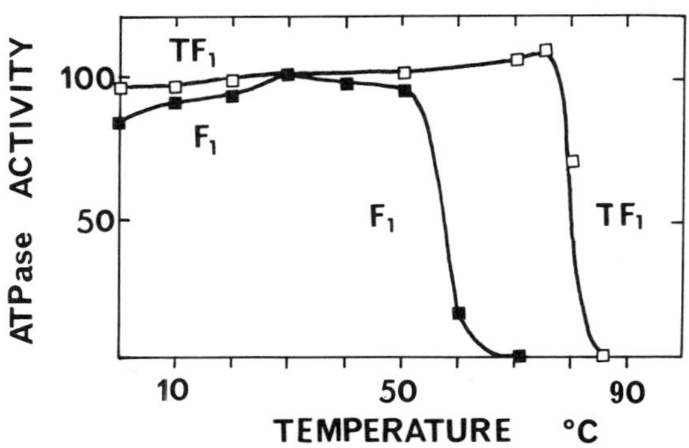

FIGURE 3. Effect of temperature on ATPase activity of beef heart mitochondrial F_1 (F_1) and thermophilic F_1(TF_1). The activity at 30°C is taken as 100%.

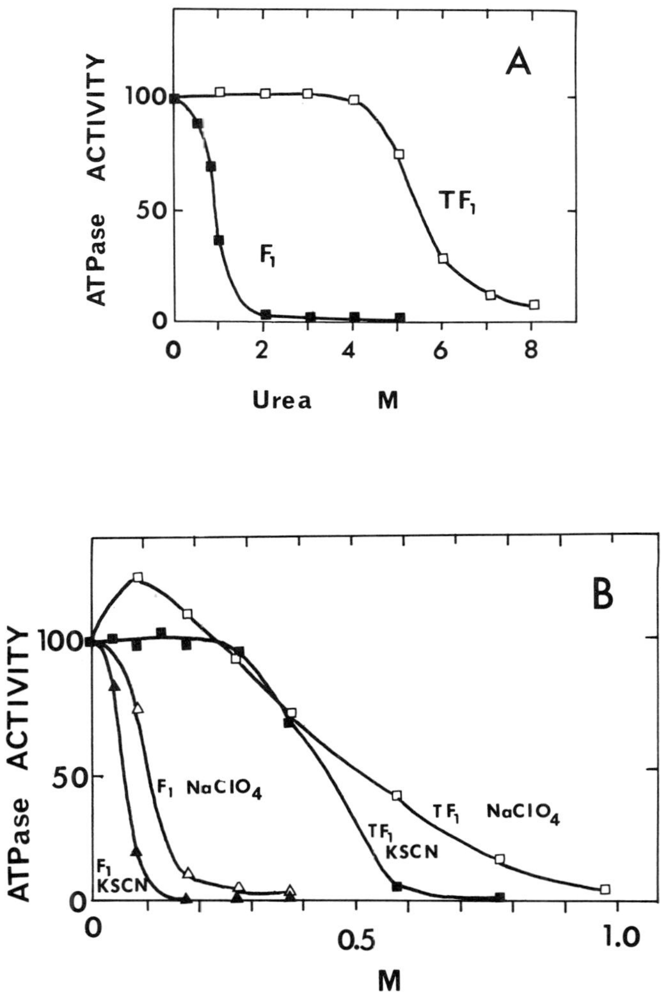

FIGURE 4. The effects of various dissociating agents on ATPase activity of F_1 and TF_1.

A: Effect of urea.

B: Effect of chaotropic agents.

C: Effect of sodium dodecylsulfate (SDS).

D: Effects of organic solvents.

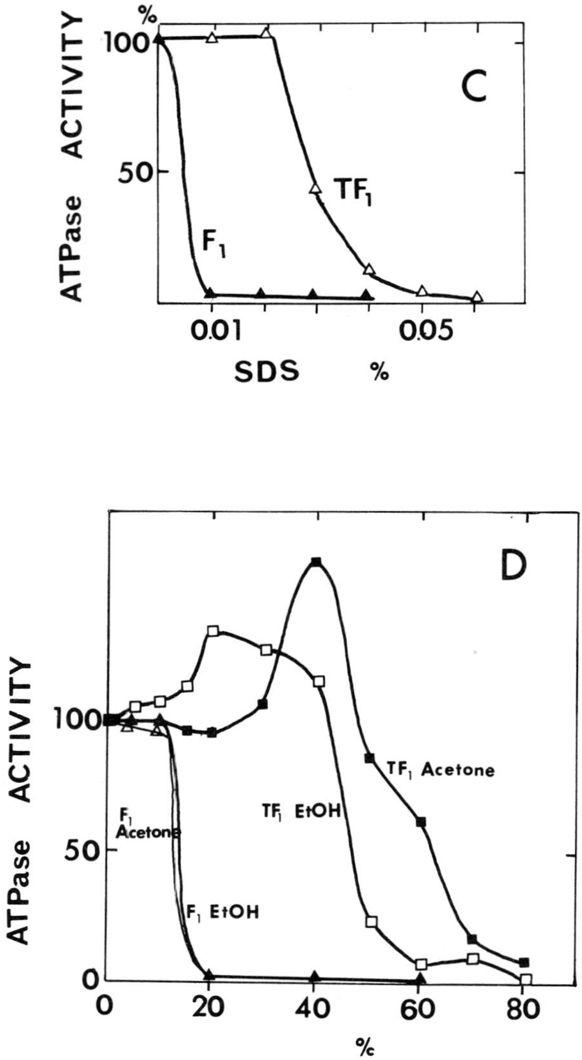
%
100
50
ATPase ACTIVITY
C
TF1
F1
0.01
0.05
SDS %
D
100
50
0
ATPase ACTIVITY
TF1 Acetone
F1
Acetone
TF1 EtOH
F1 EtOH
0
20
40
60
80
%

FIGURE 4c, d.

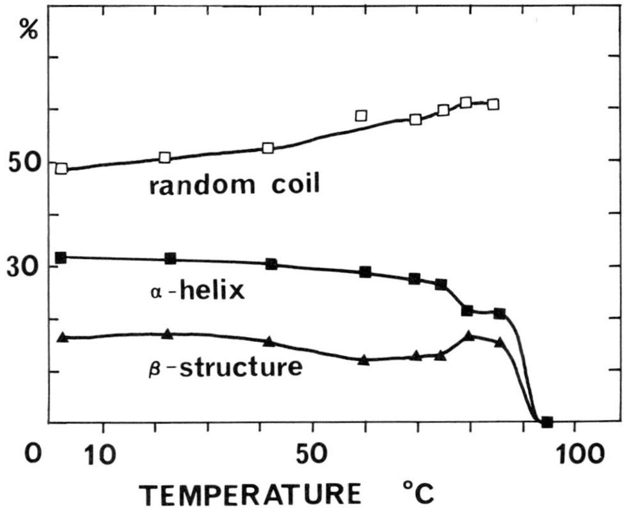

FIGURE 5. The effect of temperature on the secondary structure of TF_1.

MOLECULAR STRUCTURE OF F_1

The molecular structure of F_1 is still controversial mainly because of the dissociable property of the F_1 oligomer. On the other hand, TF_1 is hardly dissociated and its molecular weight has been determined to be 390,000 (with Dr. N. Ui, to be published). The subunits of H^+ATPase (composed of F_o and F_1) are summarized in Table 2. F_1 contains five (α, β, γ, δ, ε) out of the 8 kinds of subunits. There are many reports supporting the subunit stoichiometry of α:β:γ:δ:ε=2:2:1-2:1-2 (Nelson, 1976; Racker, 1976; Baird and Hammes, 1976; Miller *et al.*, 1977). However, our analysis (Kagawa *et al.*, 1976) as well as others (Pedersen, 1975; Bragg and Hou, 1975) support the stoichiometry of α:β:γ:δ:ε= 3:3:1:1:1. The 3 SH groups detected in each TF_1 molecule were located solely in the α subunit. Since each α subunit contained one SH group, the presence of 3 α per oligomer was established.

In order to elucidate the structure of ATPase, crystallographic analysis is essential. However, the molecular weight of F_1 is too large to permit this approach through X-ray diffraction of a three dimensional crystal. Therefore, computerized image reconstruction from electron microscopic figures of two-dimensional crystals holds more

TABLE 2
Subunits of H^+-ATPase of Thermophilic Bacterium PS3

No.	H^+-ATPase ($F_o \cdot F_1$) F_1	F_o	Molecular weight	Copies per $F_o \cdot F_1$	SH-groups per subunit
1	(α)	(-)	56,000	3	1
2	(β)	(-)	53,000	3	0
3	(γ)	(-)	32,000	1	0
4	(-)	(+)	19,000	1	-
5	(δ)	(+)	15,500	1	0
6	(-)	(+)	13,500	2	-
7	(ε)	(-)	11,000	1	0
8	(-)	(+)	5,400	5-6	-

The total SH- (or S-S) per F_1 molecule is 3 in the case of TF_1 (12 in beef heart F_1). The molecular weight of TF_1 is 390,000 (380,000 for F_1 from beef heart). Molecular weights of subunits were determined by dodecylsulfate gel electrophoresis, and numbers of copies of subunit per $F_o \cdot F_1$ were determined by the incorporation of ^{14}C-amino acids (Kagawa et al.*, 1976).*

promise. Two-dimensional paracrystals of beef heart F_1 have already been obtained, but they were unstable (Kagawa, 1974). As shown in Fig. 6, triangular two-dimensional crystals were obtained by ammonium sulfate precipitation of TF_1 in the presence of 25 mM phosphate buffer containing 5 mM $MgSO_4$ and 1 mM AMPPNP. Images of these crystals were often overlapped with those originating from contaminating salts. This interference could be removed by an optical filtration procedure. A parallel beam from a He-Ne-gas laser (NEC, GLG 2009) was projected on the electron micrograph and the diffraction pattern was observed at the focal plane of the lens behind the micrograph. Optical filtering was carried out as follows. A mask made of an opaque plate with holes permitting the selected beams to pass through was placed at the diffraction plane. The lens behind the diffraction plane produced the filtered image (Fig. 7).

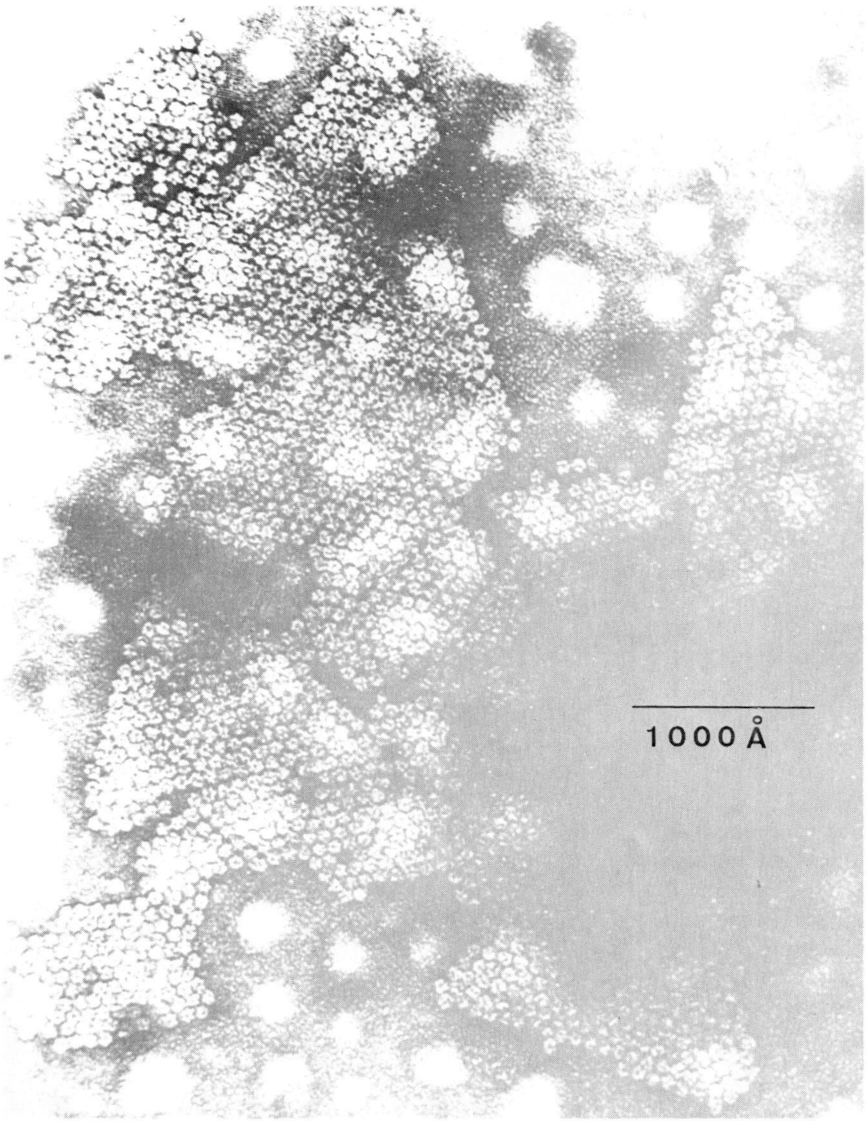

FIGURE 6. Two-dimensional crystals (triangular) of TF_1 negatively stained with uranyl acetate. The electron microscope used was Hitachi HU-12, with original magnification of 50,000 fold.

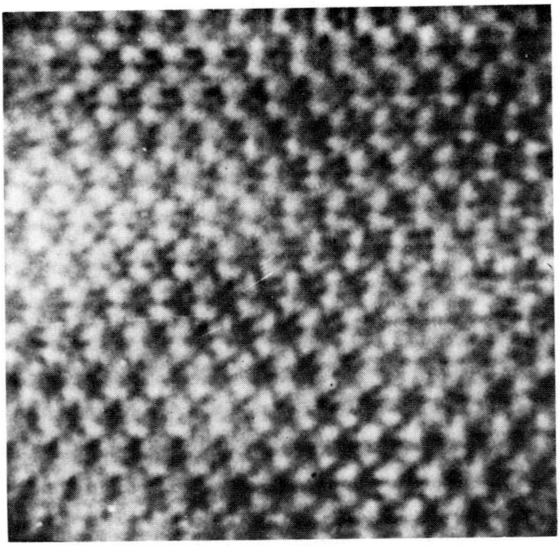

FIGURE 7. Optically filtered image of a two-dimensional crystal of TF_1. The methods used are described in the text.

Then, numerical Fourier analysis was applied to the images selected by the optical diffraction method. The optical density on the film was digitized by a computer-linked microdensitometer equipped with a moving stage scanner. The resulting digitized image was Fourier-transformed numerically. The translationally filtered image was produced by numerical Fourier synthesis by combining only the Fourier components which were approximately consistent with the translational symmetry of the crystal. The resulting digital image was then displayed as a line-printer output and contours were drawn around peaks in density (Fig. 8). The shape of TF_1 is hexagonal and has pseudo 6-fold and 3-fold symmetry, the spacing along the a-axis and b-axis being 90 Å. In addition, there is a hole near the center of the TF_1 molecule (Wakabayashi *et al.*, 1977).

MOLECULAR SPECIES OF PHOSPHOLIPIDS OF THERMOPHILIC BACTERIUM PS3

The phospholipids of PS3 were found to consist of 57.6% phosphatidylethanolamine, 19.2% cardiolipin, 12.9% phosphatidylglycerol, and 10.3% unknown materials (Kagawa

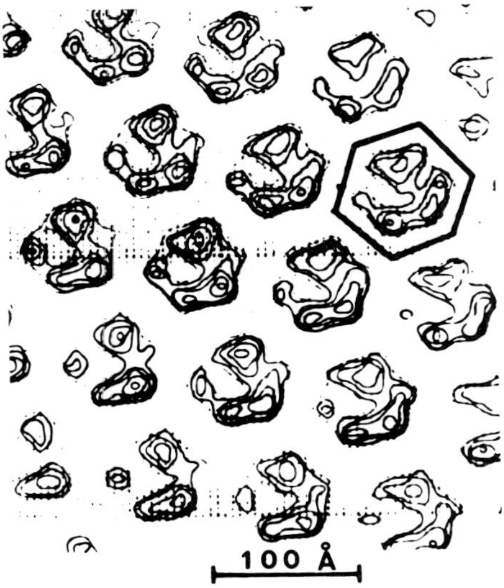

FIGURE 8. Computerized image reconstruction of TF_1 crystal. Translationally filtered image.

et al., 1976). Gas chromatographic analysis of the methylesters of the phospholipid mixture gave 14:0-iso (0.3%), 15:0-iso (34.1%), 16:0-iso (10.8%), 16:0 (5.7%), 17:0-iso (34.0%), 17:0 anteiso (12.9%), and 18:0-iso (1.0%) (Sone *et al.*, 1975). The fatty acyl groups of these phospholipids, unlike those of the usual phospholipids of mesophiles, were all saturated and were mixtures of even and odd numbered carbon chains; their $[M\text{-}C_nH_{2n+1}\ COOCH_2]^+$ and $[M\text{-}C_{n+1}H_{2n+3}\ COO]^+$ peaks overlapped. Thus, the acetyldiglycerides derived from these phospholipids were subjected to a combination of mass chromatography using chemical ionization with NH_3 and treatment of the phospholipids with phosphilipase A_2 (EC 3.1.1.4). The monoacetyldiglycerides were introduced into the column (3 mm x 3 m) filled with 1% OV-1 adsorbed on chromosorb W, at 280°C and were carried through with He at a glow rate of 26 ml/min. The monoacetyldiglycerides were separated from He in a separator at 300°C, and fragmented by chemical ionization in the presence of NH_3, at an emission current of 500 μA, electron energy of 500 eV, and accelerating voltage of 3.5 kV. An example of characteristic fragmentation mass spectrum of 1-15-methylmexadecanoyl-2-13-methyltetradecanoyl-*sn*-glycerol-3-phosphrylethanolamine (volatile form is its 3-acetyl derivative) is shown in Fig. 9. Every intensity of all the fragments in all

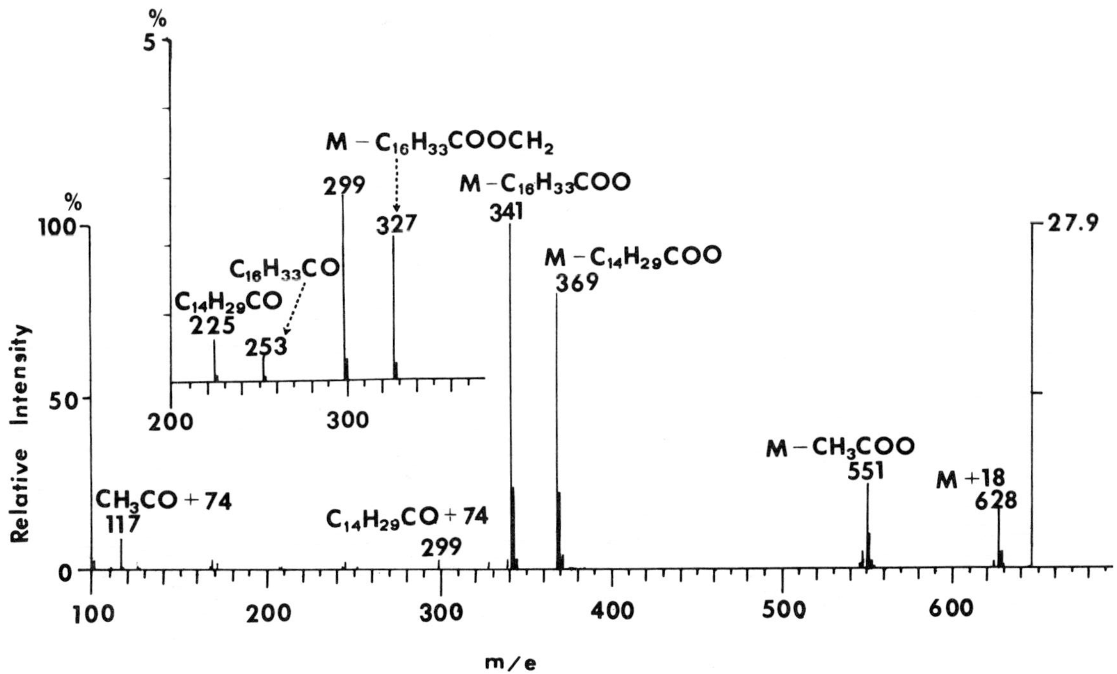

FIGURE 9. Mass spectrum of monoacetyldiglyceride obtained from phosphatidylethanolamine of thermophilic bacterium PS3. The conditions of mass spectrometry are described in the text. Because of the use of the chemical ionization technique, m/e values from 0 to 100 are omitted, and the molecular ion peak appears at $[M + NH_4]^+$. Taken from Kagawa and Ariga (1977) and reproduced with permission from The Japanese Biochemical Society.

molecular species of derived acetyldiglycerides were stored and computed with a GC-MASPAC 300 (Shimadzu, Kyoto, Japan) using an Okitac 4300 computer (16K bits, Oki Denki, Tokyo, Japan) connected to Tektronix 4010-1, Oki-Type and plotter units. Mass Chromatography of monoacetyldiglycerides obtained from phosphatidylglycerol is shown in Fig. 10, and the results are summarized in Table 3 (Kagawa and Ariga, 1977).

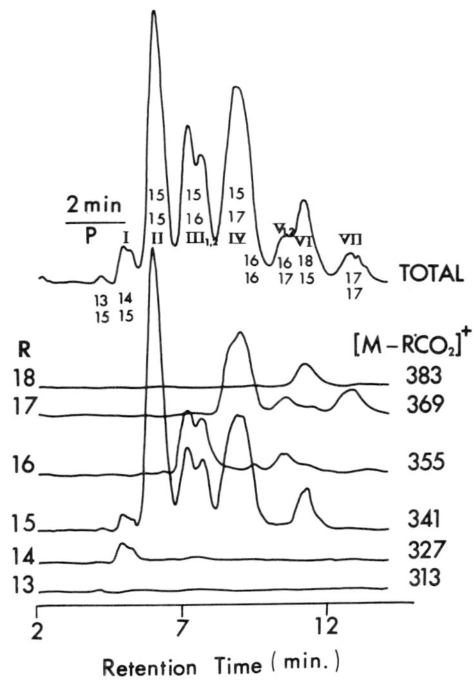

FIGURE 10. Mass chromatography of monoacetyl-diglycerides obtained from phosphatidyl glycerol of thermophilic bacterium PS3. Peaks (P in the figure) I to VII correspond to monoacetyl-diglyceride containing acyl groups (R in the figure) of C-15+14 to C-17+17. The intensities of the signals of $[M\text{-}RCOO]^+$ *ions are plotted against the retention time separately according to the molecular weight or the carbon number of the acyl group. Taken for Kagawa and Ariga (1977) and reproduced with permission from The Japanese Biochemical Society.*

TABLE 3
Molecular Species of Phospholipids of Thermophilic Bacterium PS3

Molecular	R_1	*C15*	*C15*	*C16*	*C17*	*C17*	*C18*	*C17*	*C18*
species	R_2	*C14*	*C15*	*C15*	*C15*	*C16*	*C15*	*C17*	*C16*
Phosphatidylethanolamine		--	24.9	20.4	31.5	5.9	13.2	2.9	---[a]
Phosphatidylglycerol		3.2	31.9	16.3	31.7	6.3	7.1	3.6	---[a]
Cardiolipin		3.9	22.5	18.5	28.5	8.8	10.1	6.2	1.0

[a] *$R_{1,2}$ = C14 was detected as about 1% of total.*

Values are expressed as relative amount, %.

ISOLATION OF AMINO ACID TRANSLOCATOR

Translocators mediating movements of solutes across the membrane have not been isolated except for a few cases. Since this thermophilic bacterium grows well in amino acid mixtures without glucose, and very low concentrations of alanine and glycine were detected in the original hot spring water (Kagawa, 1976), there may be a translocator of these amino acids which can accumulate them. In fact, uptake of alanine dependent on the oxidation of NADH or ascorbate-phenazine methosulfate was demonstrated in washed membrane vesicles of the bacteria (Fig. 11). This transport reaction was stimulated by the addition of DCCD, which blocked leakage of H^+.

The translocator was solubilized from the membrane (Hirata *et al.*, 1976b) and purified (Hirata *et al.*, 1977). The washed membrane was extracted with a mixture of 2% cholate and 1% deoxycholate. These detergents were then replaced with a nonionic detergent, 0.25% Triton X-100, in order to fractionate by ion exchange chromatography using a DEAE-cellulose column. The DEAE eluate was fractionated with ammonium sulfate in the presence of 1% cholate. The resulting precipitate was again dissolved in a mixture of 0.5% Triton X-100 and 6 M urea and the supernatant fraction was applied to a DEAE-cellulose column equilibrated with 10 mM Tris-sulfate (pH 8.0) containing 0.25% Triton X-100 and 4 M urea. The unabsorbed fractions were immediately dialyzed against 25 mM Tris-sulfate (pH 8) containing 0.25 mM EDTA. The dialyzate was subjected to ammonium sulfate fractionation in the presence of 1% cholate. The precipitate was again dissolved in the same Triton-urea mixture and applied to a CM-cellulose column equilibrated with 20 mM Tris-Cl (pH 7.4) containing 0.25% Triton X-100 and 4 M urea. The unadsorbed fractions were collected and precipitated with ammonium sulfate in the presence of 1% cholate.

The apparent molecular weight of this translocator is 150,000 as determined by gel filtration in the presence of 0.5% Triton X-100. This may be an oligomer, since polyacrylamide gel electrophoresis in the presence of dodecylsulfate gave two bands, the molecular weights of which were 36,000 and 9,400 (Hirata *et al.*, 1977).

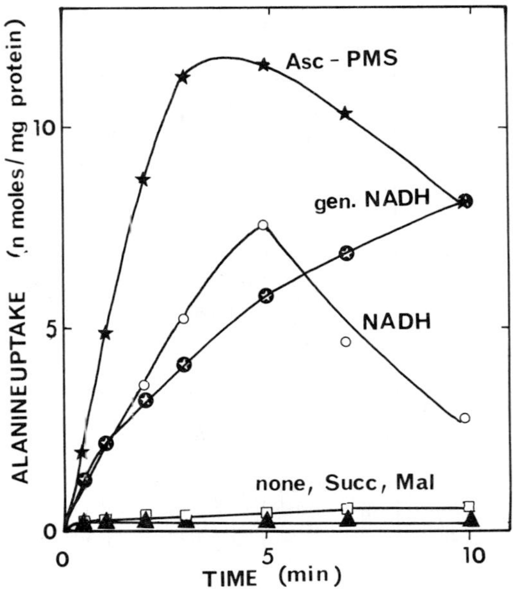

FIGURE 11. Alanine uptake dependent on respiration. The membrane vesicles were suspended in 0.1 ml of solution containing 0.1 M Tris-maleate, pH 7.0, 0.01 M $MgSO_4$, 16 μM L-[U-^{14}C] alanine (0.1 μCi) and membrane vesicles with 0.08 mg of protein. After incubation for 2 min. at 42°C, reactions were started by addition of respiratory substrates: no addition (triangle); 2 μmoles succinate (square); 2 μmoles sodium ascorbate and 0.05 μmoles of phenazine methosulfate (star); 0.4 μmoles of ethanol, 0.01 μmoles of semicarbazide, 0.1 μmole of NAD and 5 μg of alcohol dehydrogenase [EC 1.1.1.1] (star in circle); 0.5 μmole of NADH (circle). Taken from Hirata et al. *(1976a) and reproduced with permission from The Japanese Biochemical Society.*

DISSOCIATION OF H^+ ATPASE INTO SUBUNITS AND RECONSTITUTION OF H^+ ATPASE FROM SUBUNITS

H^+ATPase was dissociated into eight kinds of polypeptides as shown in Table 2. Five kinds of polypeptides are derived from F_1 and three from F_o. F_1 (TF_1) was dissociated with 8 M guanidine hydrochloride solution containing 20 mM Tris-sulfate, 1 mM dithiothreitol and 1 mM EDTA and dialyzed overnight against the same solution to complete dissociation. The solvent of the denatured protein was then displaced by a solution containing 8 M urea, 10 mM Tris-chloride (pH 7.8), 0.1 mM dithiothreitol and 0.1 mM EDTA (urea buffer). The solution was applied to a column of CM-cellulose equilibrated with the urea buffer. Subunits α, β, δ and ε were unadsorbed and subunit γ was then eluted with 0.06 M NaCl in the urea buffer. The unadsorbed fraction was applied to a DEAE-cellulose column equilibrated with the urea buffer. Subunit ε was unadsorbed and the linear gradient elution with NaCl (0 to 0.08 M) in the urea buffer resulted in the separation of δ, α and β subunits in this order (Yoshida *et al*., 1977a). All of the isolated subunits lost their secondary structure in the urea buffer.

There are 31 possible combinations of isolated subunits to reconstitute ATPase (Table 4). Of these combinations, those containing βγ or αβδ complexes were active. Assemblies containing both β and γ had the same mobility as the native TF_1 molecule on gel electrophoresis, whereas assemblies lacking the γ subunit moved more rapidly towards the anode. The ATPase activity reconstituted from α+β+δ was thermolabile and insensitive to NaN_3, whereas the activities obtained from mixtures containing β and γ were thermostable and sensitive to NaN_3, thus resembling native TF_1. This reconstitution could be achieved efficiently by mixing the subunits at temperatures from 20°C to 45°C, but not at 0°C or 60°C (Fig. 12). On transfer of the mixture from 0°C to 30°C, reconstitution was immediately initiated, whereas on transfer from 60°C to 30°C, no further reconstitution occurred (Yoshida *et al*., 1977a).

These results are interesting from the standpoint of thermostability of oligomers. Interaction between subunits β and γ strengthens the conformation of assemblies. A recent report on salt bridges between subunits of D-glyceraldehyde-3-phosphate dehydrogenase (EC 1.2.1.12) isolated from a thermophile lends support to this idea (Biesecker *et al*., 1977). In fact, Table 5 shows that the number of residues (Glu, Gln, Asp, Asn and Arg) per mole of each subunit of TF_1 are higher than (or equal to) those of F_1. Based on the isolation procedure of the subunits with ion exchange resins, subunit β is the most acidic and subunit γ is the most basic.

H^+ATPase ($F_o \cdot F_1$) was extracted from washed membranes of thermophilic bacterium PS3, after the washing with 1% cholate. The extracting medium was 2% Triton X-100 and 0.2 M Na_2SO_4. The extract was diluted with distilled water until its conductance was reduced to 1.2 mmho. Then it was applied to a DEAE-cellulose column equilibrated with a solution of 0.05 M Tris sulfate, pH 8.0 and 0.5% Triton X-100. After washing the column with 75 mM of Na_2SO_4 in the Tris buffer, H^+ATPase was eluted with 200 mM Na_2SO_4 in the

TABLE 4
Reconstitution of ATPase from Various Combinations of Subunits

	α (41μg)	β (40μg)	γ (8μg)	δ (4μg)	ε (3μg)	ATPase activity (u/mg) 40°C	60°C
1		+	+			1.48	2.38
2		+	+		+	1.44	2.58
3		+	+	+		1.36	2.60
4		+	+	+	+	1.56	3.05
5	+	+	+			1.20	1.90
6	+	+	+	+		2.14	2.10
7	+	+	+	+	+	2.33	2.39
8	+	+	+		+	1.62	2.48
9	+	+				0.15	0.08
10	+	+		+		1.21	0.35
11	+	+		+	+	1.18	0.40
12	+	+			+	0.24	0.08
13-31	Pure subunits and other combinations					--negligible---	

buffer. The eluate was fractionated with ammonium sulfate in the presence of 1% cholate and the precipitate at 210 mg ammonium sulfate/ml was dissolved in the buffer. The solution was then applied to the column of Sepharose 6B equilibrated with 0.25% Triton X-100 containing 0.25 M Na_2SO_4 and 50 mM Tris sulfate, pH 8.0 (Sone *et al.*, 1975). The H^+ channel portion of H^+ATPase is called F_o and is easily obtained by dissociating F_1 with 8 M urea from H^+ATPase. The addition of F_1 or subunits comprising F_1 to F_o restored some of the functions of H^+ ATPase. Addition of the γδε complex to F_o blocked the leakage of H^+ through F_o. This H^+-gate

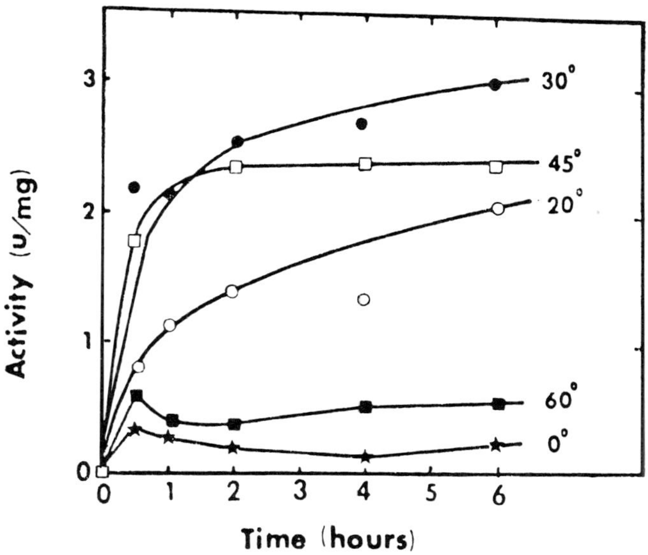

FIGURE 12. Effect of temperature on the assembly of subunits α, β and γ. Solutions containing 80 μg/ml of α subunits, 73 μg/ml of β subunits, 21 μg/ml of γ subunits, 5 mM $MgSO_4$, and 50 mM Tris/sulfate at pH 8.0 were incubated at the temperatures indicated. At appropriate times, 30 μl samples were removed for assaying ATPase activity at 60°C for 15 min. Taken from Okamoto et al. *(1977) and reproduced with permission from The American Society of Biological Chemists.*

function has been known as the structural role of F_1, since chemical or immunological treatment of the catalytic center of F_1 does not destroy the H^+-gate activity. Both δ and ε subunits are required to connect the αβγ complex to F_o (Yoshida *et al.*, 1977b). The role of H^+-gate may be the energy converter in F_1 since the H^+-flux-force characteristic is very similar to the potential barrier at the PN junction of the "Zener-diode" (Schönfeld and Newmann, 1977), and is affected by ADP concentration or energy transfer inhibitors.

In order to synthesize ATP driven by H^+-gradient, all of the five subunits are required in F_o vesicles (Yoshida *et al.*, 1977b). The components of H^+ATPase were thus functionally established.

TABLE 5
Content of Glutamic Acid, Aspartic Acid and Arginine Residues in Each ATPase Subunit

	Whole oligomer ($\alpha_3\beta_3\gamma_1\delta_1\varepsilon_1$)*		Subunit α		Subunit β		Subunit γ		Subunit δ		Subunit ε	
	F_1	TF_1	F_1	TF_1	F_1	TF_1	F_1	TF_1	F_1	TF_1	F_1	TF_1
Glu + Gln	404	471	58	64	58	66	29	40	22	30	5	11
Asp + Asn	273	307	42	44	35	41	27	29	12	15	3	8
Arg	161	202	27	27	19	27	16	17	4	16	3	7
Total number of amino acids	3,393	3,407	504	487	459	461	303	278	147	189	54	96

*The values are expressed as number of residues per mole of subunit or oligomer. The data for F_1 (beef heart F_1) are calculated from the values of Knowles and Penefsky (1972).

LIPOSOMES CONTAINING EITHER F_o OR H^+ ATPASE

When F_o was incorporated in liposomes loaded with K^+, the addition of valinomycin caused a rapid uptake of H^+ through F_o. An initial velocity of 6 H^+ uptake per sec per F_o molecule at a membrane potential of 100 mV was observed (Okamoto *et al.*, 1977). The velocity of H^+ conductance obeys the Ohmic law. The velocity corresponds to a unit conductance of 9.5 x 10^{-18} mho/F_o at pH 8.0 and the maximum unit conductance is 1.6 x 10^{-16} mho based on the titration with different pH. The effect of external pH on the passive H^+ uptake in F_o-liposomes showed that H^+, rather than OH^-, was the actual ionic species conducted. The pH-velocity curve of H^+ conductance was identical to a titration curve for a monoprotic acid of pKa = 6.76. This quantitative coincidence suggests the presence of a H^+-binding site and one to one stoichiometry of this binding. This H^+ channel is specifically inhibited by 10^{-5}M DCCD, anti- F_o antibody and assemblies of subunits of F_1 containing $\gamma\delta\varepsilon$. Recently, the smallest subunit of F_o was isolated and identified as a DCCD-binding protein.

The temperature coefficients of H^+translocation into F_o-liposomes composed of PS3 phospholipids were 22.0 kcal/mole above 24°C and 38.2 kcal/mole below 24°C (Okamoto *et al.*, 1977). Interaction of F_o with surrounding phospholipids was demonstrated also by a three-fold increase in helical content of F_o when it is added to phospholipids (Kagawa *et al.*, 1976).

Net synthesis of ATP driven by an artificially imposed electrochemical potential difference of H^+ across the membrane ($\Delta\bar{\mu}H^+$) was observed in broken chloroplasts (Jagendorf and Uribe, 1966) or submitochondrial particles (Thayer and Hinkle, 1975). However, these membranes contained many components and it was difficult to conclude that H^+ATPase was solely responsible for this reaction. ATP synthesis driven by "acid-base transition" had never been reported in reconstituted vesicles containing H^+ATPase, but this became possible through the use of stable phospholipids and H^+-ATPase obtained from thermophilic bacterium PS3 (Sone *et al.*, 1977; Kagawa *et al.*, 1977). The reconstituted H^+-ATPase liposomes were first incubated in acidic malonate buffer at pH 5.5 with valinomycin. Next, a base stage solution (Glycylglycine buffer, final pH 8.33) containing ^{32}Pi, ADP and KCl was rapidly injected. This instantaneous transition should create $\Delta\bar{\mu}H^+$, composed of both ΔpH (2.83 unit acidic inside) and $\Delta\Psi$ (125 mV positive inside) across the liposome membrane. Esterification of ^{32}Pi occurred at the velocity of 650 nmoles/mg H^+ ATPase/min.

The maximum level of Pi esterified was about 100 nmoles/

mg of H^+ATPase in the reconstituted vesicles, while less than 2.5 nmoles/mg of protein were synthesized by submitochondrial particles. Taking the molecular weight of H^+ATPase as 458,500 (Kagawa *et al.*, 1976), it was calculated that the enzyme turned over 46 times at the optimal pulse of the acid-base transition.

The H^+/ATP stoichiometry and $\Delta\bar{\mu}H^+$ required for ATP synthesis are subjects of controversy. Through the use of stable reconstituted H^+ATPase liposomes, $\Delta\bar{\mu}H^+$ in the presence of ATP was shown to be 253 mV (Sone *et al.*, 1976). This value is sufficient to explain the stoichiometry of H^+/ATP = 2.

CONCLUSIONS

An aerobic thermophilic bacterium called PS3 showed very stable membrane functions, such as oxidative phosphorylation and amino acid transport. These functions are unstable in mesophilic organisms.

These stabilities are attributed to the stable membrane proteins and phospholipids. Although the subunit structure, catalytic functions, etc., are identical in both thermophilic and mesophilic H^+ATPase, fewer (one-fourth) SH-groups and more Arg and Glu were found in the thermophilic enzyme than in the mesophilic counterparts. No unsaturated fatty acyl groups were detected in the phospholipids of thermophile membranes. The molecular species of membrane phospholipids from thermophilic bacterium PS3 were quite different than those from mesophiles.

Through the use of these proteins and phospholipids of thermophilic bacterium PS3, several unsolved problems in membrane biology have been answered as summarized in the beginning of this report.

REFERENCES

Baird, B. A. and Hammes, G. G. (1976). J. Biol. Chem. *251*, 6953.

Biesecker, G., Harris, J. I., Thierry, J. C., Walker, J. E. and Wonacott, A. J. (1977). Nature, *266*, 328.

Bragg, P. D. and Hou, C. (1975). Arch. Biochem. Biophys. *167*, 311.

Hirata, H., Sone, N., Yoshida, M. and Kagawa, Y. (1976a). J. Biochem. *79*, 1157.

Hirata, H., Sone, N., Yoshida, M., and Kagawa, Y. (1976b). Biochem. Biophys. Res. Commun. *69*, 665.

Hirata, H., Sone, N., Yoshida, M. and Kagawa, Y. (1977). J. Supramol. Structure *6*, 77.

Jagendorf, A. T. and Uribe, E. (1966). Proc. Natl. Acad. Sci. USA. *55*, 170.

Kagawa, Y. (1974). In *Methods in Membrane Biology*, (E. D. Korn, ed.), Vol. 1, p. 201, Plenum Press, New York.

Kagawa, Y. (1976). J. Cell. Physiol. *89*, 569.

Kagawa, Y., Sone, N., Yoshida, M., Hirata, H. and Okamoto, H. (1976). J. Biochem. *80*, 141.

Kagawa, Y. (1977). Adv. Biophys. *10*, 209.

Kagawa, Y. and Ariga, T. (1977). J. Biochem. *81*, 1161.

Kagawa, Y., Ohno, K., Yoshida, M., Takeuchi, Y. and Sone, N. (1977). Fed. Proc. *36*, 1815.

Knowles, A. F. and Penefsky, H. S. (1972). J. Biol. Chem. *247*, 6624.

Muller, J. L. M., Rosing, J. and Slater, E. C. (1977). Biochim. Biophys. Acta *462*, 422.

Nelson, N. (1976). Biochim. Biophys. Acta *456*, 314.

Ohta, S., Nakanishi, M., Tsuboi, M., Yoshida, M. and Kagawa, Y. (1978). Biochem. Biophys. Res. Commun. *80*, 929.

Okamoto, H., Sone, N., Hirata, H., Yoshida, M. and Kagawa, Y. (1977). J. Biol. Chem. *252*, 6125.

Pedersen, P. L. (1975). Bioenergetics *6*, 243.

Racker, E. (1976). *New Look at Mechanisms in Bioenergetics*, Academic Press, New York.

Schönfeld, M. and Newmann, J. (1977). FEBS Letters *73*, 51.

Sone, N., Yoshida, M., Hirata, H. and Kagawa, Y. (1975). J. Biol. Chem. *250*, 7917

Sone, N., Yoshida, M., Hirata, H., Okamoto, H. and Kagawa, Y. (1976). J. Membrane Biol. *30*, 121.

Sone, N., Yoshida, M., Hirata, H. and Kagawa, Y. (1977). J. Biol. Chem. *252*, 2956.

Thayer, W. S. and Hinkle, P. C. (1975). J. Biol. Chem. *250*, 5336.

Yoshida, M., Sone, N., Hirata, H. and Kagawa, Y. (1975). J. Biol. Chem. *250*, 7910.

Yoshida, M., Sone, N., Hirata, H. and Kagawa, Y. (1977a). J. Biol. Chem. *252*, 3480.

Yoshida, M., Okamoto, H., Sone, N., Hirata, H. and Kagawa, Y. (1977b). Proc. Natl. Acad. Sci. USA. *74*, 936.

Wakabayashi, T., Kubota, M., Yoshida, M. and Kagawa, Y. (1977). J. Mol. Biol. *117*, 515.

ECOLOGICAL AND GENETIC ASPECTS

THE MORPHOLOGY AND SURFACE STRUCTURE OF SOME EXTREMELY THERMOPHILIC BACTERIA FOUND IN SLIGHTLY ALKALINE HOT SPRINGS

Robert F. Ramaley
F. Rudolf Turner
Linda E. Malick
Richard B. Wilson

Department of Biochemistry
University of Nebraska Medical Center
Omaha, Nebraska

THERMUS AQUATICUS YT-1 AND *THERMUS* X-1.

In 1969, Brock and Freeze (1969) reported the isolation and characterization of a new gram negative, extremely thermophilic bacterium (*Thermus aquaticus*) ATCC-25104. The type species of this bacterium (YT-1) was isolated from a run off channel of a hot spring in Yellowstone National Park but similar isolates have been obtained from a number of slightly alkaline thermal environments including hot tap water. *T. aquaticus* YT-1 has an optimum growth temperature of 70°C, grows only in dilute organic media, and tends to grow as long filamentous cells especially on solid media.

The cell envelope structure of *T. aquaticus* YT-1 was investigated by Brock and Edwards (1970) and was shown to consist of "alternating depression, invagination and protrusion" of the outer layer giving it a scallop like appearance as may be seen in Figure 1.

In 1970, Ramaley and Hixson (1970) reported the isolation of an extremely thermophilic bacterium (X-1 isolate) ATCC-27978, that was very similar to *T. aquaticus* YT-1. However the X-1 isolate did not have the bright yellow carotenoid pigmentation of *T. aquaticus* YT-1 and also seemed to produce less slime. Thin section electron micrographs of the X-1 isolate (Figure 2) show that its call envelope structure is very similar to that of *T. aquaticus* YT-1.

ISBN 0-12-268250-5

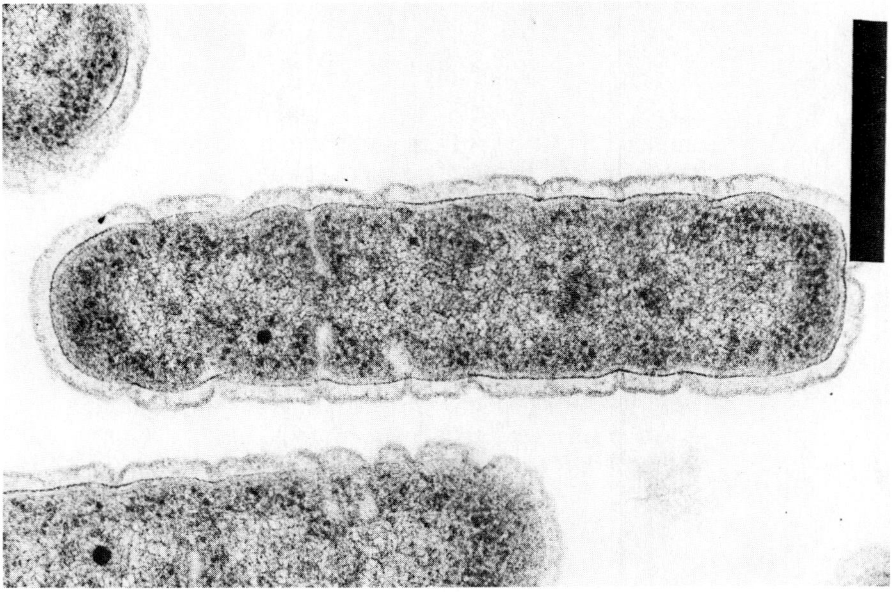

FIGURE 1. Longitudinal view of a T. aquaticus *YT-1 cell at a magnification of 89,000. The marker bar indicates 0.5 μm.*

GROWTH AND FIXATION OF CELLS FOR THIN SECTION ELECTRON MICROSCOPIC OBSERVATION

T. aquaticus YT-1 (obtained from T. D. Brock) and *Thermus* X-1 were grown aerobically at 70°C in 0.3% w/v yeast extract, 0.3% w/v tryptone in Casenholtz salts (Ramaley and Hixson, 1970) until their mid-logarithmic growth phase. The cells were chilled to 4°C and concentrated for fixation by low speed centrifugation (500 x g for 20 minutes). This procedure differs from that of Brock and Edwards (1970) in which they directly fixed the cells for electron microscopic observation without prior cooling or centrifugation. Thus, no examination of cell interaction or formation of rotund bodies (Brock and Edwards, 1970) was attempted in the present study. The fixation procedure was the same as that used by Brock and Edwards for their study and is a modification of the Ryter and Kellenberger (1958) procedure.

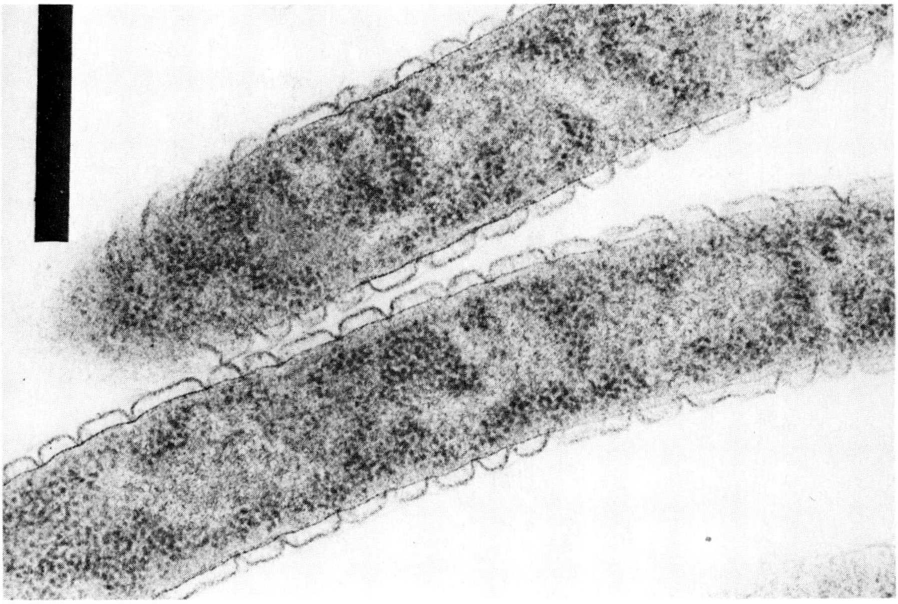

FIGURE 2. Oblique view of Thermus *X-1 cells at a magnification of 89,000. The market bar indicates 0.5* μm.

COMPARISON OF THIN SECTION ELECTRON MICROGRAPHS OF *THERMUS* X-1 WITH THOSE OBTAINED FROM OTHER *THERMUS* ISOLATES

Although *Thermus* X-1 shows the same scallop like appearance of its outer cell wall (Figure 2) as that previously reported by Brock and Edwards (1970), the definition of the outer cell wall structure is much clearer for the X-1 isolate. One explanation for this might be a reduced slime layer in *Thermus* X-1 (Ramaley and Hixson, 1970) over that previously observed with *T. aquaticus* YT-1 (Brock and Freese, 1969). However, no quantitative studies of slime production have been conducted with either *Thermus* X-1 or *T. aquaticus* YT-1.

Brock and Edwards (1970) demonstrated in grazing section of the outer wall of *T. aquaticus* YT-1 that it exists as a series of parallel bands around the circumference of the cells giving them an "annelid" appearance (Williams, 1975). Examples of this structure in *T. aquaticus* YT-1 are shown in Figure 3. *Thermus* X-1 with its greater definition of outer wall structure shows the annelid appearance even more clearly (Figure 4).

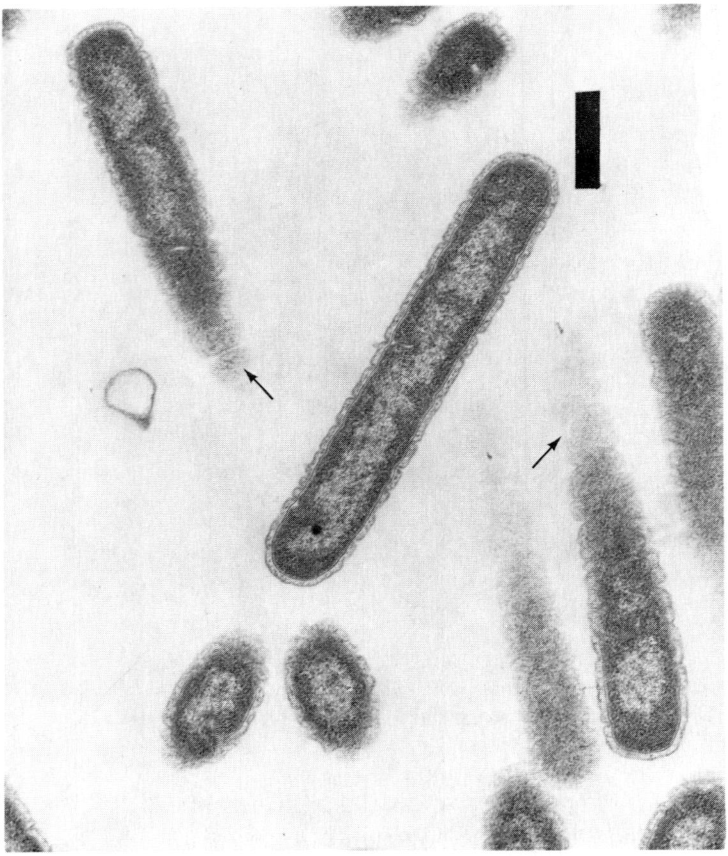

FIGURE 3. Grazing section of T. aquaticus *YT-1 at a magnification of 25,660. The marker bar indicates 0.5 µm. Note bands of outer cell wall material around circumference of cell.*

This type of outer cell wall structure appears to be common to all of the *Thermus* isolates studied to date, including *T. thermophilus* (Oshima and Imahori, 1974) and non-pigmented *Thermus* strains (NH and Dl) isolated by Pask-Hughes & Williams (1975) from hot tap water. However, the structure is not universal for all extremely thermophilic bacteria as will be shown with *Thermomicrobium* isolates.

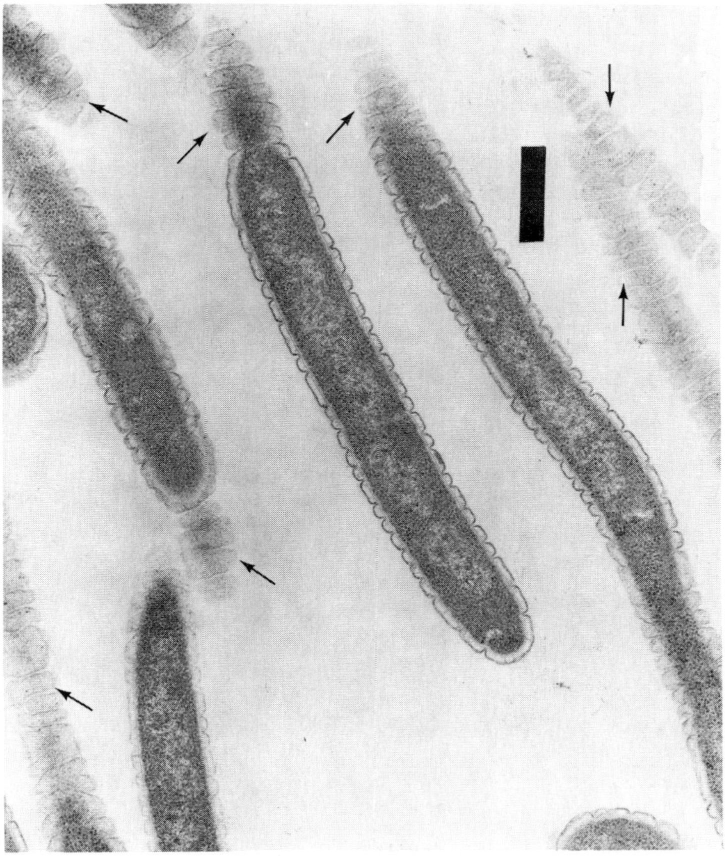

FIGURE 4. Grazing section of Thermus *X-1 at a magnification of 25,600. The marker bar indicates 0.5 μm. Note bands of outer cell material are much clearer than for* T. aquaticus *YT-1.*

SCANNING ELECTRON MICROGRAPHS OF A NEW NON-PIGMENTED *THERMUS* ISOLATE (S-1)

During a study of the distribution and types of *Thermus* isolates, Stan Johnson (an undergraduate student from Dana College working in our laboratory for his undergraduate research requirement) happened to find a non-pigmented *Thermus* type isolate growing as a pure culture in a small hot water tank in the maintenance shop at Dana College in Blair, Nebraska. This isolate attracted our attention because unlike *T. aquaticus* YT-1 or most of the previously isolated non-pigmented *Thermus* X-1-like isolates, this new isolate (*Thermus* S-1, ATCC 27979) grew only as very short rods rather than forming long filamentous cells. The shorter cell form is similar to the cell form reported for *T. thermophilus* (Oshima and Imahori, 1974) and indicates that formation of long filamentous cells is not found in all *Thermus* isolates.

Figure 5 shows a scanning electron micrograph of *Thermus* S-1 cells on agar and demonstrates the three dimensional annelid appearance suggested from the thin section micrographs.

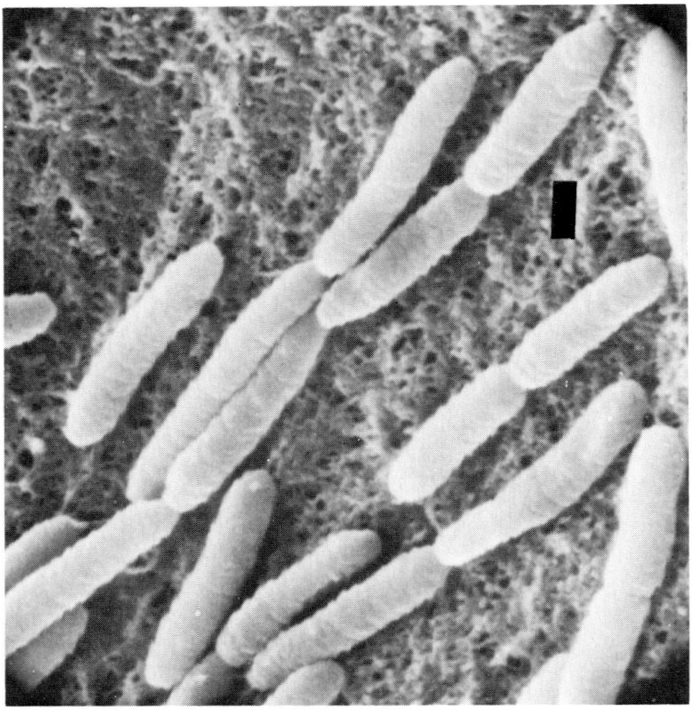

FIGURE 5. Scanning electron micrograph of Thermus *S-1 cells at a magnification of 16,000. The marker bar indicates 0.5 μm.*

SURFACE STRUCTURE OF THE PINK EXTREMELY THERMOPHILIC BACTERIUM *THERMOMICROBIUM ROSEUM*

In 1973 we described a new species of pink, extremely thermophilic bacterium that was so different from *Thermus* that we proposed its classification in a new species. This bacterium, *Thermomicrobium roseum* (Jackson, Ramaley and Meinschein, 1973), was very pleiomorphic and thin section electron micrographs of this bacterium revealed what appeared to be some type of regular repeating structure on its outer cell wall as shown in Figure 6.

Negative stained preparation of *T. roseum* have served to confirm the presence of this structure in a regular mosaic pattern as shown in Figure 7.

Recently Phillips and Perry (1976) have isolated a hydrocarbon utilizing obligate thermophilic bacterium (optimum growth temperature 60°C) which, like *T. roseum*, has a pink carotenoid pigmentation and have proposed that it be placed

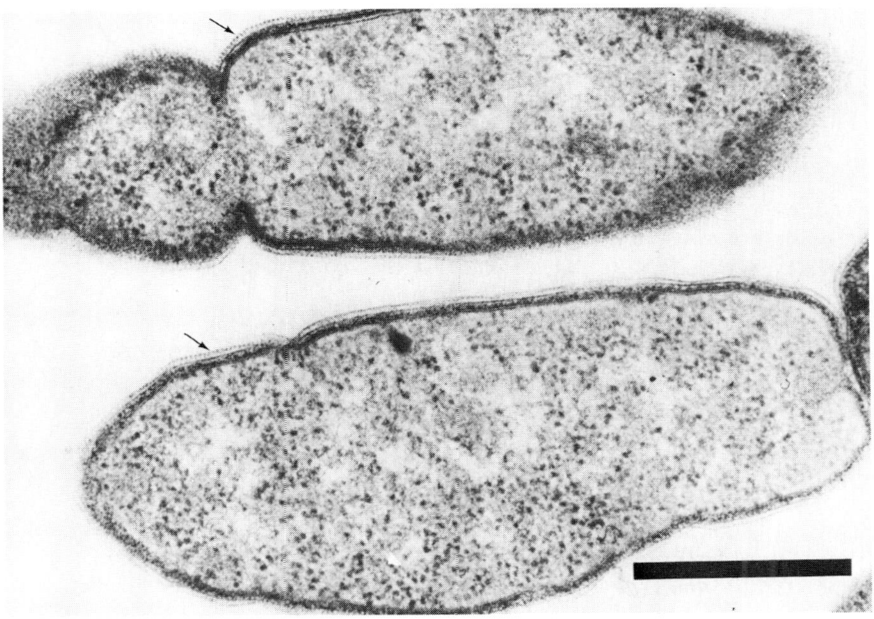

FIGURE 6. Longitudinal view of Thermomicrobium roseum *cells at a magnification of 82,000. The marker bar indicates 0.5 μm.*

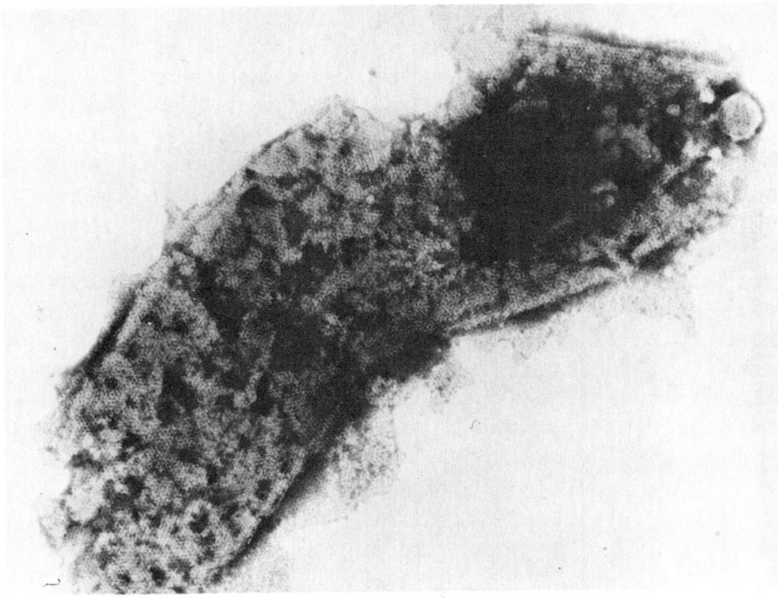

FIGURE 7. Phosphotungstate negatively stained preparation of Thermomicrobium roseum *cells (2% phosphotungstate, pH 7.2).*

in the genus *Thermomicrobium* as *T. fosteri* in honor of the late Jackson W. Foster. Some of the thin section of *T. fosteri* do look very similar to those of *T. roseum* and are certainly not similar to *Thermus* type isolates.

COMPARISON OF THE RED, MODERATELY THERMOPHILIC BACTERIUM K-2 ISOLATE WITH *THERMUS RUBER*

Although additional isolates of the pink extremely thermophilic bacterium *T. roseum* have not been obtained, a number of additional isolates of a dull red, thermophilic bacterium have been obtained since its initial isolation in May of 1972 (Ramaley and Bitzinger, 1973). This bacterium (K-2 isolate) (Ramaley *et al.*, 1975) has an optimum growth temperature of 60°C as compared with 70°C for all previous *Thermus* isolates and is found in both man made and natural thermal environments (Ramaley and Bitzinger, 1975). The growth and physiological characteristics were quite similar to both yellow pigmented and non-pigmented *Thermus*. However, because of the difference in optimum growth temperature the isolate was not proposed as a species of *Thermus*.

Loginova and Egorova (1975) have reported on the characteristics of a similar if not identical, red thermophilic bacterium isolated from hot springs in Kamachatka and Tadzhikistan (USSR) [Loginova and Khraptsova (1977)] which they have designated as *Thermus ruber*. In view of the isolation of an additional isolate of *Thermomicrobium fosteri* with its lower optimum growth temperature, we would now propose that the K-2 isolate also be designated with the species name of *ruber* recognizing its probable eventual classification as a *Thermus* species. If this is accepted, that would mean changing the recognized optimum growth temperature of the genus to include 60°C and red as one of its pigment possibilities.

SCANNING ELECTRON MICROGRAPHS OF A PINK BACTERIUM GROWING AT 80°C IN A RUN-OFF CHANNEL OF A HOT SPRING IN YELLOWSTONE NATIONAL PARK

One of the reasons that the isolation of pink, extremely natural thermophilic bacteria has been pursued with such intensity, first by Brock's group and later by ourselves was the early observation by Brock (1967) that large masses of pink bacteria were growing at 80-90°C in the upper end of one of the run-off channels of Pool A in Yellowstone National Park (now officially named as Octopus Spring). Pictures of Octopus Spring and the upper end of the run-off channel containing the bacteria are shown in Figures 8 and 9, respectively. Scanning electron micrographs of the filamentous streamers growing at 85°C (Figures 10 and 11) show that the bacteria are growing as long filamentous cells as previously reported by Brock (1967) with phase contrast microscopic observation.

These bacteria grow as long filaments and are clearly not *Thermomicrobium*. Thin section electron micrographs of these bacterial streamers have indicated that the solid appearing material that the bacteria appear to be growing on in Figure 10 is made up of these gram negative bacteria and deposited silica (M. Edwards, personal communication). However, the original electron micrographs have been lost in relocation of M. Edwards' laboratory and we are currently preparing additional thin section electron micrographs of these bacteria.

Although the presence of these bacteria growing at 85°C in the run-off channel of Pool A has been known for a decade,

FIGURE 8. View of Pool A (Octopus Spring) and upper end of run-off channel containing the pink bacteria.

FIGURE 9. Close up view of the upper end of the run-off channel showing filamentous streamers of bacteria silica deposits.

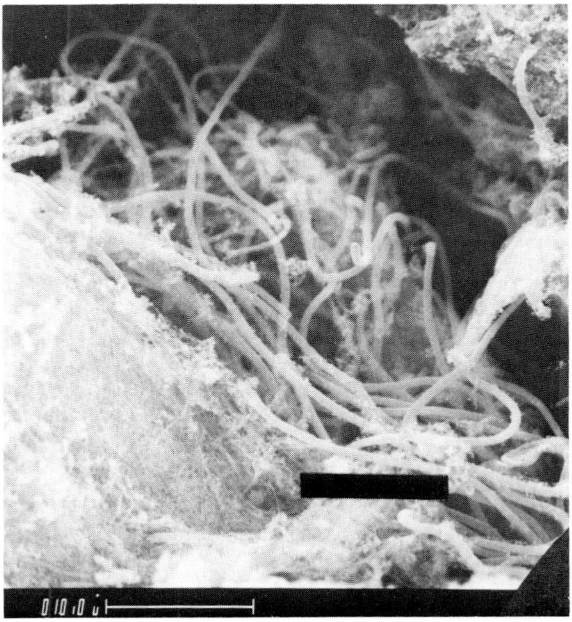

FIGURE 10. Scanning electron micrographs of pink bacterial streamers growing in the run-off channel at 85°C (680 magnification). Marker bar indicates 10 μm.

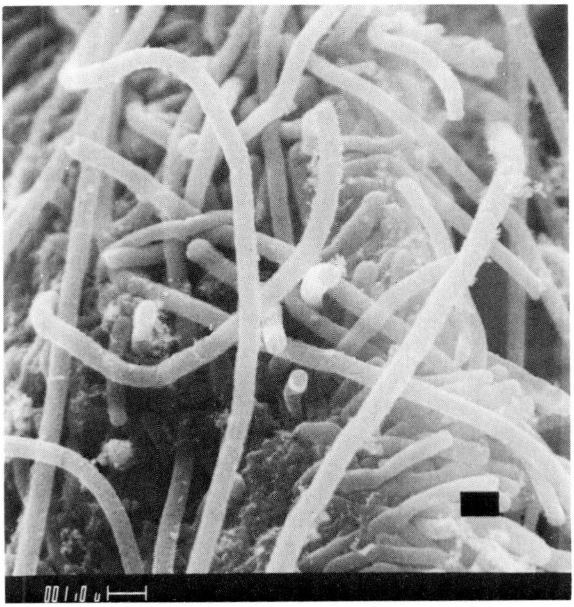

FIGURE 11. A scanning electron micrograph of the same bacterial streamers at a magnification of 4,000. Marker bar indicates 1.0 μm.

they have still not been obtained in laboratory culture. It has been possible to obtain the fatty acid (Bauman and Simmon, 1969) and hydrocarbon composition (Jackson, 1972) of the bacterial biomass and we have demonstrated that enzymes such as fumarase or phosphofructokinase can be extracted from these cells. As might be anticipated, these enzymes are extremely thermostable, exceeding the thermostability of the same enzymes from extremely thermophilic bacteria such as *T. aquaticus*. We have assigned the trivial number of PB-1 (Pink Bug No. 1) to the bacterium even though it has not been obtained in culture, and we are pursuing *in situ* physiological studies in the hot spring run-off channels which, along with continued morphological investigations, may help in gaining some insight into the nature of this pink bacterium with an aim of its eventual laboratory culture.

These studies may also aid in the comparisons made between natural ecosystems, such as the analysis of hot spring stromatolites (deposits of microorganisms embedded in layered silica deposits which resemble Precambrian bedded cherts (Jackson, 1972), and data obtained with axenic laboratory cultures of thermophilic microorganisms taken from these same thermal ecosystems.

CONCLUSIONS

Morphological studies of extremely thermophilic bacteria have shown that there are at least two very different types of outer cell wall structures that have evolved. The most common, or at least the one that is found in those bacteria that are most easily isolated (i.e., *Thermus*), has a regular scallop-like connection to the inner membrane/cell wall giving *Thermus* isolates an annelid-like appearance. This appearance is especially striking in the non-pigmented *Thermus* isolates and can even be demonstrated by scanning electron microscopic observation. In the pigmented *Thermus* isolates, which may have a thicker slime layer, the outer surface is smoother although the scallop-like connections are also present.

The other type of cell wall structure is represented by the appearance of regular knob-like structures on the outer surface of *Thermomicrobium roseum*. The presence of this mosaic was confirmed by negative stained preparation of *T. roseum*. However, nothing is currently known about the biochemical composition of this outer layer.

Biochemical studies of the cell envelope of *Thermus* and *Thermomicrobium* are currently underway in several laboratories, including those of Pask-Hughes [Ph.D. Thesis, U. of London, cited in Pask-Hughes and Williams (1977)] and Perry (personal communication) and it is anticipated that the results of these investigations should provide some very interesting answers as to how nature functions at 70°C. Perhaps similar studies of the pink bacteria growing at 85°C in the run-off channel of Pool A in Yellowstone National Park would also be very useful even though this bacteria cannot be grown in laboratory cultures.

ACKNOWLEDGEMENTS

This work was supported by a grant from the National Science Foundation (76-02707). The senior author would like to make a special acknowledgement to those undergraduate students who isolated and characterized *Thermus* and *Bacillus stearothermophilis* type isolates as part of their course work (M-473 at Indiana University) or as undergraduate research projects at Indiana University or at the University of Nebraska Medical Center.

REFERENCES

Bauman, A. J. and Simmonds, P. G. (1969). J. Bacteriol. *98*, 528.

Bitzinger, K. and Ramaley, R. F. (1973). Indiana Acad. Sci., *82*, 373.

Brock, T. D. (1967). Science *158*, 1012.

Brock, T. D. and Freeze, H. (1969). J. Bacteriol. *98*, 289.

Brock, T. D. and Edwards, M. R. (1970). J. Bacteriol. *104*, 509.

Jackson, T. J. (1972). Ph.D. Thesis: *Biogeochemistry of Thermal Environments*, Indiana University.

Loginova, L. G. and Egorova, L. A. (1975). Mikrobiologiya *44*, 661.

Loginova, L. G. and Khraptsova, G. I. (1977). Mikrobiologiya *46*, 38.

Oshima, T. and Imahori, K. (1974). Int. J. Syst. Bacteriol. *24*, 102.

Pask-Hughes, R. A. and Williams, R. A. D. (1975). J. Gen. Microbiol. *88*, 321.

Pask-Hughes, R. A. and Williams, R. A. D. (1977). J. Gen. Microbiol. *102*, 375.

Phillips, W. E. and Perry, J. J. (1976). Int. J. Syst. Bacteriol. *26*, 220.

Ramaley, R. F. and Hixson, J. (1970). J. Bacteriol. *103*, 527.

Ramaley, R. F. and Bitzinger, K. (1975). App. Microbiol. *30*, 152.

Ramaley, R. F., Bitzinger, K., Carroll, R. M., and Wilson, R. B. (1975). Int. J. Syst. Bacteriol. *25*, 357.

Ryter, A. and Kellenberger, E. (1958). Z. Naturforsch. *13b*, 597.

Williams, R. A. D. (1975). Sci. Prog. Oxf. *62*, 373.

CHANGES IN MICROBIAL FLORA, INCLUDING THERMOPHILES, DURING COMPOSTING OF ANIMAL MANURE

Takashi Saiki
Teruhiko Beppu
Kei Arima

Department of Agricultural Chemistry
University of Tokyo
Yayoi, Bunkyoku, Tokyo

Toshihiko Izawa
Kunio Morimoto

Department of Agricultural Machinery
Omiya, Saitama, Tokyo

Yasuhito Misaizu
Shigemitsu Higaki

National Institute of Animal Industry
Chiba, Tokyo

The total amount of waste material discharged in Japan is estimated to be 6-7 x 10^8 tons per year. This includes municipal animal, human and carbonaceous waste. The amount of animal feces and urine is 6.7 x 10^7 tons and that of carbonaceous waste, which can be used for solid composting, is about 3.7 x 10^7 tons per year (Table 1). Since the amount of rice production in Japan is about 1.1 x 10^7 tons per year, the importance of waste recycling can be understood easily.

The amount of fertilizer components contained in animal waste material (excreted per year) is roughly estimated to be about 70% of the chemical fertilizer consumed per year in our country (1). From the point of view of saving natural resources and of protecting the environment, one of the most effective procedures is to treat these waste as solid or

ISBN 0-12-268250-5

TABLE 1
The Amount of Waste Material in Japan

Waste material	*Tons per year*
Municipal solid waste	7.0×10^7
Animal feces and urine	6.7×10^7
Human feces and urine	4.7×10^7
Carbonaceous waste (sawdust, straw, etc.)	3.7×10^7

Taken from reference (1).

liquid compost. The products can be used as organic fertilizer as well as soil conditioners.

A variety of microorganisms, including thermophiles, are well known to play an important role in the process of thermogenetic composting. Although changes in microbial flora during the composting process has been studied for a long time and described in some excellent reviews (2, 3, 4), further details of the microfloral aspect remain to be resolved. In addition, little attention has been paid to this problem in our country and the microflora of liquid composting is quite a new problem for the future. We have, therefore, started an investigation of the microfloral changes in the composting process using well controlled composting systems. The cooperation of biochemists, ecologists, agricultural engineers and microbiologists are needed for the study of this field.

Composting may be defined as microbial degradation of organic material by aerobic respiration which passes through a thermophilic stage. There are two types of composting systems; one is solid compost and the other is liquid compost. The latter system was developed recently and can be carried out without addition of carbonaceous waste material.

The advantages of composting can be summarized as follows. From the standpoint of waste recycling: 1) saving of natural resources, and 2) improvement of soil fertility; from the standpoint of public health, namely purification or protection of environment: 1) destruction of pathogens, harmful insects and weed seeds, and 2) removal of bad odors.

APPARATUS AND PROCEDURES OF COMPOSTING

In solid composting, 400 kg of cattle or pig manure was mixed with 100 kg of sawdust or cut straw (3 cm in length) and piled in a wooden box of 1 m in width, length and height (5). The container was aerated at a rate of 5-10 volume % per min from the bottom screen. The temperatures of upper, middle and lower part of compost pile were recorded continuously.

A plastic cylindrical container (30 cm in diameter and 100 cm in height) covered with lagging was used for composting of poultry manure. The vessel contained a mixture of 40 kg of poultry manure and 10 kg of sawdust and was aerated at a rate of 6 volume % per min from the bottom screen. In the case of solid composting, water content was adjusted to about 70%.

A tank for liquid composting made of fiberglass reinforced plastic and covered with lagging has a size of 123 cm, 142 cm and 128 cm in bottom, upper diameter and height, respectively. The tank contained 500 kg of cattle manure and 400 kg of water and was agitated and aerated with a submerged air pump, which was set in the bottom of the tank, at a rate of 6 volume % per min.

Fig. 1 indicates the temperature curves observed in the solid composting process during which turning of the compost pile was not carried out. The solid line and broken line represent composting carried out in February, 1975 (cold season) and in September, 1975 (warm season), respectively.

In the experiment shown in Fig. 2, turning was carried out every day and therefore the high temperature, namely, thermophilic stage was accomplished repeatedly for a long period. It is noteworthy that at the upper part of the compost pile a temperature of 78-79°C was maintained for several hours repeatedly.

Fig. 3 shows the progress of a liquid composting system which was aerated at the rate of 6 volume % per min. In this experiment, a high temperature (near 80°C) had been maintained for about 10 days.

PROCEDURES OF BACTERIAL VIABLE COUNTS

It is very difficult to obtain exact bacterial viable counts for these ecologically complex systems. For example, we grouped bacterial colonies into 8 colony types in accordance with their size, color and some other appearances as shown in Table 2. When we prepare serial dilutions of

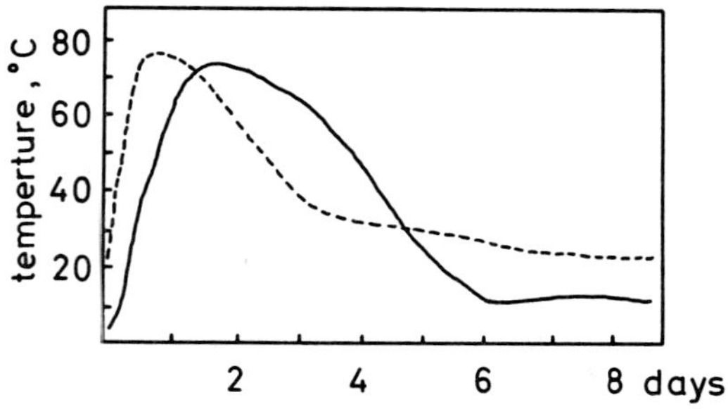

FIGURE 1. Composting Process of Cattle Manure. Cattle manure (400 kg) and sawdust (100 kg) were mixed and piled in the wooden container (1 x 1 x 1 m). The contents were aerated from the bottom screen at a rate of 6 volume per cent per min without turning. Operation in September (warm season): -----, and in February (cold season): ———

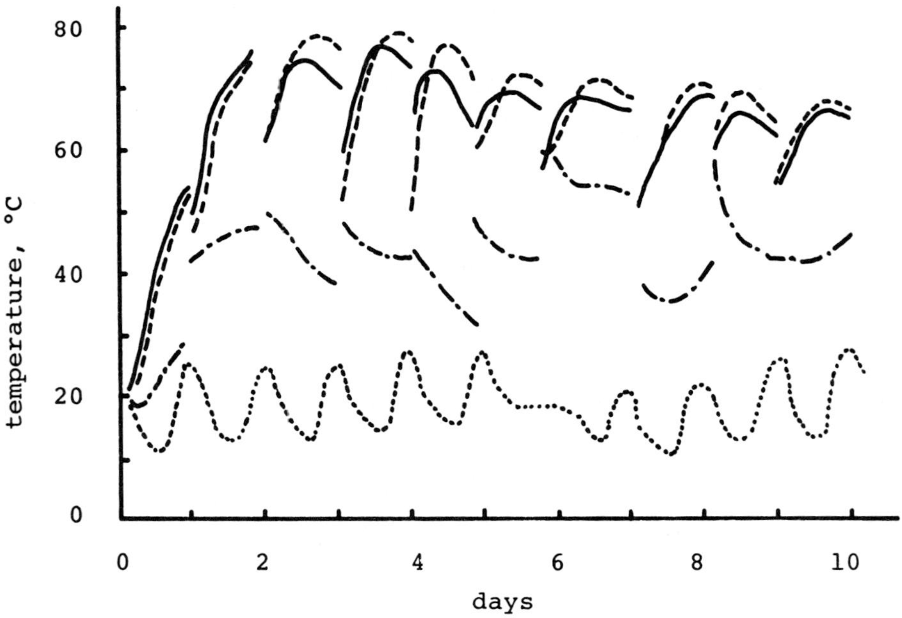

FIGURE 2. Composting Process of Cattle Manure with Turning. Experimental conditions were as in Fig. 1 except that turning was carried out every day. Temperatures of upper (-----), middle (———) and lower (–·–·–) part of compost, and atmospheric temperature (·······) were recorded continuously.

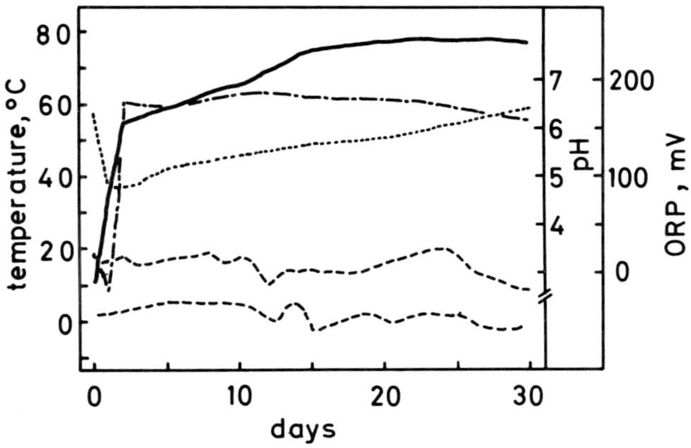

FIGURE 3. Progress of Liquid Composting. Cattle manure (500 kg) and water (400 kg) contained in a plastic tank were mixed and aerated at a rate of 6 volume per cent per min with a submersible air pump which was set at the bottom. ——— temperature of liquid compost, ····· pH, –·– oxidation-reduction potential, ---- maximum or minimum atmospheric temperature of each day.

TABLE 2
Colony Types on Bouillon Agar Medium

Colony type	*Diameter (mm)*	*Remarks*
1	5-7	white, undulate and dry
2	1-2	yellow, wet and bright
3	1-2	yellow-brown, wet and bright
4	1	gray
5	1	creamy-white
6	4-5	brownish-white, undulate
7	3-7	white, wet and flat
8	1	yellow-brown

compost suspensions in test tubes with shaking by hand or with the use of simple test tube shaker such as Vortex mixer, the viable counts fluctuated greatly as shown in Table 3. For example, 4400 colonies of type 3 appeared on petri dish No. 1, but no colonies appeared on petri dish No. 7. In this context, we first examined the procedures for preparing compost suspensions, namely, effect of buffers employed, homogenizers, time period of homogenization, and the culture medium. All procedures were performed at room temperature.

Since homogenization in a Waring blender (Power Homogenizer, Nihon Seiki Seisakusho Co. Ltd., Tokyo) from 2 to 25 min exhibited no significant influence on bacterial viable counts, we adopted 5 min as a standard homogenization time (Table 4). Among the diluents tested, distilled water gave the lowest viable counts. Although an 0.1% yeast extract solution gave the best results, it is not suitable for practical use because of the high foaming activity during homogenization. Sterilized tap water is considered to be adequate for the present purpose (Table 5).

TABLE 3
Fluctuation of Viable Counts

Colony type	*Petri dish no.*							
	1	*2*	*3*	*4*	*5*	*6*	*7*	*8*
1	87	14	27	129	0	0	0	103
2	0	277	0	0	0	1	0	0
3	4400	444	6	3	164	11	0	114
4	0	29	380	1160	104	11	0	0
5	0	0	0	0	0	0	0	0
6	2	4	11	10	8	12	322	9
7	0	0	0	0	0	0	0	0
8	0	0	0	0	7	0	0	0

Compost suspension was serially diluted using a Vortex mixer for 20 sec at each dilution. 0.1 ml of sample (10^{-8} dilution) was spread on bouillon agar and the number of colonies were counted after 48 hrs incubation at 26°C.

TABLE 4
Effect of Homogenization Time on Viable Counts

Homogenization time	*Viable counts**	
	55°C	*26°C*
2 min	--	124 ± 84
5	158 ± 67	95 ± 16
15	176 ± 14	99 ± 12
25	165 ± 43	98 ± 61

**Mean value and standard deviation from counts on 5 agar plates.*

TABLE 5
Effect of Homogenization Medium on Viable Counts

Homogenization medium	*Viable counts at 26°C**
Tap water	194 ± 74
Distilled water	94 ± 32
0.9% NaCl	212 ± 143
50 mM potassium phosphate buffer, pH 7.0	187 ± 58
0.1% yeast extract solution	248 ± 91

**Mean value and standard deviation from counts on 5 agar plates.*

Egg albumin medium (a very poor medium), consisting of egg albumin 0.025%, glucose 0.1%, K_2HPO_4 0.05%, $MgSO_4 \cdot 7H_2O$ 0.02%, $Fe(SO_3)_3$ trace and agar 2.5%, pH 7.2, was compared with bouillon medium, consisting of meat extract 0.7%, peptone 1.0%, NaCl 0.3% and agar 2.5%, for viable counts. As shown in Table 6, egg albumin medium gave 10^{-1} to 10^{-2} lower viable counts than did bouillon medium. The examinations described above were carried out with poultry manure.

A standard procedure was adopted in which the compost suspension was prepared by homogenization for 5 min in a Waring blender for the first dilution (and second dilution when necessary), and by shaking with hand for 1 min vigorously for the 2nd (or 3rd) and further dilutions. Mesophiles and thermophiles were counted on bouillon agar, pH 7.2, at 26°C (6 to 7 days incubation), and 55°C (2 to 3 days incubation), respectively. Coliform bacteria were counted on McConkey medium at 37°C (24 hr incubation). Viable counts of anerobic bacteria were performed using a Gas Pak Apparatus (Becton, Dickinson, Co.) at 26°C (6 to 7 days incubation) and at 55°C (3 days incubation). Table 7 shows one example of colony counts using the standardized procedure. Viable counts corresponding to each dilution were obtained and the degree of accuracy of the counts was considered to be satisfactory.

TABLE 6
Comparison between Egg Albumin and Bouillon Media

Sample no.	*26°C*		*55°C*	
	a	*b*	*a*	*b*
1	4×10^8	57×10^8	1×10^4	67×10^4
2			6×10^4	73×10^4
3	9×10^5	360×10^5	2×10^3	200×10^3

a: egg albumin medium, b: bouillon medium.
Viable counts are expressed as mean value of 5 plates.

TABLE 7
Viable Counts with Standardized Dilution Procedure of Compost Suspension

Plate no.	*Colony counts at 55°C dilution*			
	10^{-5}	10^{-6}	10^{-7}	10^{-8}
1	1020	229	27	5
2	880	190	21	2
3	*	294	16	4
4	*	180	21	6
5	*	225	10	1
Average	950	224	19	3.6
S.D.	70	40	4	1.5

**Count was not possible.*

CHANGES IN BACTERIAL POPULATION DURING THERMOGENETIC COMPOSTING

All solid compost samples were preserved at 3-5°C until use and subjected to analysis within 5 days after sampling. Preservation in the cold room for at least one week did not significantly influence the results.

During cattle manure composting, mesophilic counts decreased from 3 x 10^9 per gram of wet compost to 2 x 10^7, while the thermophilic counts increased from 3 x 10^6 to 2 x 10^9 (Table 8). Viable counts are expressed as values per gram of wet solid compost or per ml of liquid compost through this paper. With the pig manure compost almost the same results were obtained. It is noteworthy that coliform bacteria rapidly disappeared (Table 9).

Unexpectedly, however, the thermophilic counts did not increase in the composting of poultry manure even when the temperature increased above 70°C. The counts of anerobic thermophiles were confirmed to be of a low level also (Table 10).

The thermophilic counts of liquid compost were also observed to be low (below 10^6 per ml) even at temperatures of

TABLE 8
Changes in Bacterial Population during Composting of Cattle Manure

Day	Temperature of compost	Viable count 26°C	Viable count 55°C
0	15°C	3×10^9	3×10^6
8	73°C	2×10^7	2×10^9

TABLE 9
Changes in Bacterial Population during Composting of Pig Feces

Day	Temperature of compost	Viable count 26°C	Viable count 55°C	Viable count coliform
0	20°C	5.5×10^9	1.1×10^6	4.0×10^8
1	71°C	1.1×10^8	4.2×10^7	1.5×10^5
2	73°C	3.6×10^7	2.5×10^8	1.0×10^3

70 to 80°C (Table 11). Coliform counts decreased rapidly in this system too.

The pH of the poultry manure suspension (at 10^{-1} to 10^{-2} dilution) was observed to be alkaline (as high as 9.0-9.5) and the thermophilic counts could be increased 3-4 fold on agar medium of pH 9. Therefore, the moderately alkalophilic thermophiles are considered to be dominant in the poultry manure compost. The growth rate of one such strain at various pHs is indicated in Fig. 4. At pH 9 the strain showed good growth but at pH 7 almost no growth occurred.

TABLE 10
Changes in Bacterial Populations during Composting of Poultry Manure

Day	*Temperature of compost*	*Aerobes* 26°C	*Aerobes* 55°C	*Anaerobes* 55°C
0	15°C	10^9-10^{10}	10^5-10^6	10^4-10^5
1	52°C	10^8	10^5	10^4
2	71°C	10^6	10^6	10^3
30	15°C	10^8	10^4	10^5

TABLE 11
Changes in Bacterial Populations during Liquid Composting

Day	*Temperature of compost*	*pH*	*Viable counts* 26°C	*Viable counts* 55°C	*Viable counts* coliform
0	10°C	6.3	2×10^8	6×10^4	1×10^5
2	55°C	4.9	9×10^7	3×10^5	0
13	70°C	5.5	1×10^7	8×10^4	0
26	75°C	6.3	2×10^5	6×10^6	0

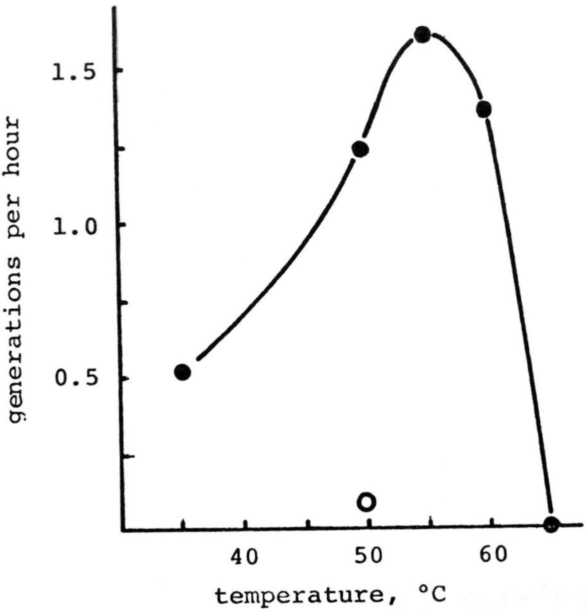

FIGURE 4. Growth Rates of a Moderately Alkalophilic Thermophile at Different pHs. Strain No. AK-12 isolated from poultry manure compost was grown on bouillon medium at pH 7 (O) and pH 9 (●) using a Monod tube shaker.

SUMMARY

1. Solid and liquid composting could successfully be carried out using the apparatus which we devised.

2. Procedures for bacterial viable counts of ecologically complex systems, such as solid and liquid compost, were established.

3. With this procedure, thermophilic counts of cattle manure and pig manure compost were estimated to be 10^8 to 10^9 per gram wet weight of compost at the thermophilic stage, respectively.

4. The thermophilic counts of poultry manure and liquid compost, however, were estimated to be below 10^6 per gram or ml of sample. These results may suggest the presence of unknown types of thermophiles.

5. Moderately alkalophilic thermophiles seem to be dominant in poultry manure compost.

6. It is important from the point of view of public health that coliform bacteria rapidly disappeared during the

course of self-heating composting.

7. Inocula or other additions were not needed in the composting of animal manure.

ACKNOWLEDGEMENT

The authors wish to express their thanks to Dr. Nobuhiro Yano and Dr. Toshiki Morichi for their valuable advice and discussions. A part of this work was supported by a grant from the Ministry of Agriculture.

REFERENCES

1. Ogura, B. and Ouchi, T. (general eds.). (1976). *Soil Fertility in Japan*. Ochanomizu Shobo Co. Ltd., Tokyo (in Japanese).
2. Golueke, C. G. (1972). *Composting*. Rodale Press, Emmaus, Pa.
3. Poincelot, R. P. (1975). Conn. Agric. Expt. Sta. Bull. (New Haven), p. 754.
4. Finstein, M. S. and Morris, M. L. (1975). Adv. Appl. Microbiol. *19*, 113.
5. Miura, T., Izawa, T. and Morimoto, K. (1977). J. Soc. Agric. Mach. (Tokyo) *39*, 93 (in Japanese).
6. Misaizu, Y. (1970). Animal Husbandry (Tokyo) *31*, 165 (in Japanese).

TRANSFORMANTS OF *BACILLUS SUBTILIS* CAPABLE OF GROWTH AT ELEVATED TEMPERATURES

S. Marvin Friedman
Tobias Mojica-a

Department of Biological Sciences
Hunter College of the City University of New York
New York, New York

INTRODUCTION

Thermophilic bacteria are a highly specialized group of procaryotic organisms that have adapted to an extreme environment created by thermal stress. A concentrated research effort over the past two decades has attempted to explain the ability of these unique cells to survive at temperatures which cause denaturation of most biological macromolecules. These studies have shown that thermophiles can synthesize intrinsically heat-stable cell components. More recently, molecular mechanisms underlying the thermal stabilities of membranes, proteins and the protein-synthesizing machinery of thermophilic bacteria have been proposed (for reviews see 1-4).

In spite of these advances which were achieved through biochemical and biophysical analysis of cellular constituents, very little is known about the genetic control of thermophily. There have been several recent reports, however, of genetic exchange systems involving bacteria that are capable of growth at elevated temperature. Thus, McDonald and Matney (5) found that strains of *Bacillis subtilis* which could not grow at 55°C were transformed to growth at this temperature by DNA from *B. subtilis* strains capable of growth at 55°C. The frequency of this transformation event was 10^{-4} and only 10-20% of the transformants retained the high level streptomycin resistance of the recipient, indicating close linkage of str^{s} and 55^{+} loci. Isono (6) reported that DNA from wild type *Bacillus stearothermophilus* transformed *amylase-*

ISBN 0-12-268250-5

deficient mutants of *B. stearothermophilus* to produce this enzyme. N. Welker (see paper in this volume) has achieved a low frequency transformation system using donor DNA from a rifampicin resistant mutant of *B. stearothermophilus* and a strain of *B. stearothermophilus* cured of prophage as recipient. Lindsay and Creaser (7) reported that DNA from the extreme thermophile, *Bacillus caldolyticus*, transformed *B. subtilis* to growth at 70°C. Cotransfer of genes mapping close to those coding for ribosomal and transfer RNA functions was observed. In addition, L-histidinol dehydrogenase isolated from a high temperature transformant was stable even at 100°C, whereas the original *B. subtilis* enzyme was inactivated above 70°C.

The two-dimensional gel electrophoretic patterns of ribosomal proteins from *B. subtilis* and *B. stearothermophilus* are quite distinct from each other (8). It seemed feasible, therefore, to identify the complement of ribosomal proteins associated with the property of thermostability by attempting to transfer genes for growth at high temperature from obligate thermophilic bacilli to the mesophile, *B. subtilis*.

MATERIALS AND METHODS

Bacterial strains: *B. subtilis* 168 *trp*$^-$ and *B. subtilis* BD-29 *arg*$^-$ *leu*$^-$ were used as recipients; *B. stearothermophilus* 2184 *str*^{r} (HC2) and *B. caldolyticus* T-10 *str*^{r} (HC11) were the sources of donor DNA.

Transforming DNA was prepared by the following procedures: phenol extraction (9), equilibrium sedimentation in an ethidium bromide-cesium chloride gradient (10) and a cleared lysate method (11).

Competent cultures were prepared according to Spizizen (12). The post-transformation protocol was that described by Lindsay and Creaser (7) except that LB broth (13) or LA agar plates were used and the selective growth temperature was 65°C.

Growth of cells (13), isolation of ribosomes and polyacrylamide gel electrophoresis of ribosomal proteins (14) and the assay for ribosome stability (15) have all been previously described.

RESULTS AND DISCUSSION

The characteristics of parental, transformant and revertant bacilli are presented in Table 1. In 69 attempts at transformation, stable putative transformants were obtained only in two experiments. During several trials, we observed colonies growing at 65°C, but upon restreaking for single cell isolation these clones did not continue to grow at 65°C. One of the stable transformants, HC31, was obtained by exposing the recipient, *B. subtilis* 168 *trp*$^-$, to 50 μg of *B. caldolyticus* (HC11) DNA, whereas the second transformant, HC14, was obtained by treating recipient cells with 50 μg of phenol-extracted *B. stearothermophilus* (HC2) DNA. No transformants were obtained when DNA was omitted from the transformation mixtures and sterility tests of the DNA preparations carried out at both 37°C and 65°C were negative.

Both transformants (HC14 and HC31) retained the *trp*$^-$ marker of the recipient, but neither transformant expressed the high level streptomycin resistance of the donor. In addition, spontaneously occurring streptomycin resistant mutants could not be isolated from either transformant although such mutants could be obtained with typical frequencies of about 5×10^{-8} from the donor and recipient strains. Both transformants, in contrast to the recipient, had lost the ability to grow at 37°C, but had acquired, in contrast to both parental donors, the capacity to grow in Spizizen minimal medium at 65°C. Transformant HC14 was the only strain which yielded spontaneous revertants capable of growth at 37°C. Clone HC33 is a representative spontaneous revertant of HC14. Attempts to revert transformant HC31 or the parental donors, HC2 and HC11, to growth at 37°C with compounds known to "cure" extrachromosomal DNA, such as acridine orange, ethidium bromide and sodium lauryl sulfate, were unsuccessful.

In order to further characterize the genetic composition of the high growth temperature transformants, DNA isolated from these strains was tested for transforming efficiency using *B. subtilis* BD-29 *arg*$^-$ *leu*$^-$ as recipient. The results presented in Table 2 show that DNA from both transformants (HC14 and HC31) and from revertant HC33 was as efficient as homologous DNA (derived from the original recipient, *B. subtilis* 168 *trp*$^-$) in transforming *B. subtilis* BD-29 from auxotrophy to prototrophy. This experiment has been repeated four times with two different sets of donor DNA preparations. Thus, the genomes of the transformants exhibit extensive homology with the *leu-arg* segment of *B. subtilis* 168 chromosomal DNA. Prototrophic recombinants were not obtained with donor DNA from the parental thermophilic strains, HC2 and

TABLE 1
Characteristics of Parental, Transformant and Revertant Strains

Strain	*Origin*	*37°C*[a]	*65°C*[a]	*MM*[b]	Str[c]	Trp[d]
168	*B. subtilis*	+	-	+	S	-
HC11	*B. caldolyticus*	-	+	-	R	+
HC31	HC11 DNA x 168	-	+	+	S	-
HC2	*B. stearothermophilus*	-	+	-	R	+
HC14	HC2 DNA x 168	-	+	+	S	-
HC33	Revertant of HC14	+	-	+	S	-

[a]*Growth (+) or lack of growth (-) in LB broth at the indicated temperature.*

[b]*Growth (+) or lack of growth (-) in Spizizen minimal medium.*

[c]*Sensitivity (S) or Resistance (R) to streptomycin (150 μg/ml).*

[d]*Tryptophan independence (+) or dependence (-). All strains were tested in Spizizen minimal medium except for HC2 and HC11 which were tested in the minimal medium described by Fodor* et al. *(16).*

HC11. Although DNA from the high growth temperature transformants was highly efficient in transforming the *arg* and *leu* markers in *B. subtilis* BD-29, no transformants capable of growth at 65°C were obtained. In addition, donor DNA derived from *B. subtilis* 168 *str*^{r} was not capable of transforming either of the high growth temperature transformants to high level streptomycin resistance (data not shown).

LaMontagne and McDonald (17) reported that bacteriophage TSP-1 formed plaques on *B. subtilis* at 53-55°C but not at 37°C. It was of interest, therefore, to test the sensitivity of donor, recipient and transformant strains to infection by TSP-1. Small drops of a high-titer stock of bacteriophage TSP-1 were spotted on LA plates seeded with the appropriate indicator strain. After incubation at

TABLE 2
Effect of Donor DNA Source on Transformation of B. subtilis *BD-29* ($arg^- leu^-$)[a]

Donor DNA Source	*Recombinants/ml*	
	arg^+	leu^+
168	0.7×10^6	2.7×10^6
HC11	0	0
HC31	1.1×10^6	2.4×10^6
HC2	0	0
HC14	0.8×10^6	2.6×10^6
HC33	0.6×10^6	2.8×10^6

[a]*Transformation mixtures (1.0 ml) contained 5 μg of phenol-extracted DNA and 2 x* 10^8 *competent cells. Incubation was at 37°C for 30 min.*

either 37°C or 55°C, the plates were examined for clear zones of lysis. Table 3 shows that *B. subtilis* 168 *trp*$^-$ was lysed by TSP-1 at 55°C but not at 37°C. Lysis at 55°C was not observed in the case of the two parental thermophiles (HC2 and HC11) and transformant HC14. When transformant HC31 was used as the indicator strain, however, a lytic response was obtained at 55°C. Further studies are required to show whether the difference between the two transformants in response to infection by TSP-1 is due to differences in receptor sites or in some post-adsorption event, such as restriction.

Ribosomes were isolated from *B. subtilis* 168 *trp*$^-$ and HC33 grown at 37°C and from HC2, HC14, HC11 and HC31 grown at 65°C. The 70S ribosome preparations were tested for heat stability by an assay which measures the residual capacity for poly U-directed C^{14}-phenylalanine incorporation at 37°C of ribosomal suspensions heated at 60°C for varying times. Table 4 shows that ribosomes from *B. subtilis* 168 and HC33 were relatively thermolabile, retaining only 6-7% of the initial protein-synthesizing activity after heating at 60°C for 30 minutes. On the other hand, ribosomes from the parental thermophiles (HC2 and HC11) and transformants HC14 and HC31 were relatively thermostable, displaying 68-107% of the

TABLE 3
Sensitivity of Parental and Transformant Strains to Bacteriophage TSP-1

Indicator strain	Lytic response	
	37°C	55°C
168	-	+
HC11	UG	-
HC31	UG	+
HC2	UG	-
HC14	UG	-

UG = indicator strain is unable to grow at this temperature. + = lysis; - = no lysis.

TABLE 4
Heat Stability of Ribosomes from Parental, Transformant and Revertant Strains[a]

Ribosome source	% Initial activity after incubation at 60°C			
	5 min	10 min	20 min	30 min
168	21	18	9	7
HC11	114	109	98	87
HC31	125	112	110	107
HC2	105	97	101	92
HC14	110	90	78	68
HC33	27	12	7	6

[a]Ribosome suspensions (1.5 ml) containing 150 A_{260}/ml were heated at 60°C. At the indicated times, 50 µl aliquots were withdrawn and assayed for poly U-directed C^{14}-phenylalanine incorporation at 37°C as described by Friedman (15).

initial protein-synthesizing activity after heating at 60°C for 30 minutes. It is of interest to note that ribosomes from transformant HC14 were somewhat less stable than those of its parental donor strain (HC2) and this difference also correlates with differing growth rates of these two strains at 65°C.

Ribosomal proteins from parental, transformant and revertant bacilli were separated by the two-dimensional polyacrylamide gel electrophoresis system of Kaltschmidt and Wittmann. Fig. 1a shows the gel pattern of *B. subtilis* 168 *trp*⁻ 70S ribosomal proteins. Figs. 1b and c show the gel patterns of 70S ribosomal proteins from the *B. caldolyticus* donor (HC11) and corresponding transformant (HC31), respectively. The gel pattern of HC11 ribosomal proteins is quite distinct from that of *B. subtilis* and furthermore, there are no immediately apparent differences between the donor and transformant patterns. Figs. 1d and e show the gel patterns of 70S ribosomal proteins from the *B. stearothermophilus* donor (HC2) and corresponding transformant (HC14), respectively. The pattern of HC2 can easily be distinguished from both those of *B. subtilis* and HC11, but once again the patterns of donor and transformant appear to be identical. Finally, the gel pattern for revertant HC33 (not shown) seems to correspond in every detail to that of *B. subtilis*. We have also separated 30S and 50S ribosomal proteins from these five strains but were unable to detect any differences between those of recipient and revertant bacilli or between those of thermophilic donors and their corresponding high growth temperature transformants (data not shown). Although the electrophoretic mobilities of ribosomal proteins from donor and transformant strains appear to be identical, biochemical analysis of similar proteins will be required to clearly establish their identity or non-identity.

This paper describes some characteristics of a genetic exchange system that can prove valuable for studying the genetic control of growth at elevated temperatures. Preliminary evidence suggests that genes coding for ribosomal proteins have been transferred from donor to recipient during transformation to thermophily. Work is in progress to determine whether these genes are integrated into the recipient genome or exist in an extrachromosomal state. In addition, the extent of homology existing between transformant and recipient genomes is being investigated. It will also be of interest to compare the structures and stabilities of additional cell components, such as enzymes and membranes, isolated from parental and transformant strains. Finally, attempts are being made to enhance the frequency of transformations to thermophily so that a larger number of

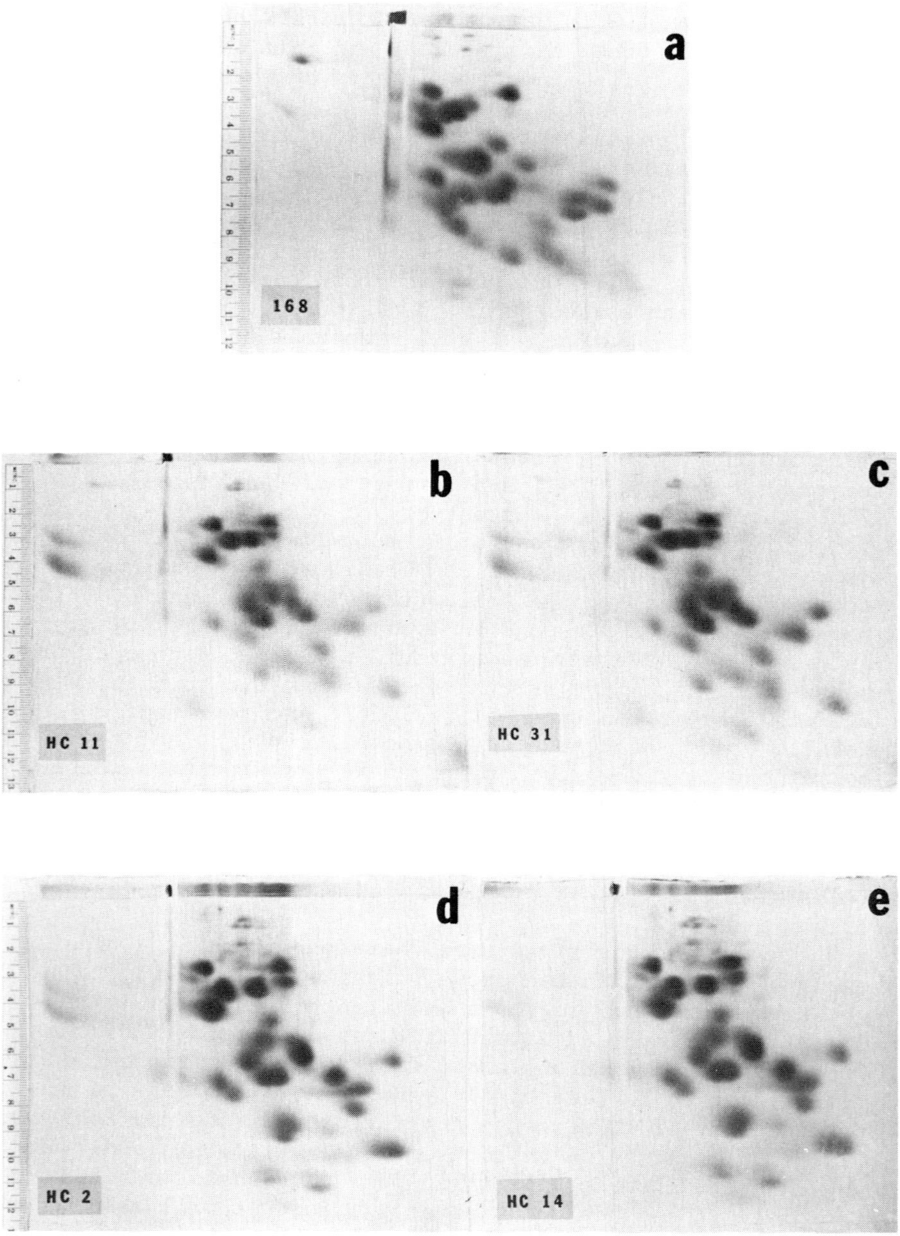

FIGURE 1. Two-dimensional polyacrylamide gel electrophoretic patterns of 70S ribosomal proteins from B. subtilis *168 (a), HC11 (b), HC31 (c), HC2 (d) and HC14 (e).*

transformants would be available for study.

Note Added in Proof: Since the report of these findings, some changes in the properties of the transformants have been observed. It is possible that reversion of certain traits has occurred during the storage of these strains in stab cultures.

ACKNOWLEDGEMENTS

We would like to thank Drs. D. Cheung and H. Murofushi for their help with gel electrophoresis. This investigation was supported by U. S. Public Health Service grant GM-22164 and BHE-PSC Faculty Research Award grant 11616.

REFERENCES

1. Friedman, S. M. (1968). Bacteriol. Rev. *32*, 27.
2. Singleton, R. and Amelunxen, R. E. (1973). Bacteriol. Rev. *37*, 320.
3. Zuber, H., ed. (1976). *Enzymes and Proteins from Thermophilic Microorganisms* (Experientia Supplement 26). Birkhauser Verlag, Basel.
4. Friedman, S. M., ed. (1978). *Biochemistry of Thermophily*. Academic Press, New York.
5. McDonald, W. C. and Matney, T. S. (1963). J. Bacteriol. *85*, 218.
6. Isono, K. (1970). J. Genet. (Tokyo) *45*, 285.
7. Lindsay, J. A. and Creaser, E. H. (1975). Nature *255*, 650.
8. Geisser, M., Tischendorf, G. W. and Stoffler, G. (1973). Mol. Gen. Genet. *127*, 129.
9. Cosloy, S. D. and Oishi, M. (1973). Mol. Gen. Genet. *124*, 1.
10. Radloff, R., Bauer, W. and Vinograd, J. (1967). Proc. Natl. Acad. Sci. USA *54*, 1514.
11. Guerry, P., LeBlanc, D. J. and Falkow, S. (1973). J. Bacteriol. *116*, 1064.
12. Spizizen, J. (1958). Proc. Natl. Acad. Sci. USA *44*, 1072.
13. Firedman, S. M. and Weinstein, I. B. (1966). Biochim. Biophys. Acta *114*, 593.
14. Miller, H. M., Friedman, S. M., Litman, D. J. and Cantor, C. R. (1976). Mol. Gen. Genet. *144*, 273.
15. Friedman, S. M. (1971). J. Bacteriol. *108*, 589.

16. Fodor, K., Hadlackzky, G. and Alfoldi, L. (1975). J. Bacteriol. *121*, 390.
17. LaMontagne, J. R. and McDonald, W. C. (1972). J. Virol. *9*, 646.

PHYSIOLOGICAL AND GENETIC FACTORS AFFECTING TRANSFECTION AND TRANSFORMATION IN *BACILLUS STEAROTHERMOPHILUS*

Neil E. Welker

Department of Biochemistry and Molecular Biology
Northwestern University
Evanston, Illinois

INTRODUCTION

Since the initial discovery of a thermophilic bacterium (1), thermophilic microorganisms have been the focus of considerable scientific and commercial interest. Over the past two decades biochemists and molecular biologists have used thermophilic microorganisms to study a variety of biological and biochemical problems. In addition, as biological systems, thermophiles have been increasingly applied to industrial technology. As a result of these studies, we now have a greater understanding of the biochemical basis of thermophily and a realization of the potential usefulness of these organisms in basic research and in applied technology. Although comparative biochemistry has provided a good deal of information concerning the biochemical basis of thermophily (for reviews see 2-6), the development of a genetic system would afford a unique opportunity to investigate the genetic as well as the biochemical properties of thermophily.

To my knowledge, there is only one published report on the exchange of genetic information in thermophilic bacteria. Isono (7) reported that deoxyribonucleic acid (DNA) from wildtype *Bacillus stearothermophilus* would transform several α-amylase-deficient mutants to produce wildtype levels of this enzyme. In this transformation system, the strain used was a derivative of *B. stearothermophilus* (ATCC 12980; an obligate thermophile) that would grow at 37°C (facultative thermophile) and transformation was carried out at 37°C. Since no additional reports have been published on this

ISBN 0-12-268250-5

transformation system, I assume that the low temperature requirement precludes its usefulness for studies on the genetic control of thermophily.

Since 1965 we have examined a number of strains for their susceptibility to DNA-mediated transformation. In these investigations, the procedures used were those which proved to be successful in establishing genetic systems in other Gram-positive bacilli (8-11). Unfortunately, DNA-mediated transformation was never detected in these studies and we abandoned this approach in favor of one which we thought had a greater probability of success.

In the *Bacillus subtilis* transformation system, it is now generally agreed that cells which are competent for transformation are also competent for transfection (see review 12) and since transfection requires neither the selection of mutants nor recombination of the newly-acquired DNA with the recipient chromosome, we used it as a screening technique for the identification of strains which are susceptible to DNA-mediated transformation.

HISTORICAL DEVELOPMENT OF TRANSFECTION IN *B. STEAROTHERMOPHILUS*

A number of strains of *B. stearothermophilus* were tested for transfection and only strain 4S was susceptible to infection with phage TP-1C DNA (13). The optimal conditions for the transfection of this competent (Com^+) strain were reported by Streips and Welker (14). In contrast to transfection in *B. subtilis* (15), a strict competence regimen such as a step-down culture or restriction of essential medium components is not necessary for the development of competence in this strain. Mid-log phase cells of cultures growing under conditions which support rapid growth (complex medium; 55-61°C) are the most susceptible to infection with phage DNA.

A cellular component (competence factor or competence-inducing factor) which enhances transfection is present in the supernatant fluids of Com^+ cultures grown to the stationary phase (14, 16). Transfection of Com^+ cells in the absence of competence factor shows that infectious centers appear between 70 and 80 min and are insensitive to deoxyribonuclease by 85 min. In the presence of competence factor, the time of the appearance of infectious centers and insensitivity to deoxyribonuclease is advanced by 40 min.

An incompetent mutant ($4SCom^-$) was isolated which is not infected with phage DNA unless competence factor is added (16). The kinetics of transfection, with exogenous competence factor, in this mutant is identical to that observed with Com+. The culture supernatant fluids of the Com− mutant do not contain competence factor.

Several mutants of Com+, resistant to phage TP-1C (Ton-r), were isolated and none of these mutants can be infected with phage DNA, even in the presence of competence factor. The Ton-r mutants, however, release relatively large amounts of competence factor into the culture medium.

The number of infectious centers is directly porportional to the concentration of phage DNA (13). These results indicate that a recipient bacterium is infected with a single phage DNA molecule.

The efficiency of transfection is the highest in cells growing at 59-62°C (14; N.E. Welker, unpublished data) and does not occur in cells growing at 67°C. The loss of transfection in cells growing at 67°C is not a result of reduced growth or impaired development of phage TP-1C because cultures of Com+ and the Com− mutant exhibit the greatest rate of growth at 67°C (doubling time of 16 min).

The average burst size of phage TP-1C by Com+ or Com− cells is 50 plaque forming units (PFU) per cell (compared to 93 PFU/cell at 55°C).

The ability of Com+ or the Com− mutant to react with competence factor is lost after these cultures have grown at 67°C for one hour (almost four doublings) or longer (16). These results indicate that a cell growing at 67°C does not contain binding 'sites' to which competence factor can bind to allow the uptake of phage DNA.

Cells competent for transfection were not competent for transformation with DNA isolated from a streptomycin-resistant (14), or a rifampicin-resistant mutant of Com+ (N.E. Welker, unpublished data).

Competence-inducing (or provoking) factors or activators had been identified in a number of transformation systems (17-20). Since competence factors were generally thought necessary for the attainment of the competent state, we turned our attention to the isolation and characterization of the competence factor.

The culture fluids of Ton-r were subjected to a variety of procedures generally used for the isolation and characterization of proteins. Although competence factor was unstable in the purified state, it appeared to have an unusually large molecular weight and did not have lytic, nuclease, or proteolytic activity (N.E. Welker, unpublished data).

Using crude competence factor from culture fluids of the Ton-r mutant, Streips and Young (21) reported that competence factor interacts with Com+ and Com− mutant (but not Ton-r) cells, protoplasts, or cytoplasmic membranes in a temperature dependent reaction. A fraction of a population of cells which were incubated with competence factor sediment more slowly in Renografin. In *B. subtilis* it was demonstrated that competent cells sediment more slowly in Renografin (22, 23). Streips and Young (21) concluded that the association between the cytoplasmic membrane and competence factor is the first and rate-limiting step in the process of transfection with phage DNA and that the cytoplasmic membrane-competence factor complex is capable of removing infectious DNA from the supernatant fluids in vitro.

Although *B. stearothermophilus* 4S Com$^-$ was known to be lysogenized with phage TP-8, this phage or some phage component was not the competence factor (16). Phage TP-8 was detected in the supernatant fluids or mitomycin C-induced lysates of Com$^+$, the Com$^-$ mutant, or the Ton-r mutant cultures using *B. stearothermophilus* 141S as the indicator.

In March of 1971, a mutant of 4S Com$^+$ was isolated which was sensitive to phage TP-8 and an additional phage present in the supernatant fluids of Com$^-$ and Ton-r mutant cultures. The new phage was designated TP-12 and the phage sensitive mutant (4S) was cured of phages TP-8 and TP-12. The isolation of the cured strain was crucial in demonstrating that phage TP-12 is the competence factor. A preliminary report of this was presented at the XV Annual Transformation Meeting in June, 1971.

The various derivatives of *B. stearothermophilus* 4S used in transfection studies are listed in Table 1. The new strain designation will be used in this report.

PHAGE TP-12 IS THE COMPETENCE FACTOR

A preliminary report demonstrating that phage TP-12 is the competence factor was presented by Eager and Welker at the Annual Meeting of the American Society for Microbiology (1972).

Competence factor is found only in the culture fluids of strains lysogenized with phage TP-12. Purified phage TP-12 suspensions (13) are more efficient than the culture fluids of 4S(8)(12) in the transfection of 4S(8). Phage TP-12 antiserum, sufficient to inactivate > 99% of a mature phage population, inhibits transfection. Competence-inducing activity copurifies with phage TP-12 plaque-forming units and incubation of a suspension of phage TP-12 with deoxyribonuclease, ribonuclease, proteolytic enzymes, lipases, and detergents (non-ionic and anionic) did not destroy competence-inducing activity. These results demonstrate that phage TP-12 and not other cellular components released along with phage TP-12 is the competence factor. Competence for transfection in

TABLE 1
Strains of Bacillus stearothermophilus *4S Used in Transfection Studies*

New designation	*Previous designation*	*Comments*
4S(8)(12)	4S[a] or 4S Com^{+}[b]	Wildtype; lysogenic for phages TP-8 and TP-12
4S(8)(12)/1[c]	4S Ton-r[b]	Spontaneous mutant of 4S(8)(12) resistant to phage TP-1C
4S(8)	4S Com^{-}[b]	Spontaneous loss of phage TP-12 from strain 4S(8)(12)
4S	-	Cured strain; nitroso-guanidine treatment of 4S(8)(12) followed by growth at 68°C
4S(12)[d]	-	Lysogenic derivative of strain 4S

[a]*Welker and Campbell (24); Streips and Welker (13).*

[b]*Streips and Welker (14).*

[c]*Resistant to phage TP-1C.*

[d]*Lees and Welker (25).*

B. stearothermophilus appears to be dependent on the helper phage (competence factor).

THE TRANSFECTION ASSAY

Mid- or late-log-phage cells of 4S(8)(12), or 4S(8), or 4S(12) are maximally susceptible to phage DNA infection (competent) when grown in a Trypticase-yeast extract-glucose (TYG; 24) medium at 55-61°C. The transfection mixture contained 5-6 x 10^6 viable competent cells, 0.0005 - 0.7 μg phage DNA, and 5-6 x 10^6 plaque-forming units of helper phage. The volume was adjusted to 2 ml with TYG medium. The mixture was incubated with aeration at 55-61°C for 80-120 min and transfection was terminated with deoxyribonuclease (0.05 μg/ml). Infectious centers were assayed on a sensitive indicator strain (13).

Transfection was also assayed on Trypticase agar (TA; 24) plates. Competent cells (0.7 - 1 x 10^8 cells), phage DNA (0.004 - 0.008 μg) and helper phage (1 - 4 x 10^7 plaque-forming units) were spread over the surface of a TA phage with a bent glass rod and incubated at 55°C for 12-16 hr.

The kinetics of phage DNA infection of cells of strain 4S(8)(12) are shown in Figure 1. Without exogenous helper phage (Figure 1, curve B), infectious centers appear initially by 63 min and are insensitive to deoxyribonuclease by 85 min. The addition of exogenous helper phage (0.5 - 3 x 10^7 plaque-forming units per ml), advances the time of the initial appearance of infectious centers and insensitivity to deoxyribonuclease by 50-60 min (Figure 1, curve A). The time between the addition of helper phage and DNA to the cells and the initial appearance of infectious centers is 3 min, even in the presence of excess helper phage. Similar kinetics are found with 4S(12) (with and without exogenous helper phase) and 4S(8) (with helper phage; Figure 1, curve A).

The time of the initial appearance of infectious centers was variable and was dependent on the number of plaque-forming units of helper phage in the culture fluids. The ratio of the number of helper phage (plaque-forming units) to the number of cells (viable cell counts) must be 1.0 or greater in order to obtain the kinetics shown in Figure 1 (curve A).

In *B. subtilis*, two classes of transfecting DNA are obtained from the various phages (12). The distinction is based on the relation between the concentration of phage DNA and the number of infectious centers (dose response). One class of phage DNA's gives a linear or near linear dose response and the other class of phage DNA's gives a dose response proportional to the second or third power of the DNA concentration. The non-linear dose response curve has been attributed to the inactivation of infectious DNA after the DNA has entered the cell or that a successful infection requires recombination between two or more phage genomes (primary recombination). The dose response curve for DNA of phages TP-1C and TP-84 (26) is shown in Figure 2. A linear dose response for each phage DNA (TP-1C, slope of 1.07; TP-84, slope of 0.97) indicates that a single phage DNA molecule is sufficient to infect a cell. The number of infectious centers produced with phage TP-84 DNA is generally ten times that observed with phage TP-1C DNA. The larger molecular weight of TP-84 DNA (28 x 10^6 daltons; 27) compared to that of TP-1C DNA (12 x 10^6

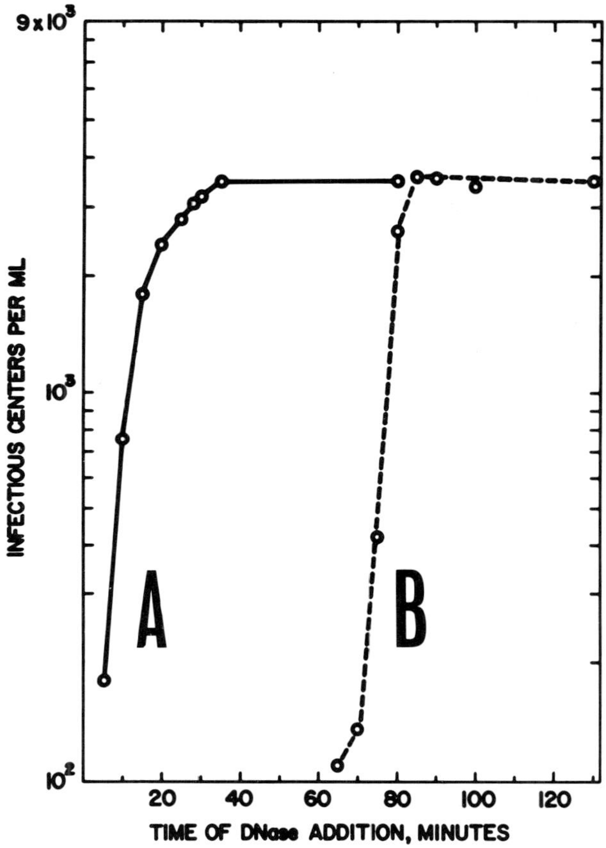

FIGURE 1. Kinetics of phage TP-1C DNA infection of B. stearothermophilus *4S(8)(12). Transfection was run in TYG medium at 55°C. At the indicated times, deoxyribonuclease was added and samples were assayed for infectious centers. Curve A (—O—), plus phage TP-12 (ratio of two phage per cell); Curve B (--O--), no exogenous phage TP-12 added.*

daltons; 28) may account for the enhanced efficiency of transfection with phage TP-84 DNA because of a greater affinity between a TP-84 DNA molecule and a competent cell.

The dose response curve for helper phage is shown in Figure 3. A linear dose response curve (slope of 1.01) indicates that one helper phage is sufficient to initiate an infection of a cell with a DNA molecule of TP-1C or TP-84. The proportion of competent cells in a population of 4S(8)

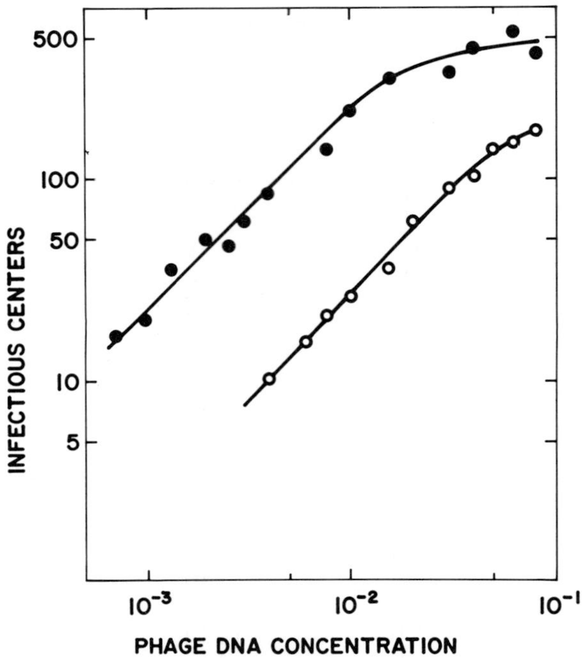

FIGURE 2. Effect of phage DNA concentration on transfection of B. stearothermophilus *45 (8). Transfection was run in TYG medium or on TA plates at 55°C. Helper phage was added at a multiplicity of three plaque-forming units per viable cell. Data are plotted as infectious centers and phage DNA concentration per 0.1 ml. (-●-), TP-84 DNA; (-O-), TP-1C DNA.*

cells can be estimated, using saturating amounts of phage DNA and helper phage and assuming that each infectious center arises from an infected cell. The frequency of transfection was found to be between 0.03 - 0.05% in TYG medium and between 0.065 - 0.085% on TA plates.

MODE OF ACTION OF HELPER PHAGE IN TRANSFECTION OF *B. STEAROTHERMOPHILUS*

Although a linear dose-response curve for transfecting DNA indicates that a primary recombination between two or more infectious DNA genomes is not required to produce a successful transfection, it is possible, however, that the helper phage genome participates in transfection.

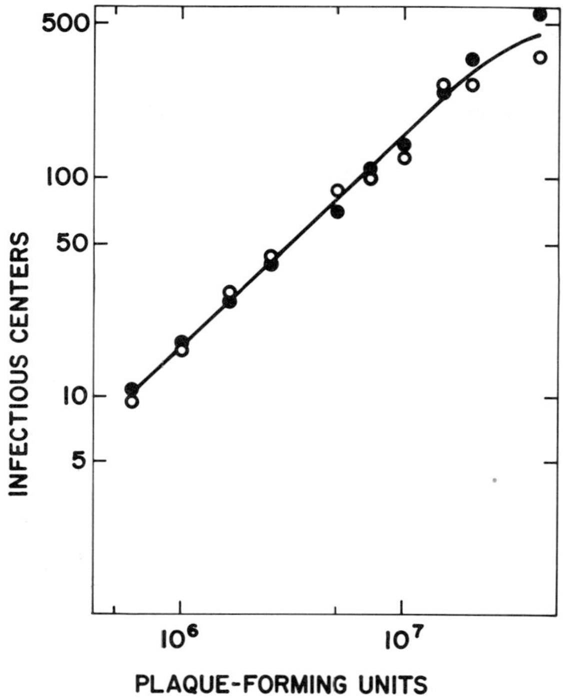

FIGURE 3. Effect of helper phage on transfection of B. stearothermophilus *45 (8). Transfection was run in TYG medium or on TA plates at 55°C. Phage TP-84 or TP-1C DNA was added at a saturating concentration of 0.15 - 0.2 µg. Data are plotted as infectious centers and plaque-forming units per 0.1 ml. (-●-), TP-84 DNA; (-O-), TP-1C DNA.*

Infection of 4S(8) with phage DNA does not occur when phage TP-12 DNA is used in place of phage TP-12. Temperature-sensitive (ts) mutants of phage TP-1C (N.D. Lees, unpublished data) and phage TP-12 (N.E. Welker, unpublished data) were used in marker rescue studies. The ts mutants of phages TP-1C and TP-12 grow normally at the permissive temperature (55°C) but failed to grow at the nonpermissive temperature (60°C). The ts mutants of phages TP-1C and TP-12 used in these studies were defective in early functions. Transfection of 4S(8) does not occur at 60°C with infectious TP-1C ts DNA and phage TP-12 and does occur with TP-1C DNA and phage TP-12 ts. The results suggest that early genes of the helper phage genome are not involved.

The effect of uv-irradiation on the ability of helper phage to initiate transfection of 4S(8) was studied by exposing helper phage to uv-irradiation for various times and determining the plaque forming units of helper phage and the ability to initiate transfection. The results are shown in Table 2. In spite of the rapid inactivation of helper phage, the number of transfectants remains constant.

Sjostrom and Philipson (29) reported that competence of *Staphylococcus aureus* for transfection and transformation is induced by phages Ø11 and 83A when present as a prophage or when used as a helper phage. Since strains lysogenic for phage TP-12 [4S(8)(12) and 4S(12)] are susceptible to infection with phage DNA it is possible that the prophage induces the competent state. A clear plaque mutant of phage TP-12 (designated TP-12C) which does not lysogenize 4S cells is as efficient as phage TP-12 in the transfection system.

The data presented indicate that in the *B. stearothermophilus* transfection system, only intact helper phage can initiate the infection of a competent cell with a phage DNA molecule. In addition, the helper phage genome does not appear to participate in the rescue of transfecting phage DNA. Since attempts to prepare intact ghosts (DNA-less phage particles) of the helper phage have been unsuccessful, we cannot rule out the possibility that the helper phage DNA may inhibit cell nucleases which otherwise would inactivate the transfecting DNA. The absence of a multipowered dose response for the helper phage and the inability of helper phage DNA to initiate transfection argue against this proposal.

Transfection does not occur in 4S(8) cells grown in a variety of media taken from different stages of growth (14), or subjected to conditions such as limited hydrolysis of the cell wall with lysozyme and suspension in a buffer containing $CaCl_2$ (N. E. Welker, unpublished data) or subjected to limited autolysis (14).

Since the requirement for an intact helper phage cannot be eliminated in the *B. stearothermophilus* transfection system, we can postulate that the adsorption of a helper phage initiates the binding uptake of infectious phage DNA.

The studies reported by Streips and Young (21) contributed significantly to our understanding of the function of the helper phage. From these studies they concluded that there is a physical association between competence factor and the cell which alters the sedimentation properties of the cell and the site for competence factor association appears to be the cytoplasmic membrane.

The demonstration that phage TP-12 is the competence factor (helper phage) in the *B. stearothermophilus* transfection system does not change or modify the conclusions presented by these authors. Using purified helper phage, we repeated the studies reported by Streips and Young (21) and obtained similar results. I will briefly describe only those

TABLE 2
Effect of UV-inactivation of Helper Phage on the Induction of Competence[a]

uv-irradiation (min)	*Surviving fraction*	*Infectious centers per ml (x 10^3)*
0	1.00	1.64
5	0.54	1.52
10	0.48	1.48
15	0.19	1.68
20	0.13	1.52
25	0.06	1.52
30	0.01	1.48

[a]*Phage TP-12 suspended in TYG medium was UV-irradiated with a General Electric germicidal lamp at a distance of 50 cm for different times. Samples of irradiated phage were assayed for plaque-forming units and used in transfection with 4S(8) cells.*

studies which contribute additional information as to the mode of action of the helper phage.

The adsorption of helper phage to 4S(8) cells growing at 55°C is shown in Table 3. Helper phage adsorb slowly with 30% of the phage population adsorbed by the time transfection is complete (insensitive to deoxyribonuclease after 35 min of incubation). Note that the extent of binding of competence factor reported by Streips and Young (values in parenthesis; 21) is identical to the values reported here, even though their data were obtained by indirect methods. On further incubation, there is a release of adsorbed phage (14-16% of the phage initially adsorbed) followed by a second and more rapid period of adsorption with 99% of the helper phage adsorbed by 180 min. The adsorption of helper phage to a TP-12 ts lysogen [4S(12ts)] was followed because of the difficulties encountered in determing adsorption of helper phage to cells of 4S(8)(12) or 4S(12) whose culture fluids contain phage TP-12. Helper phage were mixed with a culture of 4S(12ts) growing at 55°C and unadsorbed phage TP-12 were assayed on 4S at 60°C (TP-12 forms plaques but TP12ts does not at this temperature).

Phage TP-12 adsorbed to 4S(12ts) cells at a rate (data not shown) similar to that for 4S(8) cells. We can conclude therefore that adsorption of helper phage to cells of 4S(8)(12) or 4S(12) probably occurs at a similar rate.

TABLE 3
Adsorption of Helper Phage to B. stearothermophilus *4S(8)*[a]

Time (min)	*Relative number of adsorbed phage*
0	0.00
<u>3</u>[b]	0.05
5	0.11
10	0.15 (0.11)[c]
20	0.20 (0.19)
30	0.29 (0.30)
<u>35</u>[b]	0.30
40	0.32 (0.32)
50	0.25
60	0.21
70	0.16
80	0.35
110	0.58
120	0.85
180	0.99

[a]*Phage TP-12 was added to a culture of 4S(8) growing in TYG medium at 55°C. Samples were removed at the indicated times, the cells removed by centrifugation, and the supernatant fluids assayed for unadsorbed phage on strain 4S.*

[b]*Time (underlined) of the initial appearance of infectious centers and of insensitivity to deoxyribonuclease in an identical culture containing phage TP-1C DNA and helper phage.*

[c]*Values obtained from Streips and Young (21).*

The lysogenic state must somehow affect the rate of adsorption of a coimmune phage since over 99% of a population of phage TP-12 have adsorbed to cells of 4S after 3-5 min at 55°C.

At a multiplicity of two to four helper phage per cell, approximately 92-98% of the cells should adsorb one or more helper phage.

Since 99% of a helper phage population were shown to be adsorbed to cells of 4S(8) after 180 min at 55°C, we can conclude that every cell of a population is capable of adsorbing helper phage. Transfection is complete, however, after a 35 min exposure to phage DNA and helper phage. During this period only 30% of the helper phage have adsorbed. Even though every cell is capable of adsorbing a helper phage, the relative slow rate of adsorption of helper phage during the period of phage DNA uptake indicates that only a small fraction of the cells are physiologically or biochemically competent or that only a small fraction of these cells are capable of adsorbing phage.

Alternatively, it is possible that only a small fraction of the helper phage population can initiate the uptake of phage DNA. In an attempt to resolve these two alternatives, we determined whether the reversible or irreversible adsorbed phage are involved in inducing competence. The adsorption of helper phage to 4S(8) cells growing at 55°C was determined at 3, 30 and 180 min. The cells were removed by centrifugation and the number of unadsorbed phage was determined. The cells were washed exhaustively in prewarmed TYG medium and the phage in the combined supernatant fluids were determined (recovered or reversibly bound phage). The fraction of phage not accounted for by these procedures (total phage adsorbed minus reversibly bound phage) are labeled irreversibly bound (or inactivated) phage. The results are shown in Table 4. During the period of phage DNA uptake (3-30 min), the proportion of reversibly adsorbed helper phage accounts for 10-13% of the adsorbed phage, whereas by 180 min the proportion drops to less than 1%. Transfection of exhaustively washed cells, which were exposed to helper phage for 30 min, occurs at control levels (nonwashed cells plus helper phage) without additional helper phage. Helper phage recovered from exhaustively washed cells will initiate transfection at an efficiency identical to that of helper phage adsorbing for the first time. Identical results were obtained with helper phage obtained after two additional cycles of adsorption and recovery.

The results indicate that each helper phage in a population can adsorb to a competent cell and that the initiation of phage DNA uptake probably requires an irreversible adsorption of the helper phage. We have not determined if the phage genome is still within the capsid or if it has been ejected.

The relatively low frequency of transfection is probably related to the presence of a small fraction of cells in a population which are biochemically competent for transfection.

Streips and Young (21) examined the sedimentation behavios of a population of 4S(8) cells using the Renografin technique (30). Cells of 4S(8) without competence factor sedimented as a single band in the gradient at the top of the

TABLE 4
Recovery of Adsorbed Helper Phage from B. stearothermophilus *4S(8)*[a]

Exposure to helper phage (min)	*Adsorbed phage (%)*		
	Reversible	*Irreversible*	*Total*
3	0.5	4.1	5
30	3.8	23.2	29
180	1.0	98.0	>99

[a]*Phage TP-12 was added to a culture of 4S(8) growing in TYG medium at 55°C. A sample was removed at 3, 30, and 180 min, the cells removed by low speed centrifugation, and the number of unadsorbed phage determined. The cells were washed six times in prewarmed TYG medium and phage in the combined supernatant fluids determined (designated reversibly adsorbed phage). Irreversible adsorbed (or inactivated) phage were calculated by total adsorbed phase minus reversibly adsorbed phage.*

50% Renografin layer. After incubation of cells of 4S(8) with competence factor for 30 min at 55°C, a portion of the population was found at the interface of the 12.5% and 30% Renografin layers; another portion at the interface of the 30 and 40% Renografin layers and a larger portion at the top of the 50% Renografin layer. Transfection was found in cells of the upper bands without the addition of more competence factor. They were unable to quantitate the frequency of transfection because growth of cells was inhibited by 50% Renografin. We repeated these studies using purified helper phage and obtained sedimentation patterns similar to those obtained by Streips and Young (21). The number of cells at the interface of the 12.5 and 30% and the 30 and 40% Renografin layers and at the top of the 50% Renografin layer was determined by direct count with a Spencer bright line hemocytometer counting chamber. As shown in Table 5, the number of cells in the slower sedimentating bands accounts for approximately 32% and 94% of the total cell population after 30 and 180 min of incubation with helper phage respectively. These values show a direct correlation with the number of helper phage adsorbed at these times (Table 3). Exhaustive washing of cells obtained from each band revealed that only the supernatant fluids of the slower sedimentating bands contained released helper phage.

TABLE 5

Effect of Helper Phage on Sedimentation of 4S(8) Cells in a Renografin Gradient[a]

Exposure to helper phage (min)	Total number of cells at:[b]		
	Interface of the 12.5 to 30% Renografin layer	Interface of the 30 to 40% Renografin layer	Top of the 50% Renografin layer
0	ND	ND	3.4×10^{7}
30	8.7×10^{6}	2.4×10^{6}	2.0×10^{7}
180	2.4×10^{7}	8.0×10^{6}	1×10^{3}

[a]*Procedure as described by Streips and Young (21).*

[b]*Fractions (0.05 ml) were collected from the bottom of the gradient with an 18-gauge needle. Cells were enumerated with a Spencer bright line hemocytometer counting chamber.*

[c]*ND - none detected.*

Since the number of cells in the slower sedimentating bands after 30 min of incubation with helper phage is far greater than the frequency of transfection, we can conclude that although adsorption of helper phage is the first and rate limiting step in transfection (21), the cell must possess other biochemical structural characteristics in order to take up and express phage DNA.

We have demonstrated that helper phage binds to protoplasts and protoplast membranes and that the association of helper phage with protoplast membranes results in a removal of infectious phage DNA from the supernatant fluids of membrane suspensions. The extent of binding was similar to that reported by Streips and Young (21).

The adsorption of helper phage to protoplast membranes was unexpected and suggests that the receptor sites do not consist of peptidoglycan and are associated with the cytoplasmic membrane. Helper phage do not adsorb to cells or protoplast membranes of 4S(8)(12)/1 or cells of 4S(8)(12), 4S(8), or 4S(12) grown at 67°C. These results indicate that the cytoplasmic membrane contains helper phage adsorption sites.

Although the facilitation of the uptake of phage DNA as a result of adsorption of helper phage cannot be described in molecular terms, it is possible that the adsorption of helper phage causes a modification of the membrane architecture which facilitates the binding transport of infectious DNA.

Wisdom and Welker (31) reported that in another strain of *B. stearothermophilus* the thermostability of protoplasts, the protein to lipid ratio of the protoplast membrane and the membrane content of the cell increases as the growth temperature is increased. A similar mechanism could account for the observation that cultures of 4S(8)(12), 4S(8), or 4S(12) grown at 67°C do not contain receptor sites for helper phage (13, 14, 21; N. E. Welker, unpublished data) and therefore cannot be transfected.

If *B. stearothermophilus* has a unique membrane structure, a special modification of the membrane may be required in order to facilitate the uptake of infectious DNA. Since cells containing adsorbed helper phage sediment more slowly in a Renografin gradient and presumably have a lower density than cells containing no adsorbed phage, it seems reasonable to propose that the membranes of these calls may be modified in some manner.

Protoplasts were prepared from 4S(8) cells grown at 55°C in the presence of helper phage, suspended in Tris (40m)-Magnesium (10m) buffer (pH 7.5) and held at 60°C for 1 hour. The extent of protoplast rupture is shown in Table 6. There appears to be a direct correlation between the adsorption of helper phage and protoplast rupture. Compared to a control

TABLE 6
Effect of Helper Phage on Protoplast Rupture and the Membrane Composition of B. stearothermophilus *4S(8)*[a]

Incubation at 55°C (min)	*Helper phage*	*Protoplast rupture (%) at 60°C*	*Membrane*[b]	*Ratio of protein to lipid*[c]
0	+	0	16.2	3.78
	-	0	16.5	3.75
30	+	34	15.8	3.40
	-	0	16.3	3.77
180	+	70	15.1	3.14
	-	0	16.2	3.74

[a]*Cells were grown in TYG medium at 55°C in the presence of helper phage (+). The preparation of protoplasts and the quantitation of protoplast rupture were described by Wisdom and Welker (31). A control culture without helper phage (-) was treated in the same manner.*

[b]*Dry weight of the membranes was determined gravimetrically and the results are expressed as a percentage of the dry weight of the cells from which they were isolated.*

[c]*Total cell protein and lipid were determined as described by Bodman and Welker (37).*

culture without helper phage, the adsorption of helper phage results in a decrease in the membrane content of the cell and the protein content of the membrane and a decrease in the lipid content of the membrane. The increased lipid content of the membrane may account for the lower density of cells in Renografin.

The results indicate that the adsorption of helper phage has a dramatic effect on membrane structure resulting in the uptake of infectious DNA. Biochemical changes in the membrane have been reported in *E. coli* during the adsorption of phage T4 (32). The changes were shown to be independent of the expression of phage genes. Studies on the biochemical and physical properties of cells exposed to helper phage should contribute to our understanding of the function of helper phage in transfection.

TRANSFORMATION IN *B. STEAROTHERMOPHILUS*

Since bacterial DNA inhibits transfection (U. N. Streips, Ph.D. Thesis, Northwestern University, Evanston, Illinois, 1969), we assumed that cellular DNA competes with infectious DNA for binding sites on the membrane.

Using the conditions which give maximal levels of transfection, the lysogenic strains were tested for DNA-mediated transformation. Transforming DNA was isolated from rifampicin resistant (rif^r) mutants of each strain. Transformants (rif^r) were obtained only with the cured strain (Table 7). The number of rif^r transformants of strain 4S varied between 300-500 per 10^8 recipient cells and helper phage was not required.

The transfection system developed appears to be specific for infectious DNA from phages TP-1C or TP-84 with a unique and obligatory requirement for an intact helper phage (TP-12). It is obvious from these studies that the use of transfection as a screening technique for the identification of strains which are susceptible to DNA-mediated transformation does not guarantee success.

On the positive side, the characterization of the lysogens and the isolation of a cured strain of *B. stearothermophilus* 4S have been instrumental in demonstrating that the presence of a prophage or prophage induction inhibits transformation. Lysogenic strains of *B. subtilis* were reported to exhibit reduced levels or no transformation when compared to nonlysogenic strains (33). The lysogenic strains, however, remain susceptible to infectious DNA. Recent studies (34, 35, 36) indicate that the low levels of transformation result from the induction of prophage when the lysogens become competent.

It is also possible that the absence of transformation in the lysogenic strains of *B. stearothermophilus* 4S is due to an alteration of the cell membranes or other biochemical changes after lysogenization. Studies on the mode of action of the helper phage in transfection provided some evidence in favor of this proposal.

It is obvious that we must return to the beginning to either develop a competence regimen for *B. stearothermophilus* 4S or to screen other strains of *B. stearothermophilus* for transformation. Experiments using both approaches are now underway in my laboratory.

TABLE 7
DNA-Mediated Transformation of the Cured and Lysogenic Strains of B. stearothermophilus *4S*[a]

Recipient	*Rifampicin-resistant transformants per* 10^8 *cells*
4S(8)(12)	0
4S(8)	0
4S(12)	0
4S	300 - 500

[a]*Transforming DNA (isolated from rifampicin-resistant mutants) was prepared according to the procedure described by Welker and Campbell (24). Transformation mixtures (2 ml) contained 3 x* 10^8 *cells/ml and 10-20 μg of homologous DNA. Rif*^{r} *transformants were scored on TA plates containing 50 μg/ml rifampicin.*

ACKNOWLEDGEMENTS

This work was supported by Public Health Service research grant AI06382 from the National Institute of Allergy and Infectious Diseases and a Biomedical Sciences Support Grant 5 S05 RR07028 from the National Institutes of Health.

REFERENCES

1. Miguel, P. (1888). Annales de Micrographie *1*, 3.
2. Friedman, S. M. (1968). Bacteriol. Rev. *32*, 27.
3. Singleton, R. and Amelunxen, R. E. (1973). Bacteriol. Rev. *37*, 320.
4. Ljungdahl, L. G. and Sherod, D. (1976). In *Extreme Environments: Mechanisms of Microbial Adaption* (M. H. Heinrich, ed.), p. 147, Academic Press, New York.
5. Welker, N. E. (1976). In *Extreme Environments: Mechanisms of Microbial Adaption* (M. H. Heinrich, ed.), p. 229, Academic Press, New York.
6. Zuber, H., ed. (1976). *Enzymes and Proteins from Thermophilic Microorganisms* (Experientia Supplement 26). Birkauser Verlag, Basel.

7. Isono, K. (1970). J. Genet. (Tokyo) *45*, 285.
8. Anagnostopoulos, C. and Spizizen, J. J. (1971). J. Bacteriol. *81*, 741.
9. Coukoulis, H. and Campbell, L. L. (1971). J. Bacteriol. *105*, 319.
10. Gwinn, D. D. and Thorne, C. B. (1964). J. Bacteriol. *87*, 519.
11. Marmur, J. E., Seaman, E. and Levine, J. (1976). J. Bacteriol. *85*, 461.
12. Hemphill, H. E. and Whiteley, H. R. (1975). Bacteriol. Rev. *39*, 257.
13. Streips, U. N. and Welker, N. E. (1969). J. Bacteriol. *99*, 344.
14. Streips, U. N. and Welker, N. E. (1971). J. Bacteriol. *106*, 960.
15. Bott, K. F. and Wilson, G. A. (1968). Bacteriol Rev. *32*, 370.
16. Streips, U. N. and Welker, N. E. (1971). Bacteriol. *106*, 955.
17. Akrigg, A., Ayod, S. R. and Barker, B. R. (1967). Biochem. Biophys. Res. Commun. *28*, 1062.
18. Charpak, M. and Dedonder, R. (1965). C. R. H. Acad. Sci. *260*, 5638.
19. Pakula, R. and Walczak, W. (1963). J. Gen. Microbiol. *31*, 125.
20. Tomasz, A. and Hotchkiss, R. D. (1964). Proc. Natl. Acad. Sci. USA *51*, 480.
21. Streips, U. N. and Young, F. E. (1971). J. Bacteriol. *106*, 868.
22. Cahn, F. H. and Fox, M. S. (1968). J. Bacteriol. *95*, 867.
23. Hadden, C. and Nester, E. W. (1968). J. Bacteriol. *95*, 876.
24. Welker, N. E. and Campbell, L. L. (1965). J. Bacteriol. *89*, 175.
25. Lees, N. D. and Welker, N. E. (1973). J. Virology *11*, 606.
26. Saunders, G. F. and Campbell, L. L. (1966). J. Bacteriol. *91*, 340.
27. Saunders, G. F. and Campbell, L. L. (1965). Biochemistry *4*, 2836.
28. Catterall, J. F., Lees, N. D. and Welker, N. E. (1977). In *Microbiology-1976* (D. Schlessinger, ed.), p. 1110, American Society for Microbiology, Washington, D. C.
29. Sjostrom, J. E. and Philipson, L. (1974). J. Bacteriol. *119*, 19.
30. Hadden, C. and Nester, E. W. (1968). J. Bacteriol. *95*, 876.

31. Wisdom, C. and Welker, N. E. (1973). J. Bacteriol. *114*, 1336.
32. Koerner, J. F., Snustod, D. P. and Warner, H. R. (1976). In *The Enzymes of Biological Membranes* (A. Martonosi, ed.), p. 505, Plenum Press, New York.
33. Yasbin, R. E., Wilson, G. A. and Young, F. E. (1973). J. Bacteriol. *113*, 540.
34. Garro, A. J. and Law, M.-F. (1974). J. Bacteriol. *120*, 1256.
35. Yasbin, R. E., Wilson, G. A. and Young, F. E. (1975). J. Bacteriol. *121*, 296.
36. Yasbin, R. E., Wilson, G. A. and Young, F. E. (1975). J. Bacteriol. *121*, 305.
37. Bodman, H. and Welker, N. E. (1969). J. Bacteriol. *97*, 924.

PROTEIN SYNTHESIS

STUDIES ON HEAT-STABLE RIBOSOMES FROM THERMOPHILIC BACTERIA

S. Marvin Friedman

Department of Biological Sciences
Hunter College of the City University of New York
New York, New York

INTRODUCTION

Ribosomes isolated from the thermophilic bacterium, *Bacillus stearothermophilus*, are of particular interest because they differ from *Escherichia coli* ribosomes in several important properties, including heat stability (1) and the translation of polycistronic messenger RNAs (2). Efforts to explain the cistron-discriminating activity of thermophile ribosomes have led to several contradictory reports. Studies with hybrid particles reconstituted with *B. stearothermophilus* and *E. coli* ribosomal components implicated both protein S12 and 16S RNA in the failure of thermophile ribosomes to translate the coat protein and replicase cistrons of bacteriophage R17 RNA (3). Using different assay procedures and conditions, Goldberg and Seitz (4) found that the 30S protein fraction rather than the 16S RNA was the primary specificity determinant. Shine and Dalgarno (5) reported differences in the pyrimidine-rich sequence at the 3' termini of thermophile and *E. coli* 16S RNAs and proposed that a mRNA-16S rRNA interaction could account for the differences in translational specificity. In contrast, Sprague *et al.* (6) have recently shown that the 16S RNA pyrimidine tract which is complementary to R17 and Qβ RNA initiator regions appears in both *B. stearothermophilus* and *E. coli* and thus could not serve as the basis for discrimination. Isono and Isono (7) reported that the addition of *E. coli* protein S1 to *B. stearothermophilus* ribosomes programmed with f2 RNA caused a marked stimulation in the *in vitro* synthesis at 39°C of coat protein and replicase. Subsequent studies (8) indicated that the

ISBN 0-12-268250-5

thermophile ribosomes lack a protein functionally equivalent to *E. coli* S1. More recently, Visentin *et al.* (9) have isolated the S1 type ribosomal protein from *B. stearothermophilus* and its quantitative amino acid composition was strikingly similar to that of *E. coli* S1.

The molecular basis for the heat stability of thermophile ribosomes remains unknown. Resolution of this problem is complicated by the multiple factors which could contribute to thermostability, such as intrinsic stability of ribosomal proteins, the nature of protein-protein and protein-RNA interactions and secondary structure in ribosomal RNA. Studies with hybrid 30S (10) and 50S (11) subunits reconstituted from mesophile and thermophile ribosomal components revealed that thermophile ribosomal proteins play a key role in protecting the ribosome from thermal denaturation.

MATERIALS AND METHODS

Materials

Bovine pancreatic ribonuclease A (5x crystallized, protease-free) was purchased from Sigma Chemical Corp. Acridine orange obtained from National Aniline Co. was recrystallized twice as the free base according to Stone and Bradley (12). The source of the remaining materials has been described previously (13).

Preparation of Ribosomes

Cells of *B. stearothermophilus* 2184, *B. coagulans* 43P and *E. coli* B were grown at the indicated temperatures and ribosomes were isolated as previously described (14) unless otherwise specified.

Phosphate Determination

Phosphate was measured by the method of Ames and Dubin (15) using KH_2PO_4 as the standard.

Pancreatic Ribonuclease Digestion of Ribosomes

The assay procedure was a modification of that described by Santer (16). A 1.0 ml reaction mixture contained the following: 10 μmol Tris-HCl pH 7.8, 10 μmol $Mg(OAc)_2$, 0.4 μmol

KCl, 10 A_{260} units of ribosomes and 10 μg pancreatic ribonuclease A. After incubation at the designated temperatures for the specified times, 3 drops of 1 N HCl were added to terminate the reaction and precipitate the ribosomes. Following centrifugation at 20,000 x g for 20 min, the supernatant fluid was aspirated and the optical density at 260 nm was determined. Control assays containing all the components with the exception of ribonuclease were incubated at each temperature and time point. Negligible readings were obtained in all cases indicating the absence of ribosome-bound ribonuclease activity under these conditions.

Pancreatic Ribonuclease Digestion of Ribosomal RNA

A 1.0 ml reaction mixture contained the following components: 10 μmol Tris-HCl pH 6.0, 10 μmol $Mg(OAc)_2$, 0.4 μmol KCl, 5 A_{260} units of phenol-extracted (1) ribosomal RNA and 5 μg pancreatic ribonuclease A. After incubation at the designated temperatures for the specified times, the tubes were processed as described above.

Acridine Orange Titration of Ribosomes

Titrations were carried out at 24°C as described by Furano *et al.* (17). A 1 cm, glass-stoppered cuvette contained 2.4 ml of 1 mM sodium cacodylate pH 6.7 and 4.8 x 10^{-8} mol of acridine orange. Aliquots of ribosomes suspended in the cacodylate buffer at a concentration of 20 μM RNA phosphate/ml were then added stepwise. After thorough mixing, the absorbency at 504 nm was measured in a Beckman DUR spectrophotometer.

Circular Dichroism

CD spectra were obtained in a JASCO J-20 spectropolarimeter at 16°C. The measurements were carried out in a 1 cm cell at ribosome concentrations of about 1 A_{260} unit/ml in TMK buffer (10 mM Tris-HCl pH 7.8, 10 mM $Mg(OAc)_2$ and 60 mM KCl). Results are expressed as relative ellipticity per A_{260} unit.

RESULTS AND DISCUSSION

Thermostability of Ribosomes from Obligate Thermophiles

Evidence for the heat stability of ribosomes isolated from the obligate thermophile, *B. stearothermophilus*, is based upon two types of determinations. In the first case, thermal denaturation is monitored by measuring the hyperchromicity at 260 nm with increasing temperature (14). Absorbance-temperature profiles of ribosomes and ribosomal subunits from *B. stearothermophilus* 2184 and *E. coli* B are presented in Fig. 1. These data show that both the thermophile 30S (T_m = 77.0°C) and 50S (T_m = 80.7°C) subunits were more heat stable than the *E. coli* 30S (T_m = 66.0°C) and 50S (T_m = 73.5°C) subunits. T_m values for 50S subunits from both the thermophile and *E. coli* were higher than T_m values for the corresponding 30S subunits. It is of interest to note that T_m values of 50S subunits from both organisms were slightly higher than those of the corresponding 70S ribosomes. In addition, melting curves of 70S ribosomes from both organisms were biphasic, suggesting that dissociation into subunits may precede the final leg of the hyperchromic shift.

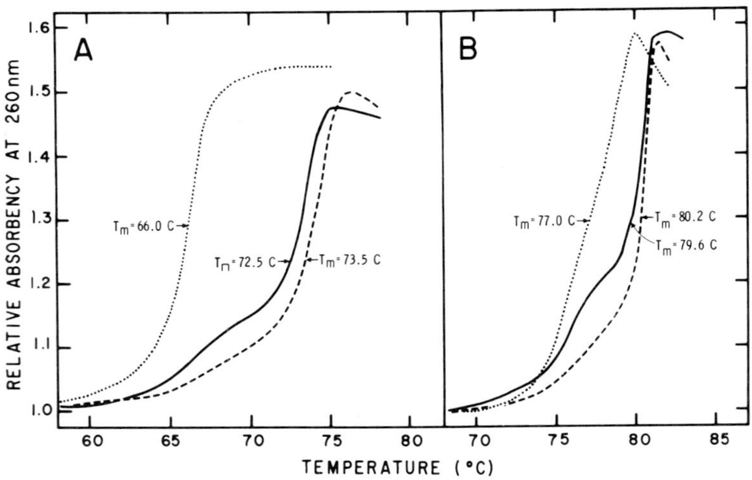

FIGURE 1. Absorbance-temperature profiles of 30S (dotted line), 50S (broken line) and 70S (solid line) ribosomal particles from E. coli *(A) and* B. stearothermophilus *(B). From Friedman (13); reproduced with permission from the American Society for Microbiology.*

These physical measurements are limited in their scope since they do not necessarily relate to the biochemical role of the ribosome in protein synthesis. We have therefore devised a functional assay which measures the residual capacity for polyuridylic acid-directed C^{14}-phenylalanine incorporation at 37°C of ribosomal suspensions heated at 65°C for varying times (1). Fig. 2 shows the heat stability of ribosomes from *B. stearothermophilus* and *E. coli* when assayed in this manner. After 5 minutes at 65°C, 94% of the initial activity was retained by thermophile ribosomes, whereas only 21% was retained by *E. coli* ribosomes. When the time interval at 65°C was extended to 15 minutes, incorporation values were 80 and 5% for thermophile and *E. coli* ribosomes, respectively.

The functional assay has also been employed to study the thermostabilities of native (nondissociated) and reassociated ribosomes (14). After 30 minutes of preincubation at

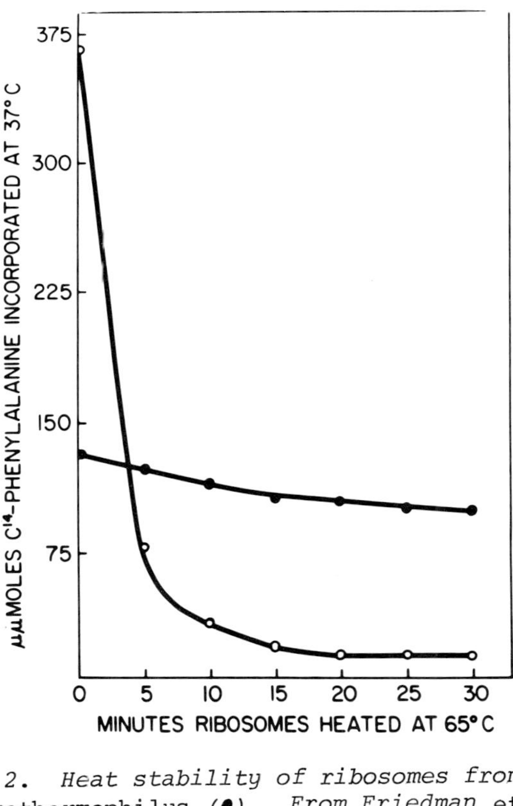

FIGURE 2. Heat stability of ribosomes from E. coli *(O) and* B. stearothermophilus *(●). From Friedman* et al. *(1); reproduced with permission from the American Society for Microbiology.*

60°C, 98% of the initial incorporation activity was retained by native *B. stearothermophilus* ribosomes, whereas only 25% was retained by nondissociated ribosomes from *E. coli*. In contrast to these results, the kinetics of thermal inactivation for reassociated ribosomes from both organisms revealed similar labilities. Thus, reassociated thermophile ribosomes retained 45% of the initial activity after preincubation at 60°C for 30 minutes, and the corresponding value for *E. coli* reassociated ribosomes was 41%. The basis for the loss of functional activity in the case of reassociated thermophile ribosomes remains unknown, but likely candidates include conformational changes and the activation of a latent ribosome-bound ribonuclease.

Thermostability of Ribosomes from Facultative Thermophiles

Over two decades ago, Campbell (18) reported that α-amylase crystallized from the facultative thermophile, *Bacillus coagulans* 43P, grown at 55°C was more thermostable than the enzyme isolated from the same organism grown at 35°C. A more detailed analysis of this phenomenon in a facultative strain of *B. stearothermophilus* revealed that the enzyme produced below 46°C was heat labile, whereas the enzyme produced above 50°C was heat stable (19). These results were confirmed by Isono (20), who studied the heat stability of amylases synthesized by a facultative derivative (BS-1) of *B. stearothermophilus* ATCC 12980 grown at 37°C and 55°C. Furthermore, Isono cited preliminary evidence for differences in the primary structures between the two enzymes. In addition to these studies, differences in fatty acid composition and membrane fluidity (21), cell wall chemistry (22) and metabolic activity (23) have been reported for facultative thermophiles grown at 37°C and 55°C.

These interesting results prompted us to investigate the effect of growth temperature on the thermostability of ribosomes isolated from a facultative thermophile. A composite absorbance-temperature profile of 70S ribosomes from *B. coagulans* 43P grown at 37°C and 55°C is shown in Fig. 3. Since the profiles were virtually identical, only a single curve is depicted. Ribosomes from cells grown at 37°C and 55°C were found to have an identical T_m value of 74°C. Furthermore, identical melting curves with a T_m of 79°C were obtained with ribosomes from *B. stearothermophilus* 2184 grown at 40°C and 65°C (data now shown). When tested under the same ionic conditions, the T_m of ribosomes from *E. coli* grown at 37°C was 72°C. These data confirm the work of Pace and Campbell (24), who found that the thermostability of ribosomes correlated with the maximal growth temperature

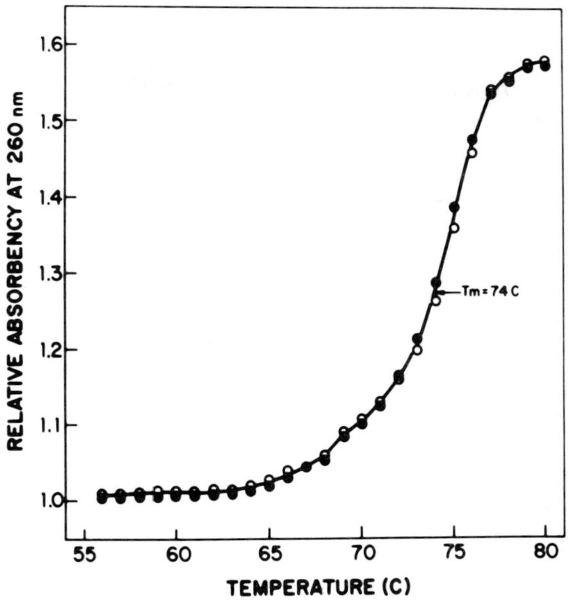

FIGURE 3. Absorbance-temperature profile of ribosomes from B. coagulans *grown at 37°C (0) and 55°C (●). 70S ribosomes were isolated from 15-30% sucrose density gradients in TMK buffer centrifuged for 16 hr at 20,000 rev/min in the Spinco SW 25.1 rotor.*

of the organism. In addition, our results demonstrate that thermophiles synthesize relatively heat stable ribosomes regardless of the growth temperature.

Ribosomal proteins were extracted from ribosomes of *B. coagulans* 43P grown at 37°C and 55°C with acetic acid (25) and separated by polyacrylamide gel electrophoresis (26). Although the gel patterns (Fig. 4) appear to be identical, these results do not necessarily exclude the possibility that the composition of some homologous proteins may differ. Nevertheless, even if differences in ribosomal proteins exist, they clearly do not influence the thermostability of the ribosome.

A third category of growth temperature effect on the stability of an enzyme synthesized by facultative thermophiles has recently been reported by Amelunxen (27, 28). Previous work showed that glyceraldehyde-3-phosphate dehydrogenase (GPDH) crystallized from the obligate thermophile, *B. stearothermophilus*, was markedly more thermostable than the rabbit muscle enzyme (29). When GPDH was crystallized from a facultative thermophile (*B. coagulans* KU) grown at

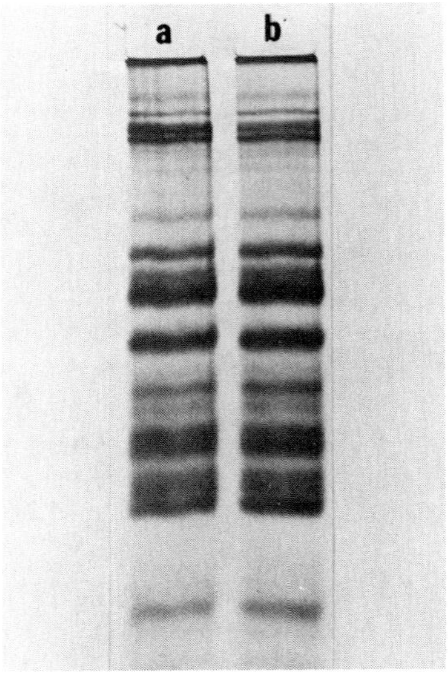

FIGURE 4. Polyacrylamide gel electrophoretic patterns of ribosomal proteins from B. coagulans *grown at 37°C (a) and 55°C (b). A 13% polyacrylamide gel slab was run from top (+) to bottom (-) at pH 4.2 in 7.5 M urea for 12.5 hr at 250 volts (70 ma).*

37°C and 55°C, two thermolabile enzymes were obtained. Although the physical properties were similar to GPDHs from other sources, a reduction in the basic residues and an apparent increase in acidic residues was detected. The enzymes could be protected against thermal inactivation at 55°C by increasing the ionic strength (μ) to 1.8 with NaCl or $(NH_4)_2SO_4$. Since these values far exceed the inorganic salt concentration found *in vivo*, the authors concluded that the intracellular charged macromolecular environment must be crucial in maintaining the functional integrity of the GPDH from *B. coagulans*. Novitzky *et al.* (22) cited unpublished evidence for the heat lability of some glycolytic enzymes from *B. coagulans* KU grown at 55°C.

These results with facultative thermophiles raise important issues concerning the role of growth temperature in determining the intrinsic stability of macromolecules and/or the stabilizing potential within the cell. The question of which genes are turned on or off as a function of temperature and how the critical gene products achieve net thermostability presents a formidable challenge for future research.

Properties of Ribosomal Ribonucleic Acid

Ribosomal ribonucleic acid (rRNA) extracted from *B. stearothermophilus* ribosomes has a somewhat higher guanine plus cytosine (G+C) content than that of isolated *Escherichia coli* rRNA (30), but the relationship between conformations of free rRNA and rRNA within the ribosome remains a controversial issue. Cotter *et al.* (31) concluded that the conformations of rRNA in yeast ribosomes and in the free state were similar, whereas Furano *et al.* (17) proposed that isolated *E. coli* rRNA contained considerable secondary structure and rRNA in the ribosome was primarily single-stranded.

Under certain ionic conditions, there was little or no difference between the T_m of ribosomes and isolated rRNA from *E. coli* (1, 32). Thermophile ribosomes, on the other hand, always displayed an appreciably higher T_m than that of free rRNA (1, 32). In addition, the slope of the hyperchromic shift was less steep with thermophile ribosomes than with *E. coli* ribosomes (1, 32). These observations have been interpreted to suggest that the thermal stability of the ribosome lies primarily in the nature of the RNA-protein interaction (24). It should be noted, however, that some of the hyperchromic effects noted above may have been due to the hydrolysis of rRNA in the ribosome by latent ribonucleases rather than to the melting of secondary structure. Thus, when a mutant of *E. coli* deficient in ribosome-bound ribonuclease I was employed, the melting curve was much sharper and the T_m was higher for the ribosomes than for isolated rRNA (33).

In order to learn more about the properties of thermophile rRNA *in situ*, we have examined the effect of incubation temperature on the kinetics of pancreatic ribonuclease A digestion of *B. stearothermophilus* and *E. coli* ribosomes (Fig. 5). More acid-soluble nucleotides were released from *E. coli* ribosomes than from thermophile ribosomes at each temperature studied. This effect was amplified as the incubation temperature was increased from 25°C to 45°C. Melting curves of *E. coli* and *B. stearothermophilus* ribosomes suspended in the digestion buffer revealed that the onset of the hyperchromic shift at 260 nm takes place at 61°C and 65°C, respectively. Thus, the extensive release of acid-soluble nucleotides from *E. coli* ribosomes catalyzed by pancreatic ribonuclease at 45°C occurs at a temperature which well precedes that coinciding with the initial disruption of stacked base pair sequences in rRNA.

The effect of temperature on the kinetics of pancreatic ribonuclease digestion of isolated rRNA is shown in Fig. 6. Although the enzyme:RNA ratio was the same as that used for the ribonuclease digestion of ribosomes, isolated rRNA was attacked much more rapidly than was rRNA contained in the ribosome. At early time points, there was more extensive degradation of *E. coli* rRNA than thermophile rRNA at both 37°C and 45°C. These results are consistent with the somewhat higher G+C content of thermophile rRNA than *E. coli* rRNA and the marked preference of pancreatic ribonuclease A for single-stranded RNA (34). At the end of the 10 minute

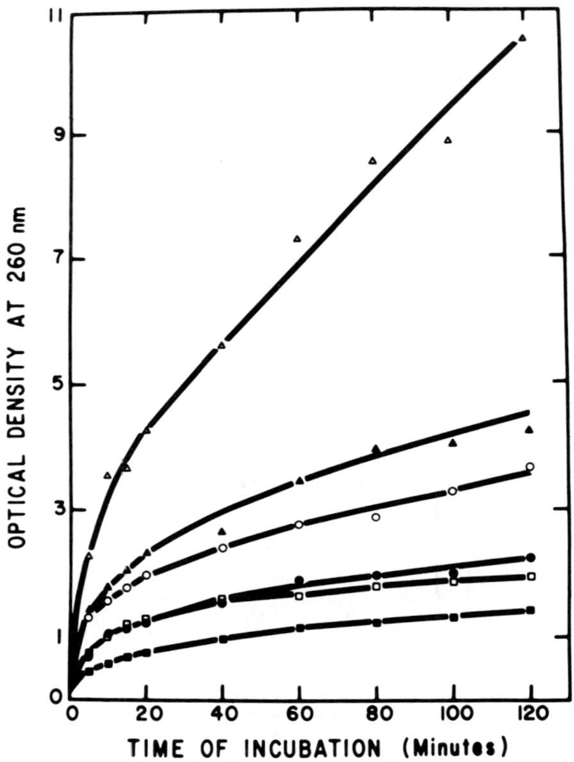

FIGURE 5. Effect of incubation temperature on the kinetics of pancreatic ribonuclease A digestion of ribosomes from B. stearothermophilus *(25°C, □; 37°C, ●; 45°C, ▲) and* E. coli *(25°C, □; 37°C, O; 45°C, Δ).*

incubation period, however, the amounts of acid-soluble nucleotides released from both preparations had reached or were approaching an equilibrium value corresponding to the initial input of 5 A_{260} units.

The greater resistance of thermophile ribosomes than *E. coli* ribosomes to digestion by ribonuclease could be due to any one or combination of the following: (i) more secondary structure in rRNA, (ii) protective protein-RNA interactions, or (iii) less available surface rRNA. In order to test the latter possibility, dye binding experiments employing acridine orange were performed. The essentially identical titration curves which were obtained with the two ribosome preparations are illustrated in Fig. 7. At the end point of the titration, defined as the intersection of the two linear limbs of the

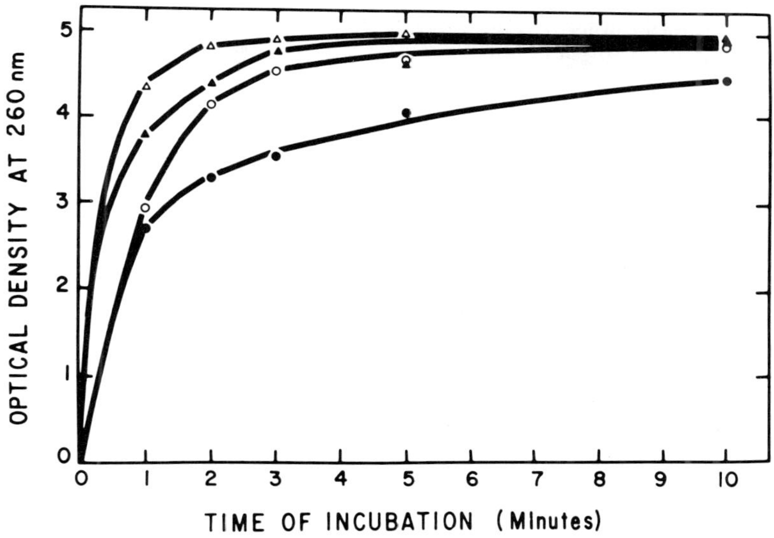

FIGURE 6. Effect of incubation temperature on the kinetics of pancreatic ribonuclease digestion of isolated rRNA from B. stearothermophilus *(37°C, ●; 45°C, ▲), and* E. coli *(37°C, O; 45°C, Δ).*

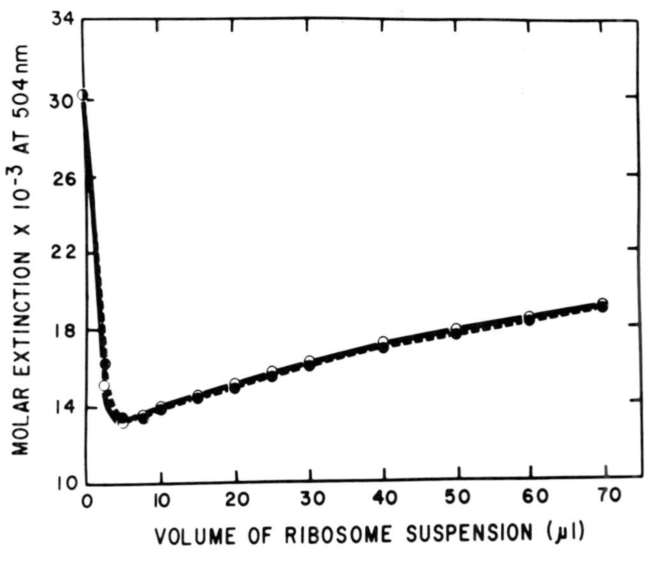

FIGURE 7. Spectrophotometric titrations of acridine orange with ribosomes from B. stearothermophilus *(--●--) and* E. coli *(— O —).*

curve (17), approximately 90% of the RNA phosphates in both *B. stearothermophilus* and *E. coli* ribosomes were titrated.

Circular dichroism (CD) spectra of ribosomes have provided useful information concerning conformations of rRNA and ribosomal proteins (35-37). Fig. 8 shows the near ultraviolet CD spectra of *B. stearothermophilus* and *E. coli* ribosomes. The spectra were similar in shape with a large, positive dichroic band centered at 265 nm and a small, negative trough at 300 nm. Thermophile ribosomes, however, displayed a small but significant increase of ellipticity at 265 nm. This difference has been consistently observed with several ribosome preparations tested over a range of magnesium concentrations. The large CD peak at 265 nm probably originates from the conformational state of RNA bases (38). Furthermore, the magnitude of ellipticity at 265 nm is regarded as an index of right handed, double-helical structure in rRNA (35).

These studies indicate that rRNA in the *B. stearothermophilus* ribosome exists in a more highly ordered conformation than does rRNA in the *E. coli* ribosome. This property undoubtedly contributes to the heat stability of thermophile

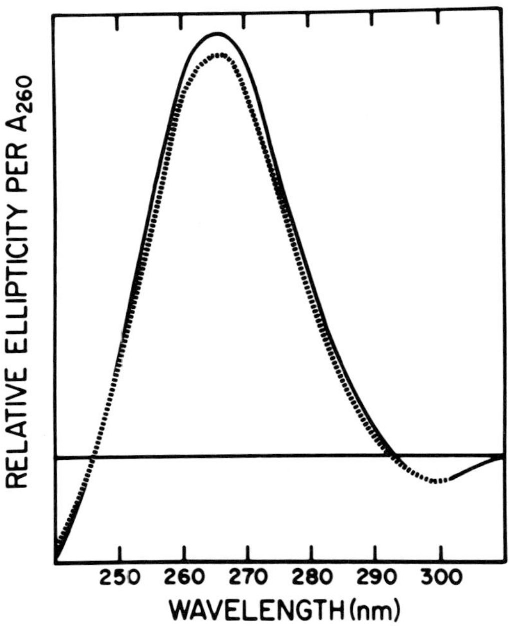

FIGURE 8. Near ultraviolet CD spectra of ribosomes from B. stearothermophilus *(——) and* E. coli *(----).*

ribosomes. The role of additional factors, such as protein-RNA interactions which are relatively resistant to thermal stress, remains to be elucidated.

Surface Topography of Ribosomal Proteins

We have investigated the surface topography of the intact 70S ribosome and free 30S and 50S subunits from *B. stearothermophilus* 2184 by lactoperoxidase-catalyzed iodination (39). Heavily iodinated 70S ribosomes and subunits retained 62-70% of the protein synthesis activity of untreated particles and sucrose gradient sedimentation profiles showed no gross conformational changes due to iodination.

Two-dimensional polyacrylamide gel electrophoresis (40) was employed to separate ribosomal proteins for analysis of their radioactivity. Fig. 9 shows the two-dimensional gel patterns and numbering system for 30S and 50S subunit proteins of *B. stearothermophilus* 2184. The radioactivity of each protein or group of proteins excised from gels was corrected for the total amount of protein that actually ran on to the gel as well as for the specific activity of each subunit. A comparison was then made between the extent of iodination of proteins in isolated subunits and the extent of iodination of proteins in subunits derived from iodinated 70S ribosomes. Analysis of this data in conjunction with the quantitative amino acid composition of ribosomal proteins (41) provides evidence for the structural organization of some proteins within the thermophile ribosome. More specifically, one can identify proteins with a large fraction of their tyrosine residues exposed on the surface, proteins with tyrosines engaged in the 30S:50S interface region and proteins with tyrosines involved in conformational changes when subunits interact.

The availability of this information enables us to make a preliminary comparison of the location of tyrosine residues in some homologous proteins from *B. stearothermophilus* and *E. coli*. Table 1 presents topography data for the two organisms which has been derived from iodination experiments carried out under similar conditions. Four pairs of homologous proteins (B-S7 and E-S7, B-S10 and E-S9, B-L2 and E-L2, B-L5 and E-L6) show multiple exposed tyrosines, suggesting a surface location. One pair of homologous proteins (B-S13 and E-S10) has tyrosines at the 30S:50S interface. Finally, tyrosines in one pair of homologous proteins (B-S9 and E-S6) are associated with conformational changes when subunits combine to form the 70S ribosome. Although further studies will be required to substantiate this point, it appears that tyrosines in only some homologous proteins from *B. stearothermophilus*

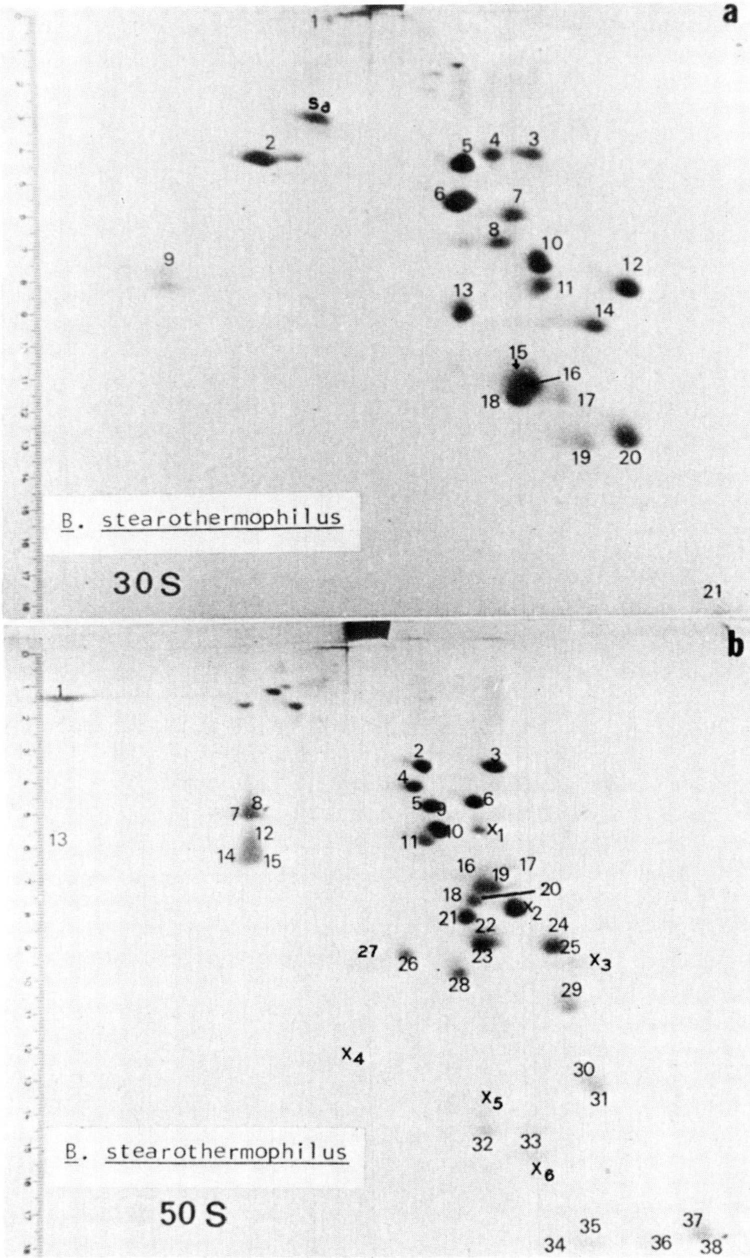

FIGURE 9. Two-dimensional electrophoretic patterns of B. stearothermophilus *strain 2184 30S (a) and 50S (b) ribosomal proteins. From Miller* et al. *(38); reproduced with permission from Springer-Verlag.*

TABLE 1

Topography of B. stearothermophilus *and* E. coli *Ribosomal Proteins as Determined by Enzymatic Iodination**

Location	B. stearothermophilus[a]	E. coli[b]
Highly exposed	S4, S7, S10, Sa L2, L4, L5/9, L6, L36	S5(6)[c], S7(7), S9(10), S18(19) L2(2)[d], L5, L6(5)[e], L10(10)[d], L11, L26
30S:50S Interface	S2, S3, S7, S13, Sa L5/9, L10, L11, L24/25	S10(13) L2, L13 (13)[e], L18(22)[e], L26, L27, L28
Conformational changes upon 70S formation	S5, S9, S12, S15/16, S18 L36	S3, S6(9), S9, S18(19) L11

*From Miller *et al.* (39); reproduced with permission from Springer-Verlag.

[a]Data from Miller *et al.* (39).

[b]Data of Litman *et al.* (42) for the 30S proteins and of Litman and Cantor (43) for the 50S proteins.

[c]Figures in parentheses indicate homologous *B. stearothermophilus* 30S proteins according to Yaguchi *et al.* (44). Homologous proteins which correspond in location are underlined.

[d]Figures in parentheses indicate homologous *B. stearothermophilus* 50S proteins according to Geisser *et al.* (45).

[e]Figures in parentheses indicate homologous *B. stearothermophilus* 50S proteins according to Horne and Erdmann (46).

and *E. coli* occupy the same position within the ribosome. One possible explanation for this finding is that tyrosine residues in homologous ribosomal proteins have not been conserved during the course of evolution. There is not sufficient amino acid sequence data available at this time to carefully evaluate this alternative. On the other hand, some homologous proteins may indeed be positioned at different loci in the ribosome. In this connection, the spatial packaging of certain key proteins could prove to be an important factor in determining the thermostability of the ribosome.

ACKNOWLEDGMENTS

I would like to express my gratitude to the following colleagues who collaborated in these studies: Drs. R. Axel, C. R. Cantor, D. J. Litman, H. M. Miller, C. Pon and I. B. Weinstein. The author's research reported here was supported by U.S. Public Health Service Grants AI-08231 and GM-22164.

REFERENCES

1. Friedman, S. M., Axel, R. and Weinstein, I. B. (1967). J. Bacteriol. *93*, 1521.
2. Lodish, H. F. (1969). Nature *224*, 867.
3. Held, W. A., Gette, W. R. and Nomura, M. (1974). Biochemistry *13*, 2115.
4. Goldberg, M. L. and Steitz, J. A. (1974). Biochemistry *13*, 2123.
5. Shine, J. and Dalgarno, L. (1975). Nature *254*, 34.
6. Sprague, K. U., Steitz, J. A., Grenley, R. M. and Stocking, C. E. (1977). Nature *267*, 462.
7. Isono, S. and Isono, K. (1975). Eur. J. Biochem. *56*, 15.
8. Isono, K. and Isono, S. (1976). Proc. Natl. Acad. Sci. USA *73*, 767.
9. Visentin, L. P., Hasnain, S., Gallin, W., Johnson, K. G., Griffith, D. W. and Wahba, A. J. (1977). FEBS Letters *79*, 258.
10. Nomura, M., Traub, P. and Bechmann, H. (1968). Nature *219*, 793.
11. Fahnestock, S. R. (1977). Arch. Biochem. Biophys. *182*, 497.
12. Stone, A. L. and Bradley, D. F. (1961). J. Am. Chem. Soc. *83*, 3627.

13. Friedman, S. M. and Weinstein, I. B. (1966). Biochim. Biophys. Acta *114*, 593.
14. Friedman, S. M. (1971). J. Bacteriol. *108*, 589.
15. Ames, B. N. and Dubin, D. T. (1960). J. Biol. Chem. *235*, 769.
16. Santer, M. (1963). Science *141*, 1049.
17. Furano, A. V., Bradley, D. F. and Childers, L. G. (1966). Biochemistry *5*, 3044.
18. Campbell, L. L. (1961). J. Am. Chem. Soc. *76*, 5256.
19. Campbell, L. L. and Pace, B. (1968). J. Appl. Bacteriol. *31*, 24.
20. Isono, K. (1970). Biochem. Biophys. Res. Commun. *41*, 852.
21. Chan, M., Virmani, Y. P., Himes, R. H. and Akagi, J. M. (1973). J. Bacteriol. *113*, 322.
22. Novitzky, T. J., Chan, M., Himes, R. H. and Akagi, J. M. (1974). J. Bacteriol. *117*, 858.
23. Jung, L., Jost, R., Stoll, E. and Zuber, H. (1974). Arch. Microbiol. *95*, 125.
24. Pace, B. and Campbell, L. L. (1967). Proc. Natl. Acad. Sci. USA *57*, 1110.
25. Hardy, S. J. S., Kurland, C. G., Voynow, P. and Mora, G. (1969). Biochemistry *8*, 2897.
26. Brown, D. G. and Abrams, A. (1970). Biochim. Biophys. Acta *200*, 522.
27. Crabb, J. W., Murdock, A. L. and Amelunxen, R. E. (1977). Abstr. Annu. Meet. Am. Soc. Microbiol. K32, p. 224.
28. McLinden, J. H., Crabb, J. W., Murdock, A. L. and Amelunxen, R. E. (1977). Abstr. Annu. Meet. Am. Soc. Microbiol. N52, p. 237.
29. Amelunxen, R. E. (1966). Biochim. Biophys. Acta *122*, 175.
30. Saunders, G. F. and Campbell, L. L. (1966). J. Bacteriol. *91*, 332.
31. Cotter, R. I., McPhie, P. and Gratzer, W. B. (1967). Nature *216*, 864.
32. Mangiantini, M. T., Tecce, G., Toschi, G. and Trentalance, A. (1965). Biochim. Biophys. Acta *103*, 252.
33. Tal, M. (1969). Biochemistry *8*, 424.
34. Geiduschek, E. P., Moohr, J. W. and Weiss, S. B. (1962). Proc. Natl. Acad. Sci. USA *48*, 1078.
35. Miall, S. H. and Walker, I. O. (1968). Biochim. Biophys. Acta *166*, 711.
36. Kabasheva, G. N., Sandakhchiev, L. S. and Sevastyanov, A. P. (1971). FEBS Letters *14*, 161.
37. Wong, K.-P. and Dunn, J. M. (1974). FEBS Letters *44*, 50.

38. Yang, J. T. and Sameijima, T. (1969). Progr. Nucleic Acid Res. *9*, 223.
39. Miller, H. M., Friedman, S. M., Litman, D. J. and Cantor, C. R. (1976). Mol. Gen. Genet. *144*, 273.
40. Kaltschmidt, E. and Wittmann, H. G. (1970). Anal. Biochem., *36*, 401.
41. Isono, S. and Isono, K. (1975). Eur. J. Biochem. *50*, 483.
42. Litman, D. J., Lee, C. C. and Cantor, C. R. (1974). FEBS Letters *47*, 268.
43. Litman, D. J. and Cantor, C. R. (1974). Biochemistry *13*, 512.
44. Yaguchi, M., Matheson, A. T. and Visentin, L. P. (1974). FEBS Letters *46*, 296.
45. Geisser, M., Tischendorf, G. W. and Stoffler, G. (1973). Mol. Gen. Genet. *127*, 129.
46. Horne, J. R. and Erdmann, V. A. (1972). Mol. Gen. Genet. *119*, 337.

STRUCTURE AND THERMAL STABILITY OF RIBOSOMAL COMPONENTS FROM THERMOPHILIC BACTERIA

M. Yaguchi
L. P. Visentin
R. N. Nazar
A. T. Matheson

Division of Biological Sciences
National Research Council of Canada
Ottawa, Ontario, Canada

INTRODUCTION

Thermophilic bacteria have evolved heat-stable ribosomes in order to live and multiply at higher temperatures (see Friedman, 1968; Stenesh, 1976). Both 30S and 50S subunits from thermophiles are more heat stable than those from mesophiles (Altenburg and Saunders, 1971; Friedman, 1971). Since heat stable subunits can be reconstituted from ribosomal components (RNA and proteins) of thermophilic bacteria, the stability of the subunits must be due to intrinsic properties of these components and the stability of their interactions (Cohlberg and Nomura, 1976; Fahnestock *et al.*, 1973; Nomura *et al.*, 1968). Although both ribosomal RNA and proteins from thermophiles were shown to be more stable than those from mesophiles, the thermal stability of hybrid ribosomes depends largely on the source of the proteins rather than RNA, suggesting that proteins play a more important role than RNA (Fahnstock, 1977b; Nomura *et al.*, 1968). Nevertheless, Pace and Campbell (1967) have previously stressed that the thermal stability of ribosomes is primarily a function of RNA-protein interactions. The molecular basis of the thermal stability of RNA-protein interactions in ribosomes is essentially unknown.

The present paper compares the structure of 5S rRNA and 30S ribosomal proteins from thermophiles with mesophiles and discusses the possible contribution of specific groups to the

ISBN 0-12-268250-5

thermal stability of isolated protein and RNA components as well as their role in stabilizing protein-nucleic acid interactions at high temperatures.

RIBOSOMAL RNA

Although the ribosomal RNAs of thermophiles have been compared with those of mesophiles for over a decade, information about their primary nucleotide sequences or structure has only recently become available. Nucleotide analyses for a number of organisms have generally indicated that the G+C content increases as a function of increasing maximum growth temperatures (Mangiantini *et al.*, 1965; Saunders and Campbell, 1966; Pace and Campbell, 1967; Stenesh and Holazo, 1967; and Zeikus *et al.*, 1970). The rRNA of the extreme thermophile, *Thermus aquaticus* for example, is 63.7 mole percent G+C compared to 56.6 mole percent for the rRNA of *Escherichia coli* (Zeikus *et al.*, 1970). Since G-C base pairs are known to be more stable than A-U pairs, this increasing G+C content was considered to be essential for the resistance to thermal denaturation which has been observed in rRNA from thermophilic organisms (see Friedman, 1968). The Tm of *Bacillus stearothermophilus* rRNA, for example, is 60°C, about 10° higher than that of *E. coli* RNA (Friedman *et al.*, 1967). Nevertheless, under the denaturation conditions used, some rRNA components were not more stable in the thermophiles and the Tm was frequently lower than the optimum growth temperature. In *T. aquaticus*, for example, 16S rRNA was not more stable over much of the denaturation profile and the Tm was 57°C, much lower than the maximum temperature (79°C) at which the organism can grow (Zeikus *et al.*, 1970). Since intact ribosomes are considerably more stable in every instance, much of their thermal stability was postulated to result from adaptative changes in the ribosomal proteins.

Because of its relatively small size and simple structure, the 5S rRNA molecule is an excellent model for studies on mechanisms by which thermophilic organisms have adapted their ribosomal RNAs to high growth temperatures. The primary nucleotide sequence of two thermophilic 5S species, *B. stearothermophilus* (a moderate thermophile) and *T. aquaticus* (an extreme thermophile) have recently been determined (Marotta *et al.*, 1976; Nazar and Matheson, 1977). When compared to other 5S rRNAs (Nazar and Matheson, 1977), the sequence homology with *T. aquaticus* 5S RNA (Fig. 1) was greater between thermophiles (approximately 69%) than with mesophilic species (e.g., approximately 65% with the closely related *B. subtilis*). The comparisons further indicate that

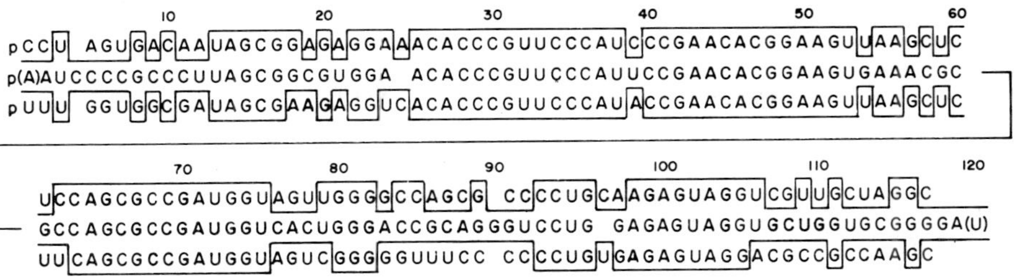

FIGURE 1. Comparison of the nucleotide sequences of Bacillus stearothermophilus *(upper line),* Thermus aquaticus *(middle line), and* Bacillus subtilis *(lower line) 5S rRNA. The solid lines enclose sequences which are identical with* T. aquaticus. *The bacilli sequences are from Marotta* et al. *(1976). The numbers correspond to nucleotide residues in* T. aquaticus *5S RNA. Taken from Nazar and Matheson (1977) and reproduced with permission from The American Society of Biological Chemists.*

among prokaryotes, eleven of the nucleotide residues (Fig. 2) are restricted to the thermophiles and may be involved in the adaptation to high growth temperatures. Based on theoretical considerations (Tinoco *et al.*, 1971; Nazar and Matheson, 1977), the generalized 5S RNA base-pairing scheme of Nishikawa and Takemura (1974) is the favored estimate of the secondary structure of *T. aquaticus* 5S rRNA (Fig. 3). Not all of the unique nucleotides add stability to this estimate. If the model is valid, these necleotides are probably involved in stabilizing the tertiary structure or in protein-RNA interactions which give stable RNA complexes within the ribosome. Little is known of these interactions and they are clearly the basis of future studies.

Thermal denaturation studies with mesophilic and thermophilic 5S rRNA have shown that in 0.1 M phosphate the Tm of *B. stearothermophilus* 5S RNA is only 1.5°C higher than that of 5S RNA from *B. subtilis* or *E. coli* (Varricchio and Marotta, 1976). Accordingly, the authors suggested that the 5S RNA makes only a minor contribution to the thermostability of *B. stearothermophilus* ribosomes. Similar observations were also made when the secondary structure of 5S rRNA of *T. aquaticus* was compared to that of *B. stearothermophilus* and *E. coli* (Nazar *et al.*, 1978). In 10 mM KCl the Tm of *T. aquaticus* 5S RNA differed by only 1°C from that of *E. coli* and the molecule was fully denatured well below the optimum growth temperature of the thermophile. The internal Na^+, K^+, and Mg^{2+} concentrations of *T. aquaticus* cells, however, were

5'	1(1)	18(17)	24(22)	62(62)	79(77)	82(84)	85(88)	89(95)	106(109)	109(112)	115(118)	3'
	A	G	A	C	U	C	C	G	U	U	G	*T. aquaticus*
	C	G	A	C	U	C	C	G	U	U	G	*B. stearothermophilus*
	U	A	U	U	U	C	U	G	A	U	G	*B. megaterium*
	U	A	U	U	C	G	U	C	A	C	A	*B. subtilis*
	U	C	G	U	G	G	U	C	G	A	C	*P. fluorescens*
	U	U	G	U	G	G	U	C	A	G	C	*Photobacterium sp. 8265*
	U	C	G	U	G	G	U	C	G	G	C	*E. coli*
	U	C	G	U	G	G	U	C	G	G	C	*S. typhimurium*
	U	C	G	U	G	G	U	C	G	G	C	*A. aerogenes*
	U	C	G	U	G	G	C	C	G	G	C	*E. ceroideae*
	U	C	G	U	G	G	U	C	A	C	C	*S. marcescens*
	U	C	G	U	G	G	U	C	G	G	C	*P. mirabilis*
	U	C	G	U	G	G	U	C	A	G	C	*Y. pestis*
	G	C	A	U	A	G	A	C	U	U	C	Human KB cells

FIGURE 2. Schematic comparison of mesophilic 5S rRNA sequences with the unique residues of thermophilic 5S rRNA. The sequences are aligned according to the method of Hori (1976). The residues are numbered to correspond to Thermus aquaticus*; the numbering scheme used by Hori (1976) follows in parentheses. Taken from Nazar and Matheson (1977) and reproduced with permission from The American Society of Biological Chemists.*

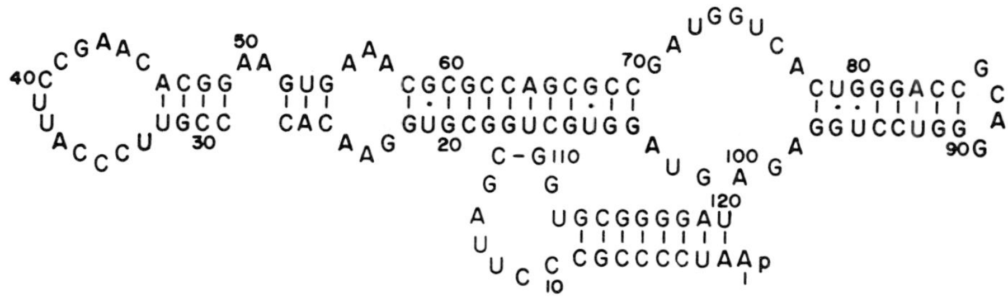

FIGURE 3. Estimate of the secondary structure of Thermus aquaticus *5S RNA. Taken from Nazar and Matheson (1977) and reproduced with permission from The American Society of Biological Chemists.*

determined to be 91 mM, 130 mM and 59 mM, respectively. Under these salt conditions, *T. aquaticus* 5S RNA was significantly more stable than *E. coli* RNA and the 5S RNA from *B. stearothermophilus* was intermediate as was its maximum growth temperature (Fig. 4). These results indicated that thermostability of 5S rRNA from thermophilic organisms were specially dependent on the internal salt concentration. Furthermore, under these salt conditions, most of the secondary structure of the RNA remained stable at the optimum growth temperature suggesting that the ribosomal RNAs of thermophilic organisms contribute more to the thermostability of the ribosomes than previously thought.

RIBOSOMAL PROTEINS

Ribosomal proteins from *B. stearothermophilus*, a moderate thermophile, are the most extensively studied ribosomal proteins among thermophiles (see Ljungdahl and Sherod, 1976). So far very little information is available on the ribosomal proteins of extreme thermophiles, such as *T. aquaticus* and *T. thermophilus* (Oshima and Imahori, 1974).

Individual proteins have been isolated from the 30S subunit from *B. stearothermophilus* (Higo *et al.*, 1973; Isono *et al.*, 1973; Isono and Isono, 1975) and characterized with respect to their amino acid composition (Isono and Isono, 1975), chromatographic behavior (Higo *et al.*, 1973; Isono *et al.*, 1973; Isono and Isono, 1975), electrophoretic mobilities (Isono *et al.*, 1973; Geisser *et al.*, 1973), immunological cross-reactivities (Higo *et al.*, 1973; Isono *et al.*, 1973; Geisser *et al.*, 1973), streptomycin mutations (Isono, 1974), functional correspondence and reconstitution (Higo *et al.*, 1973), conservation of N-terminal residues (Matheson *et al.*, 1975); N-terminal sequences (Higo and Loertscher, 1974; Higo and Nomura cited by Hori *et al.*, 1978; Vassos *et al.*, 1976; Yaguchi *et al.*, 1973; Yaguchi *et al.*, 1974), C-terminal

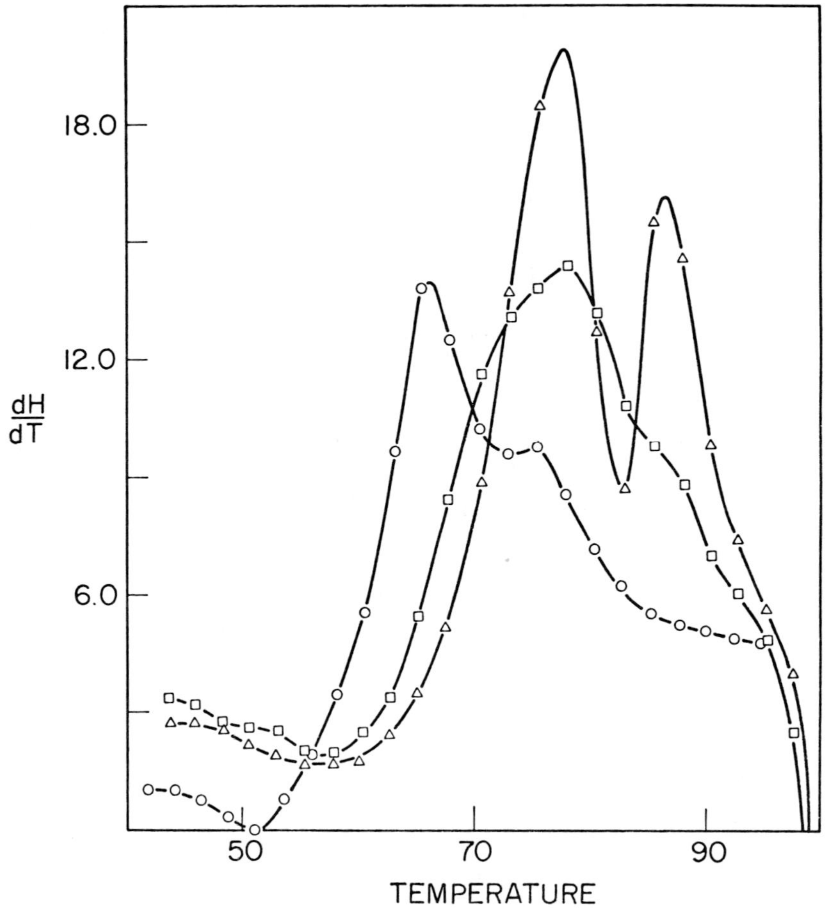

FIGURE 4. A comparison of the derivative profiles of Thermus aquaticus *(triangles),* Bacillus stearothermophilus *(squares) and* Escherichia coli *(circles) 5S rRNA under physiological salt conditions. The profiles were derived from thermal denaturation curves.*

sequences (Duggleby *et al.*, 1976), RNA binding (Chow *et al.*, 1972; Stanley and Ebel, 1977) and topography (Miller *et al.*, 1976). These results clearly show that the 20 ribosomal proteins of the 30S subunit from *B. stearothermophilus* correspond structurally and functionally to those of *E. coli* on a one to one basis. Nevertheless, there are a few exceptions. A protein corresponding to S1 of *E. coli* has not been observed in *B. stearothermophilus* (Isono and Isono, 1976). N-terminal residues of protein S5 (Wittmann-Liebold, 1973), S11 (Chen *et al.*, 1977), and S18 (Yaguchi, 1975) from *E. coli* were found to be modified, while corresponding proteins from *B. stearothermophilus* were not (Yaguchi *et al.*, 1974). Furthermore, the molecular weights and amino acid composition of protein S14 and S21 of *B. stearothermophilus* differ considerably from those of *E. coli* (Isono *et al.*, 1973; Isono and Isono, 1975), and the degree of the N-terminal sequence homology was less than that shown for other 30S ribosomal proteins (Yaguchi *et al.*, 1974). The relationship of these differences to the thermal stability of ribosomes from *B. stearothermophilus* is not known.

There is less information about 50S subunit of thermophiles. 50S ribosomal subunits of *B. stearothermophilus* have been reconstituted using either total proteins (Fahnestock *et al.*, 1973; Nomura and Erdmann, 1970) or isolated individual 50S protein components (Cohlberg and Nomura, 1976). Further characterization of some of the individual proteins from the 50S subunit of *B. stearothermophilus* also has been reported (Duggleby *et al.*, 1975; Fahnestock, 1977a; Horne and Erdmann, 1972; Miller *et al.*, 1976; Tischendorf *et al.*, 1973; Visentin and Kaplan, 1975; Visentin *et al.*, 1974).

Comparisons between the amino acid compositions of ribosomal proteins from *E. coli* and *B. stearothermophilus*, and *Clostridia* sp. (Ansley *et al.*, 1969; Friedman, 1968; Irwin *et al.*, 1973) did not show substantial differences between mesophiles and thermophiles. Apparently species related amino acid substitutions are much greater than substitutions responsible for thermal stability. Hence, the number of residues that might be responsible for the difference in the thermal stability may be relatively small and difficult to identify in any intergeneric comparison. However, it should be emphasized that variation of single functional groups have important effects on ribosome stability and function. In an *E. coli* mutant, for example, a single amino acid replacement in ribosomal protein L24 caused a defect in ribosomal subunit assembly and resulted in a change in growth temperature (Cabezon *et al.*, 1977).

Though the amino acid compositions of most ribosomal proteins from the 30S subunit of *B. stearothermophilus* are known (Isono and Isono, 1975), those for the ribosomal proteins from *B. subtilis* are limited to only three proteins: S4, S5 and S12 (Itoh *et al.*, 1975). Protein S4 is a RNA-binding protein (see Zimmermann, 1974) and essential in subunit assembly (Held *et al.*, 1974), and proteins S4 and S12 are important in the fidelity of translation (Gorini, 1974). Table 1 shows a comparison between amino acid composition of protein S4, S5, S12 pairs from *B. subtilis* and *B. stearothermophilus*, and indicate a significant increase in the

TABLE 1

Amino Acid Composition of Ribosomal Proteins from* Bacillus subtilis *(BS) and* Bacillus stearothermophilus *(BT)*

	S4		S5		S12		Average (S4, S5, S12)		
	BS	*BT*	*BS*	*BT*	*BS*	*BT*	*BS*	*BT*	*Diff.*
Asn	7.4	6.5	6.8	7.4	8.8	7.5	7.7	7.1	-0.6
Thr	3.7	4.0	4.0	4.1	6.4	5.8	4.7	4.6	-0.1
Ser	4.9	5.5	3.3	3.8	6.1	4.6	4.8	4.6	-0.2
Glx	*11.8*	*17.5*	*9.6*	*11.8*	*4.1*	*7.6*	*8.5*	*12.3*	*+3.8*
Pro	5.5	4.9	4.3	3.7	5.6	4.4	5.1	4.3	-0.8
Gly	9.5	8.7	12.2	11.7	10.9	9.3	10.9	9.9	-1.0
Ala	4.9	4.6	10.0	10.5	5.0	7.5	6.6	7.5	+0.9
Val	6.6	4.3	12.2	9.6	10.7	8.2	9.8	7.4	-2.4
Met	1.0	2.0	1.6	1.3	0.8	1.5	1.1	1.6	+0.5
Ile	4.6	4.6	6.4	7.2	5.7	5.0	5.6	5.6	0.0
Leu	11.0	9.7	9.6	8.0	6.4	6.2	9.0	8.0	-1.0

Tyr	4.4	5.3	-	0.3	3.8	4.2	2.7	3.3	+0.6
Phe	5.7	2.5	3.1	2.5	0.8	1.9	3.2	2.3	-0.9
His	2.9	2.5	1.6	2.3	3.8	2.4	2.8	2.4	-0.4
Lys	7.5	6.6	7.5	8.3	11.7	13.2	8.9	9.4	+0.5
Arg	9.2	11.4	8.1	7.5	9.5	10.3	8.9	9.7	+0.8

**Expressed as mole %.*

BS: Taken from Itoh et al. *(1975) and reproduced with permission from Springer-Verlag.*

BT: Taken from Isono and Isono (1975) and reproduced with permission from The Federation of European Biochemical Societies number of Glx (glutamic acid and glutamine) residues in *B. stearothermophilus*. Glutamic acid can stabilize RNA-protein interactions by specific interaction with guanine (Lancelot and Helene, 1977), and may stabilize protein-protein interaction by electrostatic interaction with basic groups of proteins such as the specific glutamy-arginyl electrostatic complex proposed by Helene (1977). Since most ribosomal proteins are basic, the effect of electrostatic stabilization by glutamic acid may be greater in ribosomal proteins than in other less basic proteins. Furthermore, glutamine can also stabilize RNA-protein interactions by forming strong hydrogen bonds with adenine or guanine (Helene, 1977). Other studies have shown that the regions of ribosomal proteins S4 and S7 which have been crosslinked to 16S rRNA by UV irradiation are rich not only in basic but also in acidic residues (Ehresmann *et al.*, 1975). Taken together these facts suggest that an increased glutamic acid or glutamine may play an important role in the thermostability of ribosomes.

The N-terminal sequences of 17 ribosomal proteins from the 30S subunit from *B. stearothermophilus* strain 799 (Higo and Loertscher, 1974; Vassos *et al.*, 1976; Higo and Nomura cited by Hori *et al.*, 1977) and 20 proteins from *B. stearothermophilus* strain 10 (Yaguchi *et al.*, 1973; Yaguchi *et al.*, 1974; this study) have been reported and indicate that these two strains are nearly identical. Minor differences can be attributed to strain difference; i.e., position 11 of S16 is alanine in strain 799 but threonine in strain 10 and position 4 of S20 is arginine in strain 799 but lysine in strain 10 (Higo and Loertscher, 1974; Yaguchi *et al.*, 1974). These substitutions are the result of single base changes.

In Table 2, the amino acid sequences of ribosomal proteins from *B. stearothermophilus* are compared with those from *B. subtilis* (Itoh *et al.*, 1975; Higo *et al.*, 1978; Hori *et al.*, 1978), and *E. coli* (Higo and Loertscher, 1974; Vassos *et al.*, 1976; Wittmann-Liebold, 1973; and see Stöffler and Wittmann, 1977). Amino acid residues of *B. stearothermophilus* which are identical to those of *B. subtilis* and/or *E. coli* are italicized. The overall sequence homology between *B. stearothermophilus* and *B. subtilis* is much higher (84%) than between *B. stearothermophilus* and *E. coli* (50%) and this clearly shows the importance of comparing macromolecules from the same genus (i.e., *Bacillus* sp.) as stressed by Stenesh (1976). This high degree of homology between *B. stearothermophilus* and *B. subtilis* suggests that the rate of amino acid substitution in ribosomal proteins is much slower than other cellular proteins (Higo *et al.*, 1978; Hori *et al.*, 1977) and/or that *B. stearothermophilus* is very closely related to *B. subtilis*. The former hypothesis is supported by postulates of Orgel (1963, 1970), who suggested that proteins involved in information transfer are less likely to be subjected to substitution than other cellular proteins, whose variability would not necessarily alter cell viability.

The number of amino acid residues in *B. stearothermophilus* which are different from *B. subtilis* and *E. coli* is only 17 among 171 corresponding positions examined; these amino acid substitutions are shown in Table 3. This implies that the number of positions in the sequence of *B. stearothermophilus* which are required for thermal stability is probably small, perhaps no more than 10% of the residues involved.

The amino acid substitution pattern shown in Table 3 between *B. stearothermophilus* and *B. subtilis* indicates that most of them are very conservative substitutions; six substitutions are among acids and their amides (Asp, Asn, Glu, and Gln) and five exchanges are among basic residues (His, Lys, and Arg). Since *B. stearothermophilus* has 4 glutamic acids and 1 glutamine while *B. subtilis* has 3 glutamic acids and 2 glutamines, the total glutamic acid and glutamines in both species are the same and does not show the differences suggested by the amino acid compositions of the proteins (Table 1). Apparently the positions which resulted in a net increase of glutamic acid and glutamine were not present in the N-terminal regions examined so far. Nevertheless, there is a net increase of 1 glutamic acid and 1 less glutamine in *B. stearothermophilus*. It is also noteworthy that there are two arginine residues in *B. stearothermophilus* but none in *B. subtilis*.

Hori *et al.* (1977) discussed the pattern of the amino acid substitution in ribosomal proteins between *B. stearothermophilus* and *E. coli* and Higo *et al.* (1978) between *B. stearothermophilus* and *B. subtilis*, and pointed out a significant change from phenylalanine to tyrosine in *B. stearothermophilus*. Table 4 shows nine Phe-Tyr substitutions in the N-terminal regions; eight phenylalanines in *E. coli* are changed to tyrosines in *B. stearothermophilus* and one tyrosine to phenylalanine. However, since three out of the eight corresponding positions in *B. subtilis* are known and also found to be tyrosine (Higo *et al.*, 1978), the Phe to Tyr substitution in these positions are not responsible for any difference in thermostability. The residues in other corresponding positions in *B. subtilis* are yet to be identified. The amino acid composition of three proteins shown in Table 1 tends to indicate a minor increase in tyrosine and decrease in phenylalanine in *B. stearothermophilus*. Aromatic residues such as tyrosine and tryptophan may stabilize RNA-protein complexes by stacking interaction of nucleic acid bases in single stranded RNA (Toulme and Helene, 1977). Lemieux (1974) found that the nitration of tyrosine in proteins S4, S7, and S8 of

TABLE 2

Comparison of the N-terminal Amino Acid Sequences of Ribosomal Proteins from the 30S Subunit of Escherichia coli *(EC)*,[a] Bacillus subtilis *(BS)*,[b] *and* Bacillus stearothermophilus *(BT)*[c]

						5					10					15	
S2	EC	ALA	THR	VAL	*SER*	*MET*	ARG	ASP	MET	*LEU*	LYS	ALA	*GLY*	*VAL*	*HIS*	*PHE*	*GLY*
	BT	SER	VAL	ILE	*SER*	*MET*	LYS	GLN	LEU	*LEU*	GLU	SER	*GLY*	*VAL*	*HIS*	*PHE*	*GLY*
					20												
	EC	HIS	GLN	*THR*	*ARG*												
	BT	X	GLX	*THR*	*ARG*												
						5					10					15	
S3	EC	*GLY*	*GLN*	*LYS*	*VAL*	HIS	*PRO*	ASN	*GLY*	ILE	*ARG*	LEU	*GLY*	*ILE*	VAL	LYS	PRO
	BT	*GLY*	*GLN*	*LYS*	*VAL*	ASN	*PRO*	ILE	*GLY*	LEU	*ARG*	ILE	*GLY*	*ILE*	ILE	ARG	ASN
					20					25					30		
	EC	*GLY*	ASN	*SER*	THR	GLY	PHE	ALA	ASN	THR	*LYS*	GLU	PHE	*ALA*	ASP	ASN	*LEU*
	BT	*GLY*	GLX	*SER*	ARG	TRP	TYR	SER	GLX	---	*LYS*	X	TYR	*ALA*	ASX	LEU	*LEU*
						5					10					15	
S4	EC	*ALA*	*ARG*	*TYR*	LEU	*GLY*	*PRO*	LYS	LEU	*LYS*	LEU	*SER*	*ARG*	*ARG*	GLU	*GLY*	THR
	BS	*ALA*	*ARG*	*TYR*	*THR*	*GLY*	*PRO*	X	*TRP*	*LYS*	LEU	*SER*	*ARG*	*ARG*	*LEU*	*GLY*	*ILE*
	BT	*ALA*	*ARG*	*TYR*	*THR*	*GLY*	*PRO*	MET	*TRP*	*LYS*	ILE	*SER*	*ARG*	*ARG*	*LEU*	*GLY*	*ILE*
					20					25					30		
	EC	ASP	*LEU*	PHE	LEU	LYS	*SER*	GLY	VAL	ARG	ALA	ILE	ASP	THR	LYS	CYS	*LYS*
	BS	*SER*	*LEU*	---	---	---	*SER*	*GLY*	*THR*	*GLY*	*LYS*	*GLU*					
	BT	*SER*	*LEU*	---	---	---	*SER*	*GLY*	*THR*	*GLY*	*LYS*	*GLU*	LEU	GLN	---	---	*LYS*
				35					40					45			
	EC	ILE	GLU	GLN	ALA	*PRO*	*GLY*	*GLN*	*HIS*	*GLY*	ALA	ARG	LYS	PRO	*ARG*	---	*LEU*
	BT	ARG	PRO	TYR	PRO	*PRO*	*GLY*	*GLN*	*HIS*	*GLY*	PRO	GLY	GLN	ARG	*ARG*	LYS	*LEU*

				50					55								
	EC	*SER*	ASP	*TYR*	*GLY*	VAL	*GLN*	*LEU*	ARG	*GLU*							
	BT	*SER*	GLU	*TYR*	*GLY*	LEU	*GLN*	*LEU*	X	*GLU*							
						5					10				14		
S5	BS	*MET*	*ARG*	X	*ILE*	*ASP*	*PRO*	X	*LYS*	*LEU*	*GLU*	*LEU*	*GLU*	*GLU*	PHE		
	BT	*MET*	*ARG*	ARG	*ILE*	*ASP*	*PRO*	ASN	*LYS*	*LEU*	*GLU*	*LEU*	*GLU*	*GLU*	ARG	VAL	VAL
					20					25					30		
	BT	ALA	VAL	ASN	ARG	VAL	ALA	LYS	VAL	VAL	LYS	GLY	GLY	ARG	ARG		
						5					10					15	
S6	EC	*MET*	*ARG*	HIS	*TYR*	*GLU*	*ILE*	VAL	PHE	MET	VAL	HIS	*PRO*	*ASP*	GLN	SER	GLU
	BS	*MET*	*ARG*	*LYS*	*TYR*	*GLU*	VAL	*MET*	*TYR*	*ILE*	*ILE*	*ARG*	*PRO*	ASN	ILE	*ASP*	GLU
	BT	*MET*	*ARG*	*LYS*	*TYR*	*GLU*	*ILE*	*MET*	*TYR*	*ILE*	*ILE*	*ARG*	*PRO*	*ASP*	MET	*ASP*	ASP
					20						25				29		
	EC	*GLN*	VAL	PRO	GLY	MET	ILE	---	*GLU*	*ARG*	TYR	THR	ALA	ALA	ILE		
	BT	*GLN*	ALA	ARG	GLN	SER	LEU	VAL	*GLU*	*ARG*	PHE	ASP	ASN	VAL	LEU		
						5					10					15	
S7	EC	*PRO*	*ARG*	*ARG*	ARG	VAL	ILE	GLY	GLN	*ARG*	LYS	ILE	*LEU*	*PRO*	*ASP*	*PRO*	LYS
	BS	*PRO*	*ARG*	LYS	*GLY*	*PRO*	*VAL*	*ALA*	*LYS*	*ARG*	*ASP*	*VAL*	*LEU*	*PRO*	ASX	GLY	*ILE*
	BT	*PRO*	*ARG*	*ARG*	*GLY*	*PRO*	*VAL*	*ALA*	*LYS*	*ARG*	*ASP*	*VAL*	*LEU*	*PRO*	*ASP*	*PRO*	*ILE*
					20												
	EC	PHE	GLY	*SER*	GLU	*LEU*											
	BS	*TYR*	ASX														
	BT	*TYR*	ASN	*SER*	LYS	*LEU*											
						5					10					15	
S8	EC	SER	*MET*	GLN	*ASP*	*PRO*	*ILE*	*ALA*	*ASP*	*MET*	*LEU*	*THR*	*ARG*	*ILE*	*ARG*	*ASN*	GLY
	BS	*VAL*	*MET*	*THR*	*ASP*	*PRO*	*ILE*	*ALA*	*ASP*	*MET*	*LEU*	*THR*	*ARG*	*ILE*	*ARG*	*ASN*	*ALA*
	BT	*VAL*	*MET*	*THR*	*ASP*	*PRO*	*ILE*	*ALA*	*ASP*	*MET*	*LEU*	*THR*	*ARG*	*ILE*	*ARG*	*ASN*	*ALA*

TABLE 2 - Continued

Comparison of the N-terminal Amino Acid Sequences of Ribosomal Proteins from the 30S Subunit of Escherichia coli *(EC)*,[a] Bacillus subtilis *(BS)*,[b] *and* Bacillus stearothermophilus *(BT)*[c]

					20					25					30		
S8	EC	GLN	ALA	ALA	ASN	LYS	ALA	ALA	VAL	THR	MET	*PRO*	SER	*SER*	*LYS*	LEU	
(cont'd)	BS	ASP															
	BT	ASN	MET	VAL	ARG	HIS	GLU	LYS	LEU	GLU	VAL	*PRO*	ALA	*SER*	*LYS*	ILE	*LYS*
				35					40					45			48
	EC	VAL	ALA	*ILE*	*ALA*	ASN	VAL	*LEU*	*LYS*	GLU	*GLU*	*GLY*	*PHE*	*ILE*	GLU	*ASP*	PHE
	BT	ARG	GLU	*ILE*	*ALA*	GLU	ILE	*LEU*	*LYS*	ARG	*GLU*	*GLY*	*PHE*	*ILE*	ARG	*ASP*	TYR
						5					10					15	
S9	EC	*ALA*	GLU	ASN	*GLN*	*TYR*	*TYR*	*GLY*	*THR*	*GLY*	*ARG*	*ARG*	*LYS*	*SER*	*SER*	ALA	*ALA*
	BT	*ALA*	GLN	VAL	*GLN*	*TYR*	*TYR*	*GLY*	*THR*	*GLY*	*ARG*	*ARG*	*LYS*	*SER*	*SER*	VAL	*ALA*
					20					25					30		
	EC	*ARG*	*VAL*	PHE	ILE	LYS	*PRO*	*GLY*	ASN	*GLY*	LYS	*ILE*	VAL	ILE	*ASN*	GLN	ARG
	BT	*ARG*	*VAL*	ARG	LEU	VAL	*PRO*	*GLY*	ASP	*GLY*	ARG	*ILE*	ILE	VAL	*ASN*	LYS	GLN
				35					40					45			
	EC	SER	LEU	GLU	GLN	*TYR*	PHE	GLY	ARG	*GLU*	THR	ALA	ARG	MET			
	BT	ASP	ILE	ARG	GLU	*TYR*	ILE	PRO	THR	*GLU*	ALA	LEU	ILE	GLX			
						5					10					15	
S10	EC	MET	GLN	ASN	GLN	ARG	*ILE*	*ARG*	*ILE*	*ARG*	*LEU*	*LYS*	*ALA*	PHE	*ASP*	*HIS*	*ARG*
	BS	---	*ALA*	*LYS*	GLN	*LYS*	*ILE*	*ARG*	*ILE*	*ARG*	*LEU*	*LYS*	*ALA*	*TYR*	*ASP*	*HIS*	*ARG*
	BT	---	*ALA*	*LYS*	GLU	*LYS*	*ILE*	*ARG*	*ILE*	*ARG*	*LEU*	*LYS*	*ALA*	*TYR*	*ASP*	*HIS*	*ARG*
					20					25					30		
	EC	LEU	ILE	*ASP*	*GLN*	ALA	THR	ALA	GLX	*ILE*	*VAL*	*GLU*	*THR*	*ALA*	*LYS*	GLU	THR
	BS	*ILE*	*LEU*	ASX	GLX	*SER*	*ALA*	GLX	*LYS*	*ILE*	*VAL*	GLX	*THR*	*ALA*			
	BT	*ILE*	*LEU*	*ASP*	*GLN*	*SER*	*ALA*	GLU	*LYS*	*ILE*	*VAL*	*GLU*	*THR*	*ALA*	*LYS*	ARG	SER

				35					40								
S10	EC	*GLY*	*ALA*	GLN	LEU	VAL	*GLY*	PRO	PHE								
(cont'd)	BT	*GLY*	*ALA*	LYS	VAL	X	*GLY*	X	ILE								
						5					10					15	
S11	EC	*ALA*	LYS	ALA	PRO	ILE	ARG	ALA	ARG	LYS	*ARG*	*VAL*	*ARG*	*LYS*	GLN	VAL	SER
	BT	*ALA*	ARG	ARG	THR	ASN	THR	ARG	LYS	ARG	*ARG*	*VAL*	*ARG*	*LYS*	ASN	ILE	ASP
					20					25					30		
	EC	ASP	*GLY*	VAL	*ALA*	*HIS*	*ILE*	HIS	ALA	SER	*PHE*	*ASN*	*ASN*	*THR*	*ILE*		
	BT	THR	*GLY*	ILE	*ALA*	*HIS*	*ILE*	ARG	S/C	THR	*PHE*	*ASN*	*ASN*	*THR*	*ILE*		
						5					10					15	
S12	EC	ALA	*THR*	VAL	*ASN*	*GLN*	*LEU*	*VAL*	*ARG*	*LYS*	PRO	*ARG*	ALA	ARG	*LYS*	*VAL*	ALA
	BS	*PRO*	*THR*	*ILE*	*ASN*	*GLN*	*LEU*	ILE	*ARG*	*LYS*	*GLY*	*ARG*	*GLU*	HIS	*LYS*	*VAL*	GLU
	BT	*PRO*	*THR*	*ILE*	*ASN*	*GLN*	*LEU*	*VAL*	*ARG*	*LYS*	*GLY*	*ARG*	*GLU*	HIS	*LYS*	*VAL*	VAL
					20					25					30		
	EC	*LYS*	*SER*	ASN	VAL	*PRO*	*ALA*	*LEU*	GLU	ALA	*CYS*	PRO	GLN	LYS	ARG	GLY	VAL
	BS	ASN	*SER*	LYS													
	BT	*LYS*	*SER*	LYS	SER	*PRO*	*ALA*	*LEU*	ASN	LYS	*CYS*	TYR	ASP	SER	PHE	LYS	LYS
				35													
	EC	GLY	THR	ARG	VAL	TYR											
	BT	GLX	GLX	THR	ASX	VAL											
						5					10					15	
S13	EC	*ALA*	*ARG*	*ILE*	*ALA*	*GLY*	ILE	ASN	*ILE*	*PRO*	ASP	HIS	*LYS*	HIS	ALA	*VAL*	*ILE*
	BT	*ALA*	*ARG*	*ILE*	*ALA*	*GLY*	VAL	ASP	*ILE*	*PRO*	ARG	ASP	*LYS*	ARG	VAL	*VAL*	*ILE*
					20					25							
	EC	ALA	*LEU*	*THR*	SER	*ILE*	*TYR*	*GLY*	VAL	*GLY*	*LYS*	TYR					
	BT	SER	*LEU*	*THR*	TYR	*ILE*	*TYR*	*GLY*	ILE	*GLY*	*LYS*	LYS					

TABLE 2 - Continued

Comparison of the N-terminal Amino Acid Sequences of Ribosomal Proteins from the 30S Subunit of Escherichia coli *(EC),*[a] Bacillus subtilis *(BS),*[b] *and* Bacillus stearothermophilus *(BT)*[c]

						5					10					15	
S14	EC	*ALA*	*LYS*	GLN	*SER*	*MET*	LYS	*ALA*	ARG	GLU	VAL	LYS	ARG	VAL	ALA	LEU	ALA
	BT	*ALA*	*LYS*	LYS	*SER*	*MET*	ILE	*ALA*	LYS	GLN	LYS	ARG	THR	PRO	LYS	PHE	LYS
					20					25					30		
	EC	ASP	LYS	TYR	PHE	ALA	LYS	ARG	ALA	GLU	LEU	LYS	ALA	ILE	ILE	SER	ARG
	BT	VAL	ARG	ALA	TYR	THR	ARG	THR	GLU	ARG	ARG	GLY	LYS	ARG	HIS	S/C	VAL
						5					10					15	
S15	EC	SER	*LEU*	SER	THR	*GLU*	ALA	THR	ALA	LYS	*ILE*	VAL	SER	GLU	*PHE*	GLY	ARG
	BS	*ALA*	ILE	*THR*	*GLN*	*GLU*	*ARG*	*LYS*	ASN	GLN	LEU	*ILE*	ASN	GLU	*PHE*	LYS	*VAL*
	BT	*ALA*	*LEU*	*THR*	*GLN*	*GLU*	*ARG*	*LYS*	ASX	GLU	*ILE*	*ILE*	GLU	GLN	*PHE*	HIS	*VAL*
					20					25							
	EC	ASP	ALA	*ASN*	*ASP*	*THR*	*GLY*	*SER*	THR	GLU	*VAL*	GLN	*VAL*				
	BS	HIS	*GLU*	ASX	ASX												
	BT	ARG	*GLU*	*ASN*	*ASP*	*THR*	*GLY*	*SER*	PRO	GLX	*VAL*	GLX	***VAL***				
						5					10					15	
S16	EC	MET	*VAL*	THR	*ILE*	*ARG*	*LEU*	ALA	*ARG*	HIS	*GLY*	ALA	*LYS*	*LYS*	ARG	*PRO*	*PHE*
	BT	ALA	*VAL*	LYS	*ILE*	*ARG*	*LEU*	LYS	*ARG*	MET	*GLY*	THR	*LYS*	*LYS*	LYS	*PRO*	*PHE*
					20					25					30		
	EC	*TYR*	GLN	VAL	*VAL*	*VAL*	*ALA*	*ASP*	*SER*	*ARG*	ASN	ALA	*ARG*	ASN	*GLY*	*ARG*	*PHE*
	BT	*TYR*	ARG	ILE	*VAL*	*VAL*	*ALA*	*ASP*	*SER*	*ARG*	SER	PRO	*ARG*	ASX	*GLY*	*ARG*	*PHE*
		33															
	EC	*ILE*															
	BT	*ILE*															

							5						10				
S17	EC	THR	ASP	LYS	ILE	---	*ARG*	THR	LEU	GLN	---	*GLY*	*ARG*	*VAL*	*VAL*	*SER*	*ASP*
	BT	SER	GLN	ARG	ASN	GLN	*ARG*	LYS	VAL	TYR	VAL	*GLY*	*ARG*	*VAL*	*VAL*	*SER*	*ASP*
		15					20					25					
	EC	*LYS*	*MET*	GLU	*LYS*	SER	*ILE*	*VAL*	*VAL*	ALA	ILE	GLU	ARG	PHE			
	BT	*LYS*	*MET*	ASP	*LYS*	THR	*ILE*	*VAL*	*VAL*	LEU	VAL	GLX	X	TYR			
								5							10		
S18	EC	*ALA*	---	*ARG*	---	TYR	PHE	*ARG*	---	---	*ARG*	*ARG*	*LYS*	PHE	*CYS*	ARG	*PHE*
	BT	*ALA*	GLY	*ARG*	LYS	GLY	GLY	*ARG*	GLY	LYS	*ARG*	*ARG*	*LYS*	VAL	*CYS*	TYR	*PHE*
				15					20					25			
	EC	*THR*	*ALA*	GLN	GLY	VAL	GLN	GLU	*ILE*	*ASP*	*TYR*	*LYS*	*ASP*	ILE	ALA		
	BT	*THR*	*ALA*	ASN	ASN	ILE	THR	HIS	*ILE*	*ASP*	*TYR*	*LYS*	*ASP*	VAL	ASP		
						5					10					15	
S19	EC	PRO	*ARG*	*SER*	*LEU*	*LYS*	*LYS*	*GLY*	*PRO*	*PHE*	ILE	*ASP*	LEU	*HIS*	*LEU*	LEU	*LYS*
	BS	ALA	*ARG*	*SER*	*LEU*	*LYS*	*LYS*	*GLY*	*PRO*	*PHE*	VAL	ASX	ASX	*HIS*			
	BT	GLY	*ARG*	*SER*	*LEU*	*LYS*	*LYS*	*GLY*	*PRO*	*PHE*	CYS	*ASP*	GLU	*HIS*	*LEU*	MET	*LYS*
					20					25					30		
	EC	*LYS*	VAL	*GLU*	*LYS*	ALA	VAL	GLU	SER	*GLY*	ASP	*LYS*	LYS	PRO	LEU	ARG	*THR*
	BT	*LYS*	ILE	*GLU*	*LYS*	LEU	ASN	ALA	THR	*GLY*	GLN	*LYS*	GLX	VAL	ILE	LYS	*THR*
			34														
	EC	TRP	*SER*														
	BT	X	*SER*														
						5					10					15	
S20	EC	*ALA*	*ASN*	*ILE*	*LYS*	*SER*	*ALA*	LYS	*LYS*	*ARG*	*ALA*	ILE	GLN	*SER*	*GLU*	*LYS*	ALA
	BS	*ALA*	*ASN*	*ILE*	*LYS*	*ASN*	*ALA*	*ILE*	*LYS*	*ARG*	VAL	HIS	ASN	ALA	*GLU*	ILE	LYS
	BT	*ALA*	*ASN*	*ILE*	*LYS*	*SER*	*ALA*	*ILE*	*LYS*	*ARG*	*ALA*	LYS	THR	*SER*	*GLU*	*LYS*	ARG

TABLE 2 - Continued
Comparison of the N-terminal Amino Acid Sequences of Ribosomal Proteins from the 30S Subunit of Escherichia coli *(EC)*,[a] Bacillus subtilis *(BS)*,[b] *and* Bacillus stearothermophilus *(BT)*[c]

					20					25							
S20	EC	*ARG*	LYS	*HIS*	ASN	*ALA*	*SER*	*ARG*	ARG	*SER*							
(cont'd)	BS	*ARG*	*ALA*	*HIS*	ASN	*ALA*											
	BT	*ARG*	*ALA*	*HIS*	ASX	*ALA*	*SER*	*ARG*	LYS	*SER*							
						5					10					15	
S21	EC	*PRO*	VAL	ILE	LYS	VAL	ARG	GLY	ASN	*GLU*	PRO	*PHE*	ASP	VAL	*ALA*	*LEU*	ARG
	BT	*PRO*	MET	---	---	---	---	---	---	*GLU*	VAL	*PHE*	GLU	GLN	*ALA*	*LEU*	LYS
			18														
	EC	ARG	PHE														
	BT	ASN	SER														

[a]Escherichia coli: *The complete amino acid sequences of S4, S6, S8, S9, S12, S13, S15, S16, S17, S18, S19 and S20 have been determined (see Stöffler and Wittmann, 1977) and their N-terminal sequences are shown here. The N-terminal sequences of other proteins are taken from Higo and Loertscher (1974) with permission from The American Society for Microbiology, Higo and Nomura cited by Hori* et al. *(1977) with permission from Springer-Verlag and Wittmann-Liebold (1973) with permission from the Federation of European Biochemical Societies.*

[b]Bacillus subtilis: *Taken from Itoh* et al. *(1975) with permission from Springer-Verlag, Higo* et al. *(1978) with permission from the University of Tokyo Press and Hori* et al. *(1978).*

[c]Bacillus stearothermophilus: *Taken from Higo and Loertscher (1974) with permission from the American Society for Microbiology, Higo and Nomura cited by Hori* et al. *(1977) with permission from Springer-Verlag, Vassos* et al. *(1976) with permission from Elsevier/North-Holland Biomedical Press, Yaguchi* et al. *(1973) with permission from the National Research Council of Canada and the Canadian Journal of Biochemistry, Yaguchi* et al. *(1974) with permission from the Federation of European Biochemical Societies and data presented for the first time in this paper.*

TABLE 3
Amino Acid Substitutions in the N-terminal Regions of Ribosomal Proteins from the 30S Subunit: Residues of B. stearothermophilus *which are Different from Those of* E. coli *and* B. subtilis

		Mesophile		*Thermophile*
Protein	*Position*	E. coli	B. subtilis	B. stearothermophilus
S6	16	Glu	Glu	Asp
S15	12	Ser	Asn	Glu
S19	12	Leu	Asx	Glu
S15	9	Lys	Gln	Glu
S10	4	Gln	Gln	Glu
S15	13	Glu	Glu	Gln
S20	12	Gln	Asn	Thr
S19	1	Pro	Ala	Gly
S12	16	Ala	Glu	Val
S6	14	Gln	Ile	Met
S4	10	Leu	Leu	Ile
S19	10	Ile	Val	Cys
S15	15	Gly	Lys	His
S20	11	Ile	His	Lys
S12	13	Arg	His	Lys
S15	17	Asp	His	Arg
S20	16	Ala	Lys	Arg

E. coli destroyed the binding capacity of these proteins with 16S rRNA. Taken together these facts suggest that the increased tyrosine content may also play an important role in thermophily.

CONCLUDING REMARKS

The extremely high degree of sequence homology between *B. stearothermophilus* and *B. subtilis* ribosomal RNA and proteins make these organisms very attractive for detecting nucleotide and amino acid substitutions which confer thermostability to ribosomes. On the other hand, due to great intergeneric differences between *B. stearothermophilus* and

TABLE 4
Phenylalanine and Tyrosine Substitutions in the N-terminal Regions of Ribosomal Proteins from the 30S Subunit

		Mesophile		*Thermophile*
Protein	*Position*	E. coli	B. subtilis	B. stearothermophilus
S6	8	Phe	Tyr	Tyr
S7	17	Phe	Tyr	Tyr
S10	13	Phe	Tyr	Tyr
S3	22	Phe		Tyr
S3	28	Phe		Tyr
S8	48	Phe		Tyr
S14	20	Phe		Tyr
S17	27	Phe		Tyr
S6	25	Tyr		Phe

E. coli, it is unlikely that further comparisons of these sequences from these organisms will produce definite answers to thermostability. By studying more closely related organisms, we propose that increases in specific amino acids: glutamic acid, glutamine, tyrosine, and arginine in ribosomal proteins or the unique nucleotide residues in rRNA, results in the thermostability of ribosomes of thermophilic organisms. Mechanistically, these residues may confer an increased intrinsic stability in the secondary structure of the ribosomal RNA or proteins, and/or in the stability of protein-protein and RNA-protein interactions.

ACKNOWLEDGEMENTS

We thank Dr. K. Higo and Prof. M. Nomura for permission to use some of their unpublished data cited by Hori *et al.* (1978) of the N-terminal amino acid sequences of *E. coli* and *B. stearothermophilus* 30S ribosomal proteins. We thank also Prof. S. Osawa for availability before publication of two manuscripts (Higo *et al.*, 1978; Hori *et al.*, 1978) which contain N-terminal amino acid sequences of *B. subtilis* 30S ribosomal proteins. Excellent technical assistance by F. Rollin, W. Rowsome and C. Roy is greatly appreciated. This is National Research Council of Canada publication No. 16490.

REFERENCES

Altenburg, L. G. and Saunders, G. F. (1971). J. Mol. Biol. *55*, 487.

Ansley, S. B., Campbell, L. L. and Sypherd, P. S. (1969). J. Bacteriol. *98*, 568.

Cabezon, T., Herzog, A., Petre, J., Yaguchi, M. and Bollen, A. (1977). J. Mol. Biol. *116*, 361.

Chen, R., Brosius, J., Wittmann-Liebold, B. and Schafer, W. (1977). J. Mol. Biol. *111*, 173.

Chow, C. T., Visentin, L. P., Matheson, A. T. and Yaguchi, M. (1972). Biochim. Biophys. Acta *287*, 270.

Cohlberg, J. A. and Nomura, M. (1976). J. Biol. Chem. *251*, 209.

Duggleby, R. G., Kaplan, H. and Visentin, L. P. (1975). Can. J. Biochem. *53*, 827.

Ehresmann, B., Reinbolt, J. and Ebel, J. P. (1975). FEBS Letters *58*, 106.

Fahnestock, S. R. (1977a). Arch. Biochem. Biophys. *180*, 555.

Fahnestock, S. R. (1977b). Arch. Biochem. Biophys. *182*, 497.

Fahnestock, S. R., Erdmann, V. and Nomura, M. (1973). Biochemistry *12*, 220.

Friedman, S. M. (1968). Bacteriol. Rev. *32*, 27.

Friedman, S. M., Axel, R. and Weinstein, I. B. (1967). J. Bacteriol. *93*, 1521.

Friedman, S. M. (1971). J. Bacteriol. *108*, 589.

Geisser, M., Tischendorf, G. W. and Stoffler, G. (1973). Mol. Gen. Genet. *127*, 129.

Gorini, L. (1974). In *Ribosomes* (M. Nomura, A. Tissieres and P. Lengyel, eds.), p. 791, Cold Spring Harbor Laboratory, New York.

Held, W. A., Ballou, B., Mizushima, M. and Nomura, M. (1974). J. Biol. Chem. *249*, 3103.

Helene, C. (1977). FEBS Letters *74*, 10.

Higo, K., Held, W., Kahan, L. and Nomura, M. (1973). Proc. Natl. Acad. Sci. USA *70*, 944.

Higo, K., Itoh, T. and Osawa, S. (1978). In *Proc. Intl. Symp. on Evolution of Protein Molecules*. Univ. of Tokyo Press (in press).

Higo, K. and Loertscher, K. (1974). J. Bacteriol. *118*, 180.

Hori, H. (1976). Mol. Gen. Genet. *145*, 119.

Hori, H., Higo, K. and Osawa, S. (1977). J. Mol. Evol. *9*, 191.

Hori, H., Higo, K. and Osawa, S. (1978). Proc. 2nd Taniguchi Int. Symp. on Biophysics (in press).

Horne, J. R. and Erdmann, V. A. (1972). Molec. Gen. Genet. *119*, 337.

Irwin, C. C., Akagi, J. M. and Himes, R. H. (1973). J. Bacteriol. *113*, 252.

Isono, K. (1974). Molec. Gen. Genet. *133*, 77.

Isono, K. and Isono, S. (1976). Proc. Natl. Acad. Sci. USA *73*, 767.

Isono, K., Isono, S., Stoffler, G., Visentin, L. P., Yaguchi, M. and Matheson, A. T. (1973). Molec. Gen. Genet. *127*, 191.

Isono, S. and Isono, K. (1975). Eur. J. Biochem. *50*, 483.

Itoh, T., Kosugi, H., Higo, K. and Osawa, S. (1975). Molec. Gen. Genet. *139*, 293.

Lancelot, G. and Helene, C. (1977). Proc. Natl. Acad. Sci. USA *74*, 4872.

Lemieux, G. (1974). Can. J. Biochem. *52*, 1038.

Ljungdahl, L. G. and Sherod, D. (1976). In *Extreme Environments: Mechanisms of Microbial Adaptation* (M. R. Heinrich, ed.), p. 147, Academic Press, New York.

Mangiantini, M. T., Tecce, G., Toshi, G. and Trentalance, A. (1965). Biochim. Biophys. Acta *103*, 252.

Marotta, C. A., Varricchio, F., Smith, I., Weissman, S. M., Sogin, M. L. and Pace, N. R. (1976). J. Biol. Chem. *251*, 3122.

Matheson, A. T., Yaguchi, M. and Visentin, L. P. (1975). Can. J. Biochem. *53*, 1323.

Miller, H. M., Friedman, S. M., Litman, D. J. and Cantor, C. R. (1976). Molec. Gen. Genet. *144*, 273.

Nazar, R. N. and Matheson, A. T. (1977). J. Biol. Chem. *252*, 4256.

Nazar, R. N., Sprott, G. D., Matheson, A. T. and Van, N. T. (1978), manuscript submitted.

Nishikawa, K. and Takemura, S. (1974). FEBS Letters *40*, 106.

Nomura, M. and Erdmann, V. A. (1970). Nature, *228*, 744.

Nomura, M., Traub, P. and Bechmann, H. (1968). Nature *219*, 793.

Orgel, L. E. (1963). Proc. Natl. Acad. Sci. USA *49*, 517.

Orgel, L. E. (1970). Proc. Natl. Acad. Sci. USA *67*, 1476.

Oshima, T. and Imahori, K. (1974). Int. J. Syst. Bact. *24*, 102.

Pace, B. and Campbell, L. L. (1967). Proc. Natl. Acad. Sci. USA *57*, 1110.

Saunders, G. F. and Campbell, L. L. (1966). J. Bacteriol. *91*, 332.

Stanley, J. and Ebel, J. P. (1977). Eur. J. Biochem. *77*, 357.

Stenesh, J. (1976). In *Extreme Environments: Mechamisms of Microbial Adaptation* (M. R. Heinrich, ed.), p. 85, Academic Press, New York.

Stenesh, J. and Holazo, A. A. (1967). Biochim. Biophys. Acta *138*, 286.

Stoffler, G. and Wittmann, H. G. (1977). In *Molecular Mechanisms of Protein Biosynthesis* (H. Weissbach and S. Pestka, eds.), p. 117, Academic Press, New York.

Tinoco, I., Jr., Uhlenbeck, O. C. and Levine, M. D. (1971). Nature *230*, 362.

Tischendorf, G. W., Geisser, M. and Stoffler, G. (1973). Molec. Gen. Genet. *127*, 147.

Toulme, J. J. and Helene, C. (1977). J. Biol. Chem. *252*, 244.

Varricchio, F. and Marotta, C. A. (1976). J. Bacteriol. *125*, 850.

Vassos, A., Higo, K. and Marcus, L. (1976). Biochim. Biophys. Acta *427*, 371.

Visentin, L. P. and Kaplan, H. (1975). Biochemistry *14*, 463.

Visentin, L. P., Matheson, A. T. and Yaguchi, M. (1974). FEBS Letters *41*, 310.

Wittmann-Liebold, B. (1973). FEBS Letters *36*, 247.

Yaguchi, M. (1975). FEBS Letters *59*, 217.

Yaguchi, M., Matheson, A. T. and Visentin, L. P. (1974). FEBS Letters *46*, 296.

Yaguchi, M., Roy, C., Matheson, A. T. and Visentin, L. P. (1973). Can. J. Biochem. *51*, 1215.

Zeikus, J. G., Taylor, M. W. and Brock, T. D. (1970). Biochim. Biophys. Acta *204*, 512.

Zimmermann, R. A. (1974). In *Ribosomes* (M. Nomura, A. Tissieres and P. Lengyel, eds.) p. 225, Cold Spring Harbor Laboratory, New York.

POLYPEPTIDE CHAIN ELONGATION FACTORS FROM AN EXTREME THERMOPHILE, *THERMUS THERMOPHILUS* HB8

Ken-ichi Arai
Shun Nakamura
Naoko Arai
Yoshito Kaziro

Institute of Medical Science
University of Tokyo
Takanawa, Minatoku, Tokyo

The elongation of the polypeptide chain in prokaryotes requires three complementary factors, i.e., EF-Tu, EF-Ts, and EF-G (for reviews, see Lucas-Lenard and Lipmann, 1971; Haselkorn and Rothman-Denes, 1973; Lengyel, 1974). EF-Tu and EF-Ts catalyze the GTP-dependent binding of aminoacyl-tRNA to the aminoacyl-tRNA site (A site) of ribosomes, while EF-G promotes the translocation of the peptidyl-tRNA-mRNA complex from the A site to the peptidyl-tRNA site (P site) of ribosomes with concomitant hydrolysis of GTP. These factors have been extensively purified from *E. coli*, and their molecular and catalytic properties as well as the mode of action are largely understood (for references, see above reviews). The elongation factors have also been purified from a moderate thermophile, *Bacillus stearothermophilus* (Beaud and Lengyel, 1971; Wittinghofer and Leberman, 1976).

In this paper, we describe the purification and some properties of the elongation factors from an extreme thermophile, *Thermus thermophilus* HB8. This bacterium, which was originally isolated by Oshima and Imahori (1971, 1974) from a Japanese hot spring, grows even at temperatures above 80°C. Our purpose of working on the factors from thermophilic bacteria is three-fold. First, we intend to obtain more stable

Abbreviations used: EF-Tu, EF-Ts, and EF-G, for polypeptide chain elongation factors Tu, Ts, and G, respectively; Na-DodSO$_4$ for sodium dodecyl sulfate.

ISBN 0-12-268250-5

factors for some physicochemical studies including X-ray diffraction analysis. Second, we are interested in the comparative aspects of the structure and function of elongation factors from various sources. And third, we expect to clarify the molecular basis of thermophilicity by comparing the structures of EF-Tu and EF-G from *T. thermophilus* with those from mesophilic and moderately thermophilic organisms.

OUTLINE OF PURIFICATION PROCEDURE FOR *T. THERMOPHILUS* ELONGATION FACTORS

Fig. 1 shows the outline of the purification of the elongation factors from *T. thermophilus* HB8. All of the elongation factors are purified together through the steps of aqueous two phase separation and ammonium sulfate fractionation. In DEAE-Sephadex column chromatography, they are separated into two peaks: Peak I, which appears from the column at 0.23 M NH_4Cl, contains the EF-Tu·EF-Ts complex and Peak II, which comes out from the column at 0.25 M NH_4Cl, is composed of EF-Tu·GDP and EF-G. The latter two factors are subsequently separated by Sephadex G-100 column chromato-

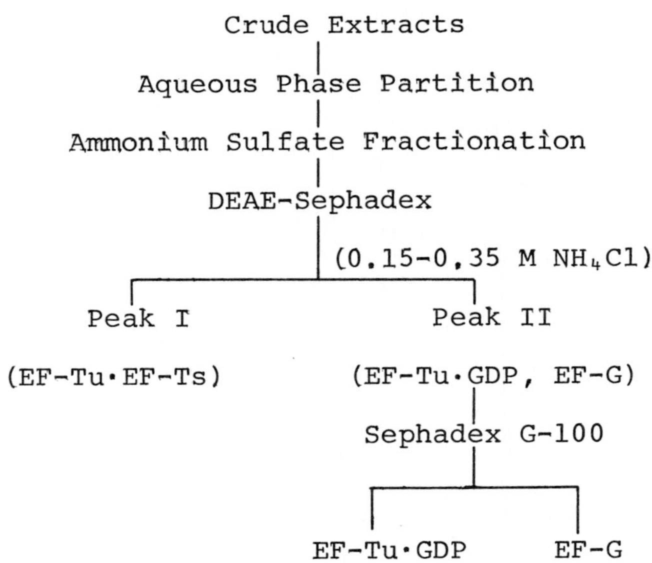

FIGURE 1. Outline of purification of elongation factors from Thermus thermophilus *HB8.*

graphy. Finally, the EF-Tu·EF-Ts complex, EF-Tu·GDP, and EF-G are purified separately to homogeneous states, and crystallized. The details of purification are described elsewhere (Arai, K., Ota, Y., Nakamura, S., Henneke, C. Oshima, T., and Kaziro, Y., submitted). The EF-Tu·EF-Ts complex is resolved into EF-Tu and EF-Ts by chromatography on a Sephadex G-200 column in the presence of 8 M guanidine-HCl. The complex is reconstituted from renatured EF-Tu·GDP and EF-Ts.

REQUIREMENT OF EF-Tu, EF-Ts, AND EF-G FOR POLYPHENYLALANINE SYNTHESIS IN A CELL FREE SYSTEM FROM *T. THERMOPHILUS*

The incorporation of [^{14}C]Phe-tRNA into polyphenylalanine in the presence of poly(U) and *T. thermophilus* ribosomes requires three elongation factors, EF-Tu, EF-Ts, and EF-G, purified from *T. thermophilus*. The reaction is completely dependent on added EF-Tu·EF-Ts in the presence of excess EF-G. In the presence of excess EF-Tu·GDP and EF-G, the reaction is markedly stimulated by the addition of increasing amounts of EF-Ts. Essentially similar results are obtained when *T. thermophilus* ribosomes are replaced by *E. coli* ribosomes.

MOLECULAR WEIGHTS

The $S_{20,w}$ values and molecular weights of *T. thermophilus* elongation factors determined by several different procedures are listed in Table 1. The molecular weights of EF-Tu·GDP and EF-G are almost comparable to those of the corresponding factors from *E. coli* (47,000 for EF-Tu, and 83,000 for EF-G; Arai, K. *et al.*, 1973). On the other hand, the molecular weight of *T. thermophilus* EF-Ts under the native condition (64,000) is about twice that of *E. coli* EF-Ts (36,000), and also that of *T. thermophilus* EF-Ts determined under denaturing conditions (31,000 and 27,000 by sedimentation equilibrium and Na-DodSO$_4$ gel electrophoresis, respectively). Therefore, *T. thermophilus* EF-Ts appears to exist in a dimeric structure under the native condition.

The molecular weight of the EF-Tu·EF-Ts complex from *T. thermophilus* determined by sedimentation equilibrium (142,000) and gel filtration (220,000), is significantly higher than that expected from the sum of the molecular weights of EF-Tu and EF-Ts determined by sedimentation equilibrium (about 80,000). Since the molar ratio of EF-Tu

TABLE 1
Semimentation Coefficients and Molecular Weights of T. thermophilus *EF-Tu·GDP, EF-Ts, EF-TU·EF-Ts, and EF-G*

	Sedimentation coefficients ($s_{20,w}$)	*Molecular weights*		
		Sedimentation equilibrium	*Sephadex G-150 gel filtration*	*Na-DodSO$_4$ gel electrophoresis*
EF-Tu·GDP	3.4	49,000	51,000	51,000
EF-Ts	3.4	31,000*	64,000	27,000
EF-Tu·EF-Ts complex	6.4	142,000	220,000	--
EF-G	---	85,000	80,000	78,000

Determined in 0.01 M NH_4OH and 0.1 M NH_4Cl

to EF-Ts in the EF-Tu·EF-Ts complex, determined by densitometric tracings of Na-DodSO_4 gel electrophoretogram, is approximately one to one, it is concluded that the complex contains 2 moles each of EF-Tu and EF-Ts, and exists probably in the form of EF-Tu·$(EF\text{-}Ts)_2$·EF-Tu.

From heterologous combination experiments, it was shown that *T. thermophilus* EF-Tu interacted only with its homologous EF-Ts to form a high molecular weight EF-Tu·EF-Ts complex (M. W. about 150,000), but did not form any complex with the heterologous *E. coli* EF-Ts. On the other hand, *E. coli* EF-Tu interacted not only with its homologous EF-Ts, but also with *T. thermophilus* EF-Ts to form an EF-Tu(*E. coli*)·EF-Ts (*T. thermophilus*) complex which appears to exist also in a dimeric form as in the case of the *T. thermophilus* EF-Tu·$(EF\text{-}Ts)_2$·EF-Tu complex.

INTERACTION WITH GUANINE NUCLEOTIDES

Interaction of EF-Tu and the EF-Tu·EF-Ts Complex with Guanine Nucleotides

T. thermophilus EF-Tu interacts with guanine nucleotides in a similar manner to *E. coli* EF-Tu. The dissociation constants of *T. thermophilus* EF-Tu·GDP and EF-Tu·GTP are 1.1×10^{-9} M and 5.9×10^{-8} M, respectively, and are similar to the values for *E. coli* EF-Tu·GDP and EF-Tu·GTP, which are 4.9×10^{-9} M and 3.6×10^{-7} M, respectively (K. Arai *et al.*, 1974a). On the other hand, the interaction with guanine nucleotides of the EF-Tu·EF-Ts complex from *T. thermophilus* and *E. coli* are quite different. Although the EF-Tu·Ts complex isolated from *E. coli* was readily dissociated by the addition of GDP to form EF-Tu·GDP and EF-Ts (K. Arai *et al.*, 1972), the same complex from *T. thermophilus* could not be dissociated by GDP. In the *E. coli* system, EF-Ts displaces the GDP moiety of EF-Tu·GDP forming an EF-Tu·EF-Ts complex (Miller and Weissbach, 1969; Miller and Weissbach, 1970; Weissbach *et al*, 1970), whereas EF-Ts from either *T. thermophilus* or *E. coli* fails to displace GDP from *T. thermophilus* EF-Tu·GDP even when added at more than 10-fold molar excess to EF-Tu·GDP. These results suggested that the EF-Tu·EF-Ts complex from *T. thermophilus* may be converted to a ternary GDP·EF-Tu·EF-Ts complex on addition of GDP. The formation of an [^{3}H]GDP·EF-Tu·EF-Ts complex was more directly demonstrated by gel filtration on a Sephadex G-150 column equilibrated with [^{3}H]GDP. As shown in Fig. 2, the peak of ^{3}H-radioactivity coincided with a protein peak eluted at the position of the EF-Tu·EF-Ts complex, followed by a plateau

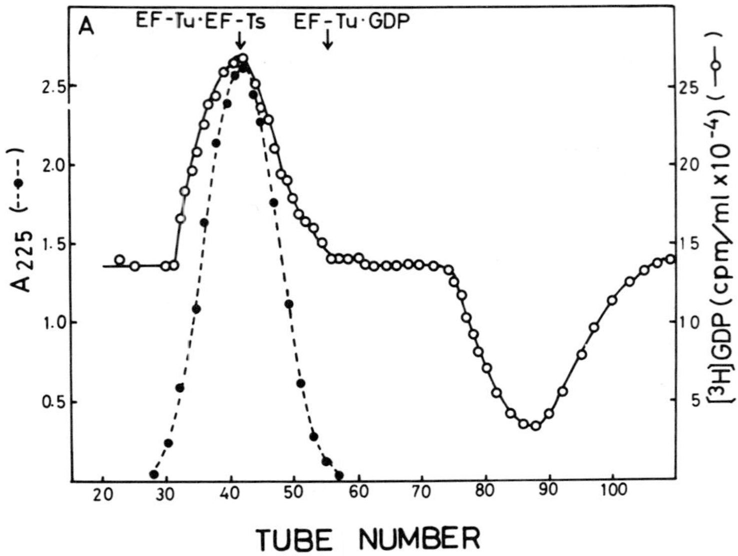

FIGURE 2. Chromatography of a GDP·EF-Tu·EF-Ts complex on a Sephadex G-150 column equilibrated with 3.6 μM [^{3}H]GDP. ^{3}H-radioactivity (-O-); protein concentration (-●-).

and then by a trough of radioactivity. Neither protein nor racioactivity was found at the position corresponding to EF-Tu·GDP, indicating that little, if any dissociation of the EF-Tu·EF-Ts complex occurred in the presence of GDP. From the sum of the radioactivity of the peak corresponding to the EF-Tu·EF-Ts complex, it was calculated that approximately 1.9 mol of [^{3}H]GDP was bound per 150,000 g of protein. Thus, it was concluded that the *T. thermophilus* EF-Tu·EF-Ts complex, existing in a dimeric form $(EF\text{-}Tu\cdot EF\text{-}Ts)_2$, interacted with GDP to form a $(GDP\cdot EF\text{-}Tu\cdot EF\text{-}Ts)_2$ complex. The dissociation constant for the complex of EF-Tu·EF-Ts with GDP was 6.1×10^{-7} M, a value about 600-fold higher than that for EF-Tu·GDP (1.1×10^{-9} M). On the other hand, the similar complex with GTP had a Kd value of 1.9×10^{-6} M, which was higher than that for EF-Tu·GTP (5.8×10^{-8} M) by about 30-fold. Therefore, it appears that *T. thermophilus* EF-Ts reduces the affinity of EF-Tu for guanine nucleotides and facilitates the exchange of GDP bound to EF-Tu with external GTP.

From these and other observations, we proposed a hypothetical reaction sequences for the *T. thermophilus* EF-Tu·EF-Ts complex as depicted in Fig. 3. We are now tempted to believe that the same mechanism is also true in the *E. coli*

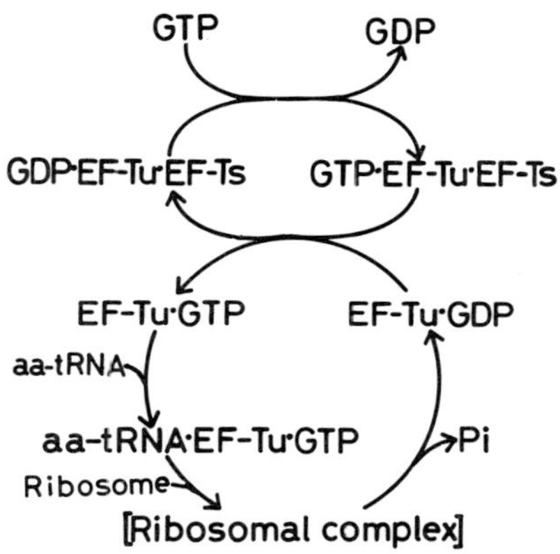

FIGURE 3. Reaction scheme for T. thermophilus *EF-Tu and EF-Ts. Based on the observation that the EF-Tu moiety of the EF-Tu·EF-Ts complex from* T. thermophilus *exchanges rapidly with free external EF-Tu, it is postulated that EF-Tu·GDP released from ribosomes reacts with a GTP·EF-Tu·EF-Ts complex to form a GDP·EF-Tu·EF-Ts complex with a concomitant displacement of EF-Tu·GTP. The latter then binds with an aminoacyl-tRNA to form a ternary aminoacyl-tRNA·EF-Tu·GTP complex to be transferred to the ribosomal A site. GDP bound to the EF-Tu·EF-Ts complex can be exchanged with external GTP to regenerate the GTP·EF-Tu·EF-Ts complex.*

system, i.e., a ternary GDP·EF-Tu·EF-Ts complex might be formed transiently in which the affinity between GDP and EF-Tu may be weakened so that it rapidly exchanges its GDP moiety with GTP. Presumably, interaction of EF-Ts with EF-TU·GDP would induce a conformational change in EF-Tu around the GDP binding site, thus promoting the dissociation of GDP. This contention was supported by the recent observation that kirromycin, a potent inhibitor of *E. coli* EF-Tu (Wolf *et al.*, 1974) which prevents the formation of the EF-Tu·EF-Ts complex, stimulates the exchange of GDP bound to EF-Tu with external [^{3}H]GDP (Brown and Blumenthal, 1976; Blumenthal *et al.*, 1977). These results suggest that kirromycin may occupy the binding site for EF-Ts and mimic the action of EF-Ts. In this respect, it is noteworthy that a kirromycin-induced conformational change in EF-Tu has recently been demonstrated by Wilson and Cohn (1977) near the binding site for guanine nucleotides.

Binding of Guanine Nucleotides to EF-G

As reported previously (N. Arai *et al.*, 1975; N. Arai *et al.*, 1977a; Rohrbach and Bodley, 1976; Baca *et al.*, 1976), the binary EF-G·GTP or EF-G·GDP complex is rather unstable and can not be isolated by the nitrocellulose membrane filter procedure. In contrast, *T. thermophilus* EF-G formed a relatively stable EF-G·GDP binary complex which can be isolated on the nitrocellulose membrane filter. The dissociation constant (Kd) of EF-G·GDP and the number of binding sites for GDP were calculated as 6.7×10^{-7} M and one, respectively. On the other hand, the interaction of EF-G with GTP was rather weak, and the Kd value for EF-G·GTP determined by the equilibrium dialysis method was 1.2×10^{-5} M. The ternary EF-G·GDP·ribosome complex was much more stable than the same complex involving *E. coli* EF-G, and could be isolated on the nitrocellulose membrane filter in the absence of fusidic acid. *T. thermophilus* EF-G catalyzes, as does *E. coli* EF-G, the uncoupled GTPase reaction in the presence of either *T. thermophilus* or *E. coli* ribosomes. However, the specific activity of the most purified *T. thermophilus* EF-G used in this assay was about one-third to one-fourth of the specific activity of *E. coli* EF-G. Since the rate-limiting step in the uncoupled GTPase reaction is the dissociation of the EF-G·GDP·ribosome complex, stabilization of the complex would reduce the rate of the reaction. These properties of *T. thermophilus* EF-G are more similar to those of the eukaryotic EF-2 from pig liver (Mizumoto *et al.*, 1974) than to those of *E. coli* EF-G (N. Arai *et al.*, 1977a).

SULFHYDRYL GROUPS

In the reactions catalyzed by EF-Tu and EF-G from *E. coli*, it was proposed that the conformational transitions in EF-Tu and EF-G induced by the ligand change from GDP to GTP, or *vice versa*, play the central role (Kaziro, 1973, 1977; Yokosawa *et al.*, 1973). The local conformational change around the sulfhydryl group of the proteins induced by the change of nucleotide ligands was demonstrated by means of hydrophobic (Crane and Miller, 1974; N. Arai *et al.*, 1975; K. Arai *et al.*, 1975; N. Arai *et al.*, 1977a) and spin label

(K. Arai *et al.*, 1974b; K. Arai *et al.*, 1976; N. Arai *et al.*, 1976) probes, and also by tritium (Prinz and Miller, 1973) and deuterium (Ohta *et al.*, 1977) exchange experiments. Therefore, the involvement of SH groups in *T. thermophilus* EF-Tu, EF-Ts, and EF-G activities has been investigated.

Lack of Involvement of Sulfhydryl Groups in EF-Tu and EF-Ts Activities

In *E. coli* EF-Tu, two reactive sulfhydryl groups are present, of which one (SH_1) is essential for the binding of guanine nucleotides while the other (SH_2) is required for interaction with aminoacyl-tRNA and ribosomes (Miller *et al.*, 1971; K. Arai *et al.*, 1974c). On the other hand, *E. coli* EF-Ts contains two sulfhydryl groups, one reactive and the other non-reactive, and the reactive sulfhydryl is required for interaction with EF-Tu (K. Arai *et al.*, 1974c).

In contrast to EF-Tu and EF-Ts from *E. coli,* the factors from *T. thermophilus* are not inactivated by treatment with N-ethylmaleimide, even after 20 min at 60°C. The amount of sulfhydryl and disulfide groups in EF-Tu and EF-Ts from *T. thermophilus* was determined by titration with *p*-chloromercuri-[^{14}C]benzoate (Table 2). Approximately 1 mol of sulfhydryl group was titrated per mol of EF-Tu after treatment of EF-Tu·GDP with sodium sulfite in 8M guanidine hydrochloride for 60 min at 37°C, whereas no sulfhydryl groups could be titrated in the samples incubated without sodium sulfite. The reduced EF-Tu retained full activity and displayed a heat stability similar to that of nonreduced EF-Tu. These results indicate that *T. thermophilus* EF-Tu contains one disulfide bond, but no sulfhydryl groups (Table 2). In EF-Ts, no sulfhydryl groups could be titrated under native conditions. However, 0.8 and 1.4 mol of sulfhydryl group was titrated per mol of protein in the presence of protein denaturants, before and after treatment with sodium sulfite, respectively.

There are two possible interpretations for the difference in involvement of the sulfhydryl group for the activity between *T. thermophilus* and *E. coli* factors. First, the structure of the active center of these proteins may be different in *T. thermophilus* and *E. coli*. The second alternative is that, in *E. coli* factors, the sulfhydryl group(s) may not function directly at the catalytic center, but are only required to maintain the conformation of the active sites. Although, in general, enzymes with the same catalytic function from mesophilic and thermophilic sources are believed to share similar structures of the active site, it was recently reported that glyceraldehyde-3-phosphate dehydrogenase from *Thermus aquaticus* lacks one sulfhydryl group situated near

TABLE 2
Titration of Sulfhydryl Groups in EF-Tu, EF-Ts, and EF-G from T. thermophilus

	Conditions		
Factors	*Native*	*8M guanidine-HCl*	*8M guanidine-HCl and Na_2SO_3*
		mol/mol of protein	
EF-Tu·GDP	<0.1	<0.1	1.1
EF-Ts	<0.1	0.8	1.4
EF-G	0.8	1.0	1.85

the active site of its mesophilic counterpart (Hocking and Harris, 1976).

Involvement of Sulfhydryl Group in EF-G Activities

T. thermophilus EF-G is also completely resistant to N-ethylmaleimide at 30°C. At 60°C, however, both the uncoupled GTPase activity and the ability to form a ternary EF-G·GDP·ribosome complex are inactivated proportionally to the extent of modification of the sulfhydryl group. Under native conditions, 0.8 mol of sulfhydryl group was titrated with *p*-chloromercuri[^{14}C]benzoate after incubation for 30 min at 0°C without affecting the GDP binding activity of EF-G. These results indicate that *T. thermophilus* EF-G possesses a sulfhydryl group which is essential for interaction with ribosomes, but not for binding of guanine nucleotides as in the case of *E. coli* EF-G (N. Arai *et al.*, 1977b). It must be noted that the reaction of sulfhydryl reagents with the sulfhydryl group of *T. thermophilus* EF-G occurs much more slowly than the reaction with *E. coli* EF-G.

STABILITY AND THERMORESISTANCE OF *T. THERMOPHILUS* FACTORS

As expected, the factors from *T. thermophilus* are extremely stable against heat as compared to the corresponding factors from *E. coli*. The temperature (Tm) at which 50% of

TABLE 3
Amino Acid Compositions of EF-Tu·GDP, EF-Ts, and EF-G. The Amino Acid Compositions were determined as described previously (K. Arai, et al., *1976b) and the values are expressed as mol %.*

	EF-Tu·GDP		*EF-Ts*		*EF-G*	
	T. thermophilus	E. coli[c]	T. thermophilus	E. coli	T. thermophilus	E. coli
Lys	4.41	6.20	7.19	8.82	5.72	5.62
His	2.90	2.65	1.64	1.39	1.68	1.96
Arg	6.15	5.89	6.27	2.76	5.97	4.55
Trp[a]	0.39	0.41	0.45	0	0.79	0.86
Asp	8.59	8.99	8.96	9.99	7.54	10.36
Thr	7.54	7.92	2.29	4.53	5.95	5.50
Ser	1.81	2.85	1.80	3.82	2.85	5.25
Glu	12.57	12.45	17.30	15.51	13.96	12.95
Pro	5.67	5.10	3.51	1.95	4.31	4.50
Gly	9.90	9.97	7.67	9.06	8.49	8.40
Ala	6.90	6.85	12.03	13.56	7.72	9.61
Half-Cys[b]	0.23	0.92	0.82	0.87	0.38	0.79
Val	12.20	8.29	7.77	9.52	8.86	8.08
Met	2.77	1.75	2.51	1.51	2.91	2.69
Ile	5.56	7.02	6.02	6.57	8.55	6.79
Leu	6.75	7.24	9.06	5.88	7.91	6.45
Tyr	2.69	2.12	2.04	0.83	3.57	2.38
Phe	2.97	3.38	2.67	3.43	3.00	3.31

$H\phi_{ave}$ [d]	108.2	107.1	105.5	96.5	116.0	104.5
Hydrogen bonds former [e]	34.00	34.16	29.64	27.19	33.68	34.47
Non-helix former [f]	27.34	27.00	18.70	25.31	26.59	26.41

[a] *Determined fluorometrically (Pajot, 1976).*

[b] *Determined as cysteic acid after performic acid oxidation.*

[c] *Amino acid compositions of the factors from* E. coli *were taken from K. Arai* et al. *(1973) and Kaziro* et al. *(1972).*

[d] *Average hydrophobicity defined by Bigelow (1967).*

[e] *0.45 x (Glu + Asp) + Arg + 1/2Cys + His + Met + Ser + Thr + Trp + Tyr (Barnes and Stellwagen, 1973).*

[f] *Ser + Thr + Val + Ile + Cys.*

the activity is inactivated after heating for 5 min is as follows: EF-Ts (88°C), EF-Tu·GDP (78°C), free Ef-Tu (75°C), and EF-G (85°C). These values are to be compared to those of *E. coli* factors: EF-Ts (65°C), EF-Tu·GDP (50°C), free EF-Tu (33°C), and EF-G (59°C). The most remarkable difference in Tm (42°C) was observed in the case of free EF-Tu.

It was observed that *T. thermophilus* EF-Tu·GDP was far more stable than *E. coli* EF-Tu·GDP against various protein denaturants. *T. thermophilus* EF-Tu·GDP exhibited full activity in 5.5 M urea, 1.5 M guanidine-HCl, and 4 M LiCl, whereas *E. coli* EF-Tu·GDP started to lose its activity in 2 M urea, 0.5 M guanidine-HCl, and 1.5 M LiCl. *T. thermophilus* EF-Tu·GDP, which had been treated with 8 M guanidine-HCl, exhibited complete restoration of activity when diluted sufficiently to decrease the concentration of the denaturant. The ability of *T. thermophilus* EF-G to interact with GDP was fully preserved in 4 M urea, but inactivated by about 50% in 7 M urea, 2 M guanidine-HCl, or 0.03% Na-DodSO_4.

AMINO ACID COMPOSITIONS

Amino acid compositions of EF-Tu·GDP, EF-Ts, EF-G from *T. thermophilus* and *E. coli* are shown in Table 3. As can be seen from the table, amino acid compositions of EF-Tu and EF-G from both microorganisms are largely similar, except for the contents of half-cystine. On the other hand, considerable differences are found between the amino acid compositions of EF-Ts from *T. thermophilus* and *E. coli*. The former contains tryptophan which is absent in the latter. The content of non-helix formers are significantly different. Average hydrophobicity, $H\phi_{ave}$, is estimated for each of the factors from thermophilic and mesophilic sources based on the data on hydrophobicity of amino acid side chains according to Bigelow (1967). $H\phi_{ave}$ for *T. thermophilus* EF-Tu is similar to that of *E. coli*, whereas $H\phi_{ave}$ for *T. thermophilus* EF-Ts and EF-G are larger than those of *E. coli* EF-Ts and EF-G. The latter results suggest the presence of larger intramolecular hydrophobic interactions in *T. thermophilus* EF-Ts and EF-G which may play an important role in its conformational stability as described below.

In spite of the functional interchangeability, no immunological cross reaction could be detected between the factors from *T. thermophilus* and the antibodies to the corresponding *E. coli* factors in Ouchterlony double diffusion tests.

ORDERED STRUCTURES

As in the case of *E. coli* EF-Tu·GDP (K. Arai *et al.*, 1976b; Nakamura *et al.*, 1977), *T. thermophilus* EF-Tu·GDP is hydrolyzed with trypsin through two distinct stages. In the first stage, *T. thermophilus* EF-Tu·GDP is transformed into a fragment with a molecular weight of 40,000 (Fragment A) within 1 min at 0°C. The ability of EF-Tu to interact with aminoacyl-tRNA or ribosomes is destroyed at this stage. In the second stage of digestion, Fragment A is further cleaved into two subfragments (Fragments B and C) with molecular weights of 23,000 and 12,000. Fragments B and C exist in a complexed form as judged by disc polyacrylamide gel electrophoresis under native conditions, and the ability to interact with GDP or EF-Ts is fully retained. The molecular weights of individual subfragments and the kinetics of their appearance are essentially analogous to what was observed with *E. coli* EF-Tu·GDP. Furthermore, the recent report by Gast *et al.* (1976) indicated that EF-Tu from *B. stearothermophilus* is also degraded by trypsin in a similar manner as *E. coli* EF-Tu·GDP. Presumably, the EF-Tu molecules from mesophilic, moderately thermophilic, and extremely thermophilic bacteria may share a common a tertiary structure with comparable sites susceptible for trypsin digestion.

In this respect, it is noteworthy that CD spectra of EF-Tu·GDP from *T. thermophilus* and *E. coli* are remarkably similar. The contents of α-helix, β-structure, and unordered form are estimated as 24, 10, and 66%, and as 25, 12, and 64%, respectively, for *E. coli* and *T. thermophilus* EF-Tu·GDP. However, the conversion of Fragment A into Fragments B and C proceeded more slowly in the case of EF-Tu from *T. thermophilus* and *B. stearothermophilus* as compared to *E. coli* EF-Tu. This may be related to a low degree of motility of the protein from the extreme thermophile, as observed by hydrogen-deuterium exchange (Ohta *et al.*, 1977).

Skar *et al.* (1975) have shown that *E. coli* EF-G is split by trypsin into two fragments of molecular weights of 71,000 and 3,000 without loss of activity. The former fragment was then cleaved into two fragments of molecular weights of 47,000 and 29,000, resulting in decreased activity. In contrast, *T. thermophilus* EF-G is found to be strongly resistant to trypsin. The activity of *T. thermophilus* EF-G to form a ternary EF-G·GDP·ribosome complex is hardly affected by exposure to trypsin for 1 hr at 30°C. Analysis by Na-DodSO$_4$ gel electrophoresis has revealed that *T. thermophilus* EF-G, originally with a molecular weight of 78,000, remains almost undigested for the initial 10 min of incubation, but is thereafter gradually transformed into a polypeptide of 75,000 molecular weight. The remarkable stability against tryptic digestion and the extremely low reactivity towards sulfhydryl reagents of *T. thermophilus* EF-G may be due to the rigid structure of the protein. Measurements of CD spectra indicated that *T. thermophilus* EF-G contains more ordered structure than *E. coli* EF-G. The contents of α-helix, β-structure, and unordered form of *T. thermophilus* EF-G was 43, 35, and 22%, respectively. The corresponding values of *E. coli* EF-G were 37, 26, and 36%, respectively.

From the studies on the primary and ordered structures of the elongation factors from *T. thermophilus* described in this paper, it is still difficult to speculate on the molecular basis for thermophilicity. In the case of EF-G, for example, it is likely that the increased intramolecular hydrophobic interactions, and the greater extent of secondary structure contribute to the conformational stability of the protein. However, similar studies with EF-Tu have revealed no significant differences in these values between *E. coli* and *T. thermophilus*. Therefore, it appears to be rather difficult to assume, at present, any general rule to explain the structural basis for thermophilicity. Since the elongation factors are available from various sources in large amounts and in highly purified states, and have relatively simple structures with highly elaborated functions, they appear to be suitable for studies on the structural basis for thermophilicity.

SUMMARY

The polypeptide chain elongation factors EF-Tu, EF-TU·EF-Ts, and EF-G have been purified from *Thermus thermophilus* HB8, and their properties are compared with those of *E. coli* factors. The factors from both organisms possess similar molecular and catalytic properties and function essentially in an analogous manner in promoting polypeptide chain elongation. However, the following differences in their properties have been observed:

1. *T. thermophilus* EF-Ts appears to exist in a dimeric form, $(EF\text{-}Ts)_2$, and the EF-Tu·EF-Ts complex also exists as a dimer, presumably in the form of EF-Tu·$(EF\text{-}Ts)_2$·EF-Tu.

2. Unlike the *E. coli* EF-Tu·EF-Ts complex, *T. thermophilus* EF-Tu·EF-Ts does not dissociate in the presence of GDP but forms a ternary complex containing 2 mol each of EF-Ts, EF-Tu, and GDP. The binding of EF-Ts to EF-Tu·GDP seems to lower the affinity of EF-Tu toward GDP, thus facilitating the nucleotide exchange reaction.

3. As expected, *T. thermophilus* factors are extremely

stable against heat, acid, alkali, and other protein denaturants. Even free EF-Tu which was extremely labile in *E. coli* was fairly stable in *T. thermophilus*.

4. *T. thermophilus* EF-Tu did not contain any free sulfhydryl groups but did contain one disulfide bond. The sole sulfhydryl present in EF-Ts could not be titrated in the absence of a denaturant. These observations are in contrast to *E. coli* EF-Tu and EF-Ts, whose reactive sulfhydryls appear to play an essential role in their catalytic function.

5. EF-G from *T. thermophilus* forms a stable binary complex with GDP which could be isolated on nitrocellulose membranes. It also forms a stable ternary complex with ribosomes and GDP in the absence of fusidic acid. These properties are more similar to EF-2, a eukaryotic counterpart of EF-G, rather than to *E. coli* EF-G.

ACKNOWLEDGEMENTS

We would like to thank Dr. Tairo Oshima, the Mitsubishi-Kasei Institute of Life Sciences, Tokyo, for kindly supplying *T. thermophilus* cells and for valuable advice.

REFERENCES

Arai, K., Kawakita, M. and Kaziro, Y. (1972). J. Biol. Chem. *247*, 7029.

Arai, K., Kawakita, M., Kaziro, Y., Kondo, T. and Ui, N. (1973). J. Biochem. *73*, 1095.

Arai, K., Kawakita, M. and Kaziro, Y. (1974a). J. Biochem. *76*, 293.

Arai, K., Kawakita, M., Maeda, T., Ohnishi, S. and Kaziro, Y. (1974b). J. Biol. Chem. *249*, 3311.

Arai, K., Kawakita, M., Nakamura, S., Ishikawa, I. and Kaziro, Y. (1974c). J. Biochem. *76*, 523.

Arai, K., Arai, T., Kawakita, M. and Kaziro, Y. (1975). J. Biochem. *77*, 1095.

Arai, K., Maeda, T., Kawakita, M., Ohnishi, S. and Kaziro, Y. (1976a). J. Biochem. *80*, 1057.

Arai, K., Nakamura, S., Arai, T., Kawakita, M. and Kaziro, Y. (1976b). J. Biochem. *79*, 69.

Arai, N., Arai, K. and Kaziro, Y. (1975). J. Biochem. *78*, 243.

Arai, N., Arai, K., Maeda, T., Ohnishi, S. and Kaziro. Y. (1976). J. Biochem. *80*, 1057.

Arai, N., Arai, K. and Kaziro, Y. (1977a). J. Biochem. *82*, 687.

Arai, N., Arai, K., Nakamura, S. and Kaziro, Y. (1977b). J. Biochem. *82*, 695.

Baca, O. G., Rohrbach, M. S. and Bodley, J. W. (1976). Biochemistry *15*, 4570.

Barnes, L. D. and Stellwagen, E. (1973). Biochemistry *12*, 1559.

Beaud, G. and Lengyel, P. (1971). Biochemistry *10*, 4899.

Bigelow, C. C. (1967). J. Theor. Biol. *16*, 187.

Blumenthal, T., Douglass, J. and Smith, D. (1977). Proc. Natl. Acad. Sci. USA *74*, 3264.

Brown, S. and Blumenthal, T. (1976). J. Biol. Chem. *251*, 2749.

Crane, L. J. and Miller, D. L. (1974). Biochemistry *13*, 933.

Gast, W. H., Leberman, R., Schulz, G. E. and Wittinghofer, A. (1976). J. Mol. Biol. *106*, 943.

Haselkorn, R. and Rothman-Denes, L. B. (1973). Ann. Rev. Biochem. *41*, 397.

Hocking, J. D. and Harris, J. I. (1976). In *Enzymes and Proteins from Thermophilic Microorganisms* (H. Zuber, ed.) p. 121, Birkhäuser-Verlag, Basel.

Kaziro, Y., Inoue-Yokosawa, N. and Kawakita, M. (1972). J. Biochem. (Tokyo) *72*, 853.

Kaziro, Y. (1973). In *Organization of Energy Transducing Membranes* (M. Nakao and L. Packer, eds.), p. 187, University of Tokyo Press, Tokyo.

Kaziro, Y. (1977). In *Reflections on Biochemistry* (A. Kornberg, B. L. Horecker, L. Cornudella and J. Oro, eds.), p. 85, Pergamon Press, Oxford, New York, Toronto, Sydney.

Lengyel, P. (1974). In *Ribosomes* (M. Nomura, A. Tissieres and P. Lengyel, eds.), p. 13, Cold Spring Harbor Laboratory, Cold Spring Harbor, N. Y.

Lucas-Lenard, J. and Lipmann, F. (1971). Ann. Rev. Biochem. *40*, 409.

Miller, D. L., Hachmann, J. and Weissbach, H. (1971). Arch. Biochem. Biophys. *144*, 115.

Miller, D. L. and Weissbach, H. (1969). Arch. Biochem. Biophys. *132*, 146.

Miller, D. L. and Weissbach, H. (1970). Biochem. Biophys. Res. Comm. *38*, 1017.

Mizumoto, K., Iwasaki, K. and Kaziro, Y. (1974). J. Biochem. (Tokyo) *76*, 1269.

Nakamura, S., Arai, K., Takahashi, K. and Kaziro, Y. (1977). Biochem. Biophys. Res. Comm. *77*, 1418.

Ohta, S., Nakanishi, M., Tsuboi, M., Arai, K. and Kaziro. Y. (1977). Eur. J. Biochem. *78*, 599.

Oshima, T. and Imahori, K. (1971). J. Gen. Appl. Microbiol. *17*, 513.

Oshima, T. and Imahori, K. (1974). Int. J. Syst. Bacteriol. *24*, 102.

Pajot, P. (1976). Eur. J. Biochem. *63*, 263.

Prinz, M. P. and Miller, D. L. (1973). Biochem. Biophys. Res. Comm. *53*, 149.

Rohrbach, M. S. and Bodley, J. W. (1976). Biochemistry *15*, 4565.

Skar, D. D., Rohrbach, M. S. and Bodley, J. W. (1975). Biochemistry *14*, 3922.

Weissbach, H., Miller, D. L. and Hachmann, J. (1970). Arch. Biochem. Biophys. *137*, 262.

Wilson, G. E., Jr., and Cohn, M. (1977). J. Biol. Chem. *252*, 2004.

Wittinghofer, A. and Leberman, R. (1976). Eur. J. Biochem. *62*, 373.

Wolf, J., Chinali, G. and Parmeggiani, A. (1974). Proc. Natl. Acad. Sci. USA *71*, 4910.

Yokosawa, H., Inoue-Yokosawa, N., Arai, K., Kawakita, M. and Kaziro, Y. (1973). J. Biol. Chem. *248*, 375.

NOVEL POLYAMINES OF EXTREMELY THERMOPHILIC BACTERIA

Tairo Oshima

Mitsubishi-Kasei Institute of Life Sciences
Machida, Tokyo

IMPLICATION OF POLYAMINES IN PROTEIN SYNTHESIS

Polyamines and diamines have become the subject of intense research interest in life sciences in recent years (Bachrach, 1970, 1973; Cohen, 1971, 1972; Herbst and Bachrach, 1970; Russell, 1973a, 1973b; Smith, 1972; Tabor and Tabor, 1972). They are ubiquitously distributed in animal and plant tissues and microbial cells, and thought to play important roles in diverse cellular activities such as stabilization of nucleic acids, protein biosynthesis, DNA and RNA synthesis, cell growth, cell proliferation, some enzyme actions, cellular adaptation to environmental stresses, and the like. In most cases, however, the molecular mechanisms of polyamine action in cellular activities have not been clarified yet.

In the course of biochemical studies on protein biosynthesis catalyzed by a cell-free extract of an extremely thermophilic bacteria, *Thermus thermophilus*, it was found that only very low activity was observed when the reaction was carried out at temperatures higher than 50°C (Ohno-Iwashita *et al.*, 1975; 1976a, 1976b), whereas the thermophile is not capable of growing at temperatures lower than 50°C (Oshima and Imahori, 1974). The components involved in protein synthesis such as soluble enzymes, factors, tRNA and ribosomes were stable to heat and the low incorporation observed at higher temperature was not due to irreversible denaturation of some components in the reaction mixture.

It was found that there were two ways to restore protein synthesis at higher temperatures: (i) preincubation of

ISBN 0-12-268250-5

a mixture containing the messenger, aminoacyl-tRNA and ribosomes at lower temperature such as 10°C prior to the reaction at higher temperature, or (ii) the addition of spermine to the reaction mixture. Polyamines such as spermine may be an essential cofactor *in vivo* protein synthesis at the physiological temperature of the bacterium, because the cold preincubation seems to be a condition far from natural environments of the bacterium. The microorganism is able to proliferate at higher temperatures without preincubation in the cold.

More detailed works revealed that polyamines affected on two stages in the Poly(U) directed poly(Phe) synthesis by the cell-free extract. The presence of polyamine was essential for the formation of an active ternary complex between 30S ribosomal subunit, the messenger and aminoacyl-tRNA at high temperature. The polyamine requirement for the reaction was absolute and specific, and spermine was the most active among the polyamines and diamines tested. Diamines such as putrescine showed inhibition for the activation by the addition of spermine. The addition of polyamines also enhanced the elongation stage. However polyamine was not essential and polyamine specificity was not observed at this stage. The effect of polyamine observed are one of rare cases in which the action mechanism of polyamine is able to be interpreted at the molecular level.

The observations described above strongly suggest implications of polyamines in thermal stability and regulation of *in vivo* protein synthesis. To identify polyamines involved *in vivo* protein synthesis, extracts of *T. thermophilus* were analyzed.

Except for diamines such as putrescine and cadaverine, naturally occurring polyamines are generally limited to spermine and spermidine. However, it was found that *T. thermophilus* cells contain at least four novel polyamines, whose presence in nature has not been known or has been very rare. They are also found in other extreme thermophiles belonging to the genus Thermus, and may play essential roles in life at high temperatures.

POLYAMINE ANALYSIS

A simple assay method for determination of polyamine content in crude extracts was devised. A column (0.8 cm diameter x 7 cm long) of cation exchange resin (the same resin used in a commercial amino acid analyzer) was connected to an automatic amino acid analyzer and eluted with the buffer described below at a flow rate of 0.5 ml/min at 60°C. The buffer was prepared by dissolving 38.1 g $CH_3COONa \cdot 3H_2O$, 7.2 g acetic acid, 146.1 g sodium chloride, 50 ml n-propanol, 0.4 g Brij 35, 0.18 g 1,2-cyclohexane-diamine-N,N, N',N'-tetraacetic acid and 1.0 ml 1.0 N NaOH solution in a final volume of 1000 ml.

The cell suspension or extract was added to an equal volume of 10% trichloroacetic acid and the precipitate formed was removed by centrifugation. An aliquot of the supernatant was diluted with H_2O to a polyamine concentration of 10-200 nmoles per ml. The diluted sample (1 ml) was applied directly to the column without removal of trichloroacetic acid. The column was repeatedly used without buffer change or NaOH wash. The samples were applied every two hours and the column was continuously eluted with the buffer.

A mixture of standard polyamines, namely putrescine, spermidine and spermine, could be analyzed within 80 min. The method can be used for determining the polyamine compositions of other microorganisms, animal tissues, and urine.

ISOLATION OF POLYAMINES

Fig. 1 shows the typical polyamine composition for *T. thermophilus* grown in a polypeptone-yeast extract medium at 75°C. The cells synthesize two tetraamines, a and b, as major polyamine components, and three triamines, c, d and e, as minor components along with trace amounts of diamines and other polyamines, f, g, h and i.

To determine chemical structures, polyamines a-e were isolated and purified as follows: Polyamines extracted in 5% trichloroacetic acid were applied on a Dowex 50 x 2-H^+ column (1.7 x 24 cm). The column was eluted with a linear gradient of HCl from 0 to 6 N. Fractions containing triamines and tetraamines were pooled separately. Triamines c, d and e were isolated by using CK-10S ion exchange resin column (0.8 x 40 cm) chromatography eluted with 0.4 M sodium acetate buffer containing 2.2 N NaCl, pH 4.9 (final concentration of sodium was 2.48 M). Tetraamines a and b were isolated in a similar manner using 0.12 M potassium citrate -HCl buffer, pH 5.64, containing 2 M KCl (final concentration of potassium was 2.12 M). If necessary, the CK-10S column chromatography was repeated until the preparation was free from contamination of other polyamines.

The final preparations were dissolved in dilute HCl solution after salts were removed, and then excess HCl was removed by evaporation. The residues were dissolved in the minutest quantity of water and crystals of the hydrochloride salt of the purified polyamine were obtained by adding an

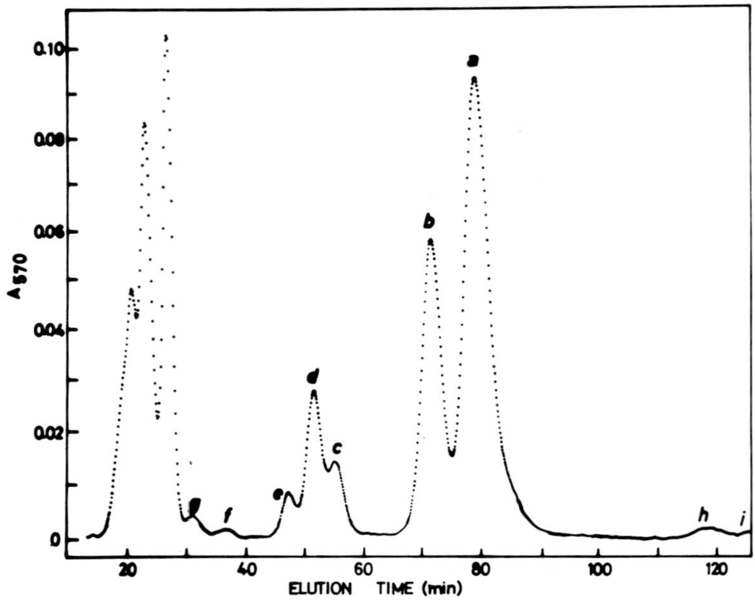

FIGURE 1. Polyamine composition of T. thermophilus. *Cells grown at 75°C and harvested in the middle log phase were homogenized in 5% trichloroacetic acid and an aliquot of the supernatant was subjected to analytical column chromatography.*

excess amount of methanol:ethanol (1:1) mixture to the solution. The chemical structures of isolated polyamines thus obtained were investigated by recording their IR, proton-NMR, ^{13}C-NMR and mass spectra.

THERMINE AND THERMOSPERMINE

Polyamine b was determined to be 1,11-diamino-4,8-diazaundecane, $NH_2(CH_2)_3NH(CH_2)_3NH(CH_2)_3NH_2$, a new polyamine as reported in a previous paper (Oshima, 1975). Thermine was proposed for the trivial name of the new polyamine. The chemical structure of thermine was confirmed by comparison of IR, proton NMR and mass spectra with those of an authentic preparation.

Polyamine a had been thought to be spermine (Oshima, 1975) because retention times of the compound on cation exchange resin column chromatography with several different buffer systems were identical with those of authentic spermine, and Rf values on thin layer chromatography were also the same as those for authentic spermine. Recently, however, it was found that the IR spectrum of the isolated polyamine differs from that of spermine.

Proton-NMR (60 M Hz) spectrum of polyamine a was similar to that of spermine, suggesting that carbon atoms of polyamine a consist of two propyl groups and one butyl group. ^{13}C-NMR studies revealed that unlike spermine, the arrangement of carbon atoms in the polyamine a molecule is asymmetric. These data led the author to conclude that polyamine a is also a new polyamine, 1,12-diamino-4,8-diazadodecane, $NH_2(CH_2)_3NH(CH_2)_3NH(CH_3)_4NH_2$, which is an isomer of spermine.

This conclusion was confirmed by recording the mass spectrum of polyamine a as shown in Fig. 2. A strong peak at mass number 149 was seen only for polyamine a, but not for spermine. This fragment can be produced by removing $NH_2CH_2CH_2$- from the aminobutyl group in one end of the molecule. It would be difficult to obtain this fragment from spermine, in which the butyl group is located in the center of the molecule.

The author would like to propose to give this polyamine the trivial name of thermospermine. If spermine is present

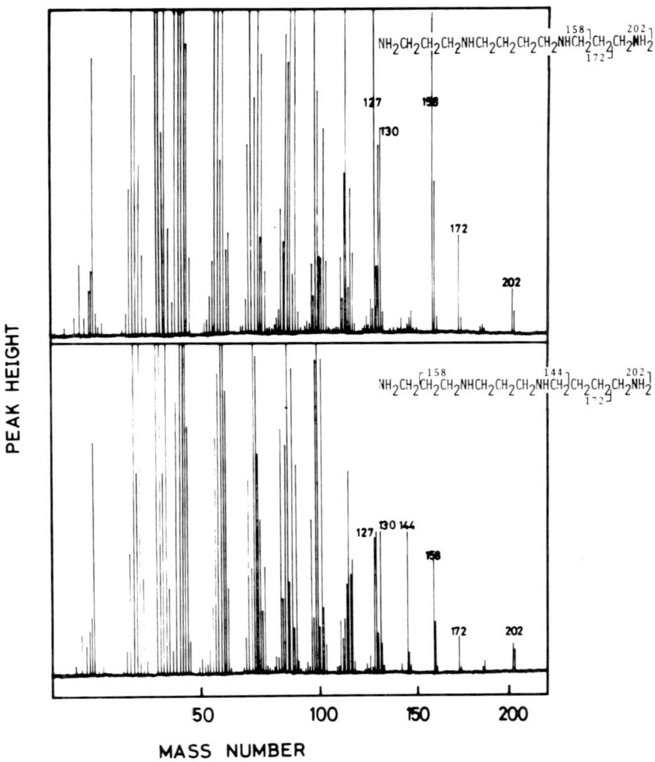

FIGURE 2. Comparison of the mass spectrum of thermospermine (lower) with that of spermine (upper).

in *T. thermophilus*, the polyamine may contaminate the isolated thermospermine preparation, since they have an identical retention time on any kind of chromatography so far investigated. However ^{13}C-NMR spectra of the purified preparation did not show any spike which is characteristic of the ^{13}C-NMR spectrum of spermine. This finding suggests that no or only a small amount of spermine is present in the thermophile.

TRIAMINES

Polyamine d was confirmed to be spermidine. Polyamine e was also confirmed to be 1,7-diamino-4-azaheptane, $NH_2(CH_2)_3NH(CH_2)_3NH_2$ (caldine, DeRosa *et al.*, 1976) by comparison with the synthetic preparation (Oshima *et al.*, manuscript in preparation).

Polyamine c was confirmed to be 1,9-diamino-5-azanonane, $NH_2(CH_2)_4NH(CH_2)_4NH_2$, by measuring its IR, mass, proton- and ^{13}C-NMR spectra. Details will be published elsewhere. The polyamine, *sym*-homospermidine, was reported to be present in sandal leaves (Kuttan *et al.*, 1971), but has not been known to be present in microorganisms, animals and other plants.

Table 1 summarizes the chemical names of the five polyamines, a-e (in Fig. 1), found in *T. thermophilus*.

ROLES OF NOVEL POLYAMINES

In vitro poly(Phe) synthesis catalyzed by an extract of *T. thermophilus* in the presence of poly(U) at 65°C was induced by the addition of spermine, thermospermine or thermine as shown in Fig. 3. It is noteworthy that spermine and thermospermine showed the highest activity although spermine seems to be absent in the thermophile. Phenylalanine incorporation in the presence of thermine was about two-thirds of that in the presence of thermospermine. *In vivo* protein synthesis may be regulated by the concentrations of thermospermine and thermine.

Among the triamines, spermidine and caldine may serve as precursors of the tetraamines, thermospermine and thermine, respectively. Polyamine c, *sym*-homospermidine, might not be a precursor for a tetraamine since the corresponding tetraamine was not detected in cells grown under normal conditions.

It was found that a thermophilic phage to *T. thermophilus*, YS40 (Sakaki and Oshima, 1975), contains *sym*-

TABLE 1
Polyamines found in Thermus thermophilus

Designation in Fig. 1	*Systematic name*	*Trivial name*	*Distribution in nature*[1]
a	1,12-Diamino-4,8-diazadodecane	Thermospermine	Not reported
b	1,11-Diamino-4,8-diazaundecane	Thermine	Not reported
c	1,9-Diamino-5-azanonane	*sym*-Homospermidine	Only in sandalwood[2]
d	1,8-Diamino-4-azaoctane	Spermidine	Widespread
e	1,7-Diamino-4-azaheptane	Caldine	Only in turnip yellow mosaic virus[3] and white shrimp[4]

[1] *Except in thermophiles.*

[2] *Kuttan* et al. *(1971).*

[3] *A later report (Beer and Kosuge, 1970) could not confirm the presence of this polyamine as reported in the earlier literature (Johnson and Markham, 1962).*

[4] *Stillway and Walle (1977).*

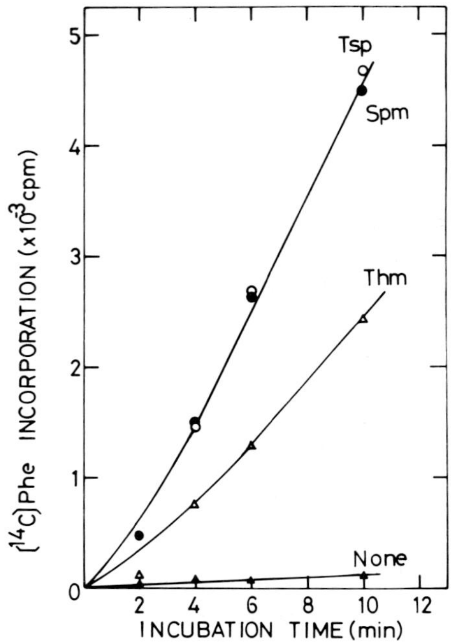

FIGURE 3. Rate of phenylalanine incorporation at 65°C in the absence and presence of polyamines. Polyamine concentrations were 2 mM and other experimental conditions were as described previously (Ohno-Iwashita et al., *1975). Thermospermine (o), spermine (●) and thermine (Δ).*

homospermidine. The polyamine composition of the phage consisted of thermospermine (or spermine) and *sym*-homospermidine as the major components along with smaller amounts of thermine and spermidine. It was also found that *sym*-homospermidine was rapidly synthesized after the thermophile cells had been infected by the phage. A more detailed mechanism for the role of *sym*-homospermidine in the replication and/or assembly of phage YS40 has not been clarified yet.

DISTRIBUTION OF NOVEL POLYAMINES

Polyamines were rich in young cells of *T. thermophilus*, and decreased in older cells. Especially thermospermine decreased rapidly when cells reached the stationary stage. Polyamine composition also depended on the growth temperature.

Triamines were rich and tetraamines were poor in cells grown at lower temperatures, such as 50°C.

Numerous efforts have been made to establish an analytical method for the determination of spermine and thermospermine in mixed solutions. Nevertheless, the author and his collaborators have not as yet succeeded in this attempt. Both polyamines always eluted at the same time on various analytical column chromatographies. Thus, the distribution of thermospermine has not been studied.

Thermine, caldine and *sym*-homospermidine were present in other *Thermus* species, such as *T. aquaticus* and *T. flavus*. No direct relationship was observed between sensitivity to phage YS40 and *sym*-homospermidine content of these *Thermus* species.

Moderate thermophiles, *Bacillus stearothermophilus* and *Bacillus acidocaldarius*, possessed spermine in addition to spermidine and putrescine as already reported in the literature (Stevens and Morrison, 1968), but no thermine was detected. It seems that the novel polyamines, thermine, caldine and *sym*-homospermidine (and probably thermospermine), are distributed only in extreme thermophiles in the bacterial world, suggesting their involvement in extreme thermophily.

REFERENCES

Bachrach, U. (1970). Ann. Rev. Microbiol. *24*, 109.

Bachrach, U. (1973). *Function of Naturally Occurring Polyamines*, Academic Press, New York.

Beer, S. V. and Kosuge, T. (1970). Virology *40*, 930.

Cohen, S. S. (1971). *Introduction to the Polyamines*, Prentice-Hall, New Jersey.

Cohen, S. S. (1972). Adv. Enzyme Regul. *10*, 207.

DeRosa, M., DeRosa, S., Gambacorta, A., Carteni-Farina, M. and Zappia, V. (1976). Biochem. Biophys. Res. Comm. *69*, 253.

Herbst, E. J. and Bachrach, U. (eds.) (1970). "Metabolism and Biological Functions of Polyamines," Ann. N.Y. Acad. Sci. *171*, 693.

Johnson, M. W. and Markham, R. (1962). Virology *17*, 276.

Kuttan, R., Radhakrishnan, A. N., Spande, T. and Witkop, B. (1971). Biochemistry *10*, 361.

Ohno-Iwashita, Y., Oshima, T. and Imahori, K. (1975). Arch. Biochem. Biophys. *171*, 490.

Ohno-Iwashita, Y., Oshima, T. and Imahori, K. (1976a). J. Biochem. *79*, 1245.

Ohno-Iwashita, Y., Oshima, T. and Imahori, K. (1976b). In *Enzymes and Proteins from Thermophilic Microorganisms*

(H. Zuber, ed.), p. 317, Birkhauser Verlag, Basel.

Oshima, T. (1975). Biochem. Biophys. Res. Comm. *63*, 1093.

Oshima, T. and Imahori, K. (1974). Intl. J. Syst. Bacteriol. *24*, 102.

Russell, D. H. (1973a). Life Sciences *13*, 1635.

Russell, D. H. (ed.) (1973b). *Polyamines in Normal and Neoplastic Growth*, Raven Press, New York.

Sakaki, Y. and Oshima, T. (1975). J. Virology *15*, 1449.

Smith, T. A. (1972). Endeavour *31*, 22.

Stevens, L. (1970). Biol. Rev. *45*, 1.

Stevens, L. and Morrison, M. R. (1968). Biochem. J. *108*, 633.

Stillway, L. W. and Walle, T. (1977). Biochem. Biophys. Res. Comm. *77*, 1103.

Tabor, H. and Tabor, C. W. (1972). Adv. Enzymol. *36*, 203.

PROTEINS

THERMOSTABILITY OF PROTEINS

Earle Stellwagen
Harvey Wilgus

Department of Biochemistry
University of Iowa
Iowa City, Iowa

During the last ten years a large number of proteins of diverse function have been purified and characterized from thermophilic microorganisms particularly *Bacillus stearothermophilus* which grows optimally at 60°C and *Thermus aquaticus* which grows optimally at about 70°C. The primary, secondary, tertiary and, where appropriate, quaternary structural parameters of thermophilic proteins are in general strikingly similar to mesophilic proteins having the same function (1). Indeed the variability of structural parameters between thermophilic and mesophilic proteins of the same function is often no greater than the variability between proteins of the same function obtained from various mesophilic organisms. In spite of their similar structural parameters, thermophilic proteins are invariably more stable to denaturing conditions than their mesophilic counterparts. This stability extends to the effects of all chemical denaturants at ambient temperatures as well as to an enhanced thermostability of 10-50°C. Examination of functional or structural parameters of thermophilic proteins with increasing temperature frequently indicates that the same native conformation is retained between ambient and the functionally optimal temperature followed by a cooperative irreversible inactivation at higher temperature.

Various investigators have proposed that the enhanced stability of thermophilic proteins results from the increased presence of amino acid residues capable of forming hydrophobic (2), hydrogen (3) or ionic bonds (4), producing conformations having larger and more densely packed protein interiors (5),

ISBN 0-12-268250-5

more or different secondary structure (6), more extensively interlaced protein surfaces (4) or more complementary intersubunit contacts (7).

Since some mesophilic proteins are stable up to 90°C (5) and since precise structures of many mesophilic proteins are available from crystallographic measurements, we initiated a comparative study of the thermostability of mesophilic proteins. We chose to detect thermal conformational changes using the $\Delta pH/\Delta T$ procedure described by Bull and Breese (8) for several reasons. Firstly, in contrast to optical measurements the $\Delta pH/\Delta T$ procedure is not perturbed by the turbidity which usually accompanies irreversible thermal inactivation. Secondly, no ligands such as substrates, cations or buffers which can complex with proteins and alter thermostability (1) need be present as in measurements of biological activity. And thirdly, both reversible and irreversible conformational changes can be detected in contrast to preincubation studies at elevated temperatures followed by cooling and measurement, a procedure that only measures irreversible changes. Comparative measurements (8) indicate that the $\Delta pH/\Delta T$ procedure usually detects a single thermal transition whose midpoint temperature corresponds to the functional and native conformational transition midpoint detected by other procedures using the relatively few proteins whose thermally inactivated conformations remain soluble.

The variation of pH with temperature together with the derivative $\Delta pH/\Delta T$ profile measured for a typical protein sample is shown in Fig. 1A. The regular $\Delta pH/\Delta T$ dependence on T both prior and subsequent to the discontinuity is considered to represent the effect of temperature on the pK values of ionizable protein groups in the native and thermally denatured protein conformations, respectively. The discontinuity bridging these two regions then represents the release of conformational perturbations which perturb the pK values of ionizable groups in the native but not thermally denatured protein. The temperature at which a maximum $\Delta pH/\Delta T$ value is observed in the thermal denaturation region is defined as a conformational melting temperature, denoted as T_M. Since irreversible aggregation is coincident with the thermal denaturation of nearly all the protein examined, the measured T_M values should not be considered to represent the temperature at which equal amounts of native and denatured protein is present. Rather the measured T_M should be considered as a scanning analog of kinetic measurements of thermal inactivation. Thus the measured T_M values should be meaningful in a relative but not an absolute sense.

The T_M values measured for 19 proteins purified from mesophilic organisms and subjected to successful high resolution crystallographic analysis are listed in Table 1 in order of increasing T_M. The $\Delta pH/\Delta T$ profiles of each of these proteins exhibited a single transition. The reported T_M is the highest value measured in the neutral pH range. Also listed in Table 1 are the logarithm of the molecular weight of the polypeptide chain(s) in each protein, the number of chains constituting the native protein, the percentage residues participating in α-helical and β-strand secondary structures, the average residue volume and its predicted melting temperature (5) and the hydrophobicity index (9). Table 1 also lists the T_M values and structural parameters for an additional 16 well characterized proteins whose structures have not been reported to high resolution. The dependence of the T_M values measured for the proteins listed in Table 1 on the logarithm of the molecular weight of their polypeptide chain is illustrated in Fig. 1B. Table 1 also lists the optimal temperature for catalysis, T_{opt}, reported for several enzymes purified from the extreme thermophiles *Thermus aquaticus* and Thermophilic bacterium PS 3 together with their structural parameters. These entries include all proteins from extreme thermophiles for which measurements of both subunit size and T_{opt} have been reported to date. The T_{opt} values and polypeptide chain sizes of these enzymes are also plotted in Fig. 1B.

Inspection of the values listed in Table 1 indicates that the measured T_M values exhibit little correlation with average residue volume hydrophobicity index, volume, or the amount of helical or β-strand structure having correlation coefficients of 0.01, -0.13, 0.11 and 0.27, respectively. Similar conclusions (10) were reached when comparing the same parameters of thermophilic proteins as a class with mesophilic proteins. As also shown in Table 1, increasing α-helical content at the expense of β-structure in response to changes in glutamate content (6) would not necessarily raise T_M values.

As shown in Fig. 1B an unexpected inverse trend appears between the T_M values of the proteins examined and the size of their constituent polypeptide chains. This trend which had a correlation coefficient of -0.67 appears to be independent of the presence of quaternary structure, strongly bound ligands such as metal cations or heme moieties, or intrachain disulfide bonds in the native protein. These considerations may in fact account for the variation from a more strict linear relationship of T_M on polypeptide chain size observed in Fig. 1B. Reported (11) measurements of the activation parameters per mole of polypeptide chain for several proteins indicate that $\Delta S^{\ddagger}$ increases with increasing polypeptide chain length more rapidly than $\Delta H^{\ddagger}$, an observation that would lower T_M as chain length increases. This observation may be understood in terms of the ratio of the accessible surface (A_s) of globular proteins to their total or packing volume (V_p) which decreases with molecular weight by the relationship (12), $A_s/V_p = (8.8 \pm 0.1)\ M^{-1/3}$. Since the hydrophobicity index appears to be independent of chain size and since the internal volume of the tertiary structure of globular structure increases with chain size, it follows that a larger fraction of internal volume of large tertiary structures must be contributed by polar residues. As more polar residues need be buried, more opportunity exists for failure to hydrogen bond all of the buried residues, a situation that would preferentially diminish $\Delta H^{\ddagger}$ for thermal inactivation.

Extrapolation of the trend illustrated in Fib. 1B for mesophilic proteins suggests that polypeptide chains having a molecular weight of about 2 x 10^5 in globular proteins would be thermolabile at 37°C. This observation provides another reason for the relative absence of such long polypeptide chains in addition to problems associated with the fidelity of transcription and translation processes involved in the biosynthesis of long chains.

Surprisingly, the inverse trend between thermostability measured as the functional temperature optimum and polypeptide chain length persists for the proteins isolated from the extreme thermophiles, as shown in Fig. 1B. While functional temperature optima are not the most reliable index of protein

Fig. 1. (A) The effect of temperature on the pH of a 2.0 ml solution of hens egg lysozyme, 2.0 mg/ml, in 0.2 M KCl adjusted to pH 6.8. The open circles indicate the corrected pH values measured at the indicated temperatures and the filled circles indicate the derivative $\Delta pH/\Delta T$ values for 1.0°C temperature intervals. The apparatus used was similar to that described by Bull and Breese (8) employing a Radiometer model 25 pH meter, a Radiometer GK 2401C combination electrode, a Yellow Springs Instrument Co. model 425C bridge and thermister probe, and a Houston Instruments Omniscribe dual pen recorder. The protein solution, pH electrode, thermister probe and a small magnetic stirring bar were inserted into a jacketted vessel surrounded by a grounded wire screen. The temperature of the protein solution was controlled by a circulating water bath programmed to increase by 0.83°C per minute. The observed pH values were corrected for temperature as described by Goodno et al. (13). (B) Dependence of measured T_M or T_{opt} values on the size of the largest constituent polypeptide chain in the native protein. The circles indicate mesophilic proteins and the triangles indicate thermophilic proteins. The number beside each value indicates the protein as encoded in Table 1.

thermostability as noted above, the T_M measured by the $\Delta pH/\Delta T$ procedure for one enzyme obtained from a thermophilic organism exhibited only a modest difference from its functional optimal temperature (6). In view of this correspondence and the prodigious task of purifying a number of proteins from extreme thermophiles for $\Delta pH/\Delta T$ in measurements, we believe that the reported T_{opt} represent adequate preliminary measure of the relative thermostability of these proteins. Since thermophilic and mesophilic proteins generally have very similar structures, it had been anticipated that the formation of a small number of additional noncovalent interactions in the thermophilic structure or optimization of existent interactions in the mesophilic structure to increase the $\Delta H^{\ddagger}$ of the thermophilic protein without increasing chain length, $\Delta S^{\ddagger}$, would obscure the inverse dependence noted with the mesophilic proteins. However, appearance of a parallel

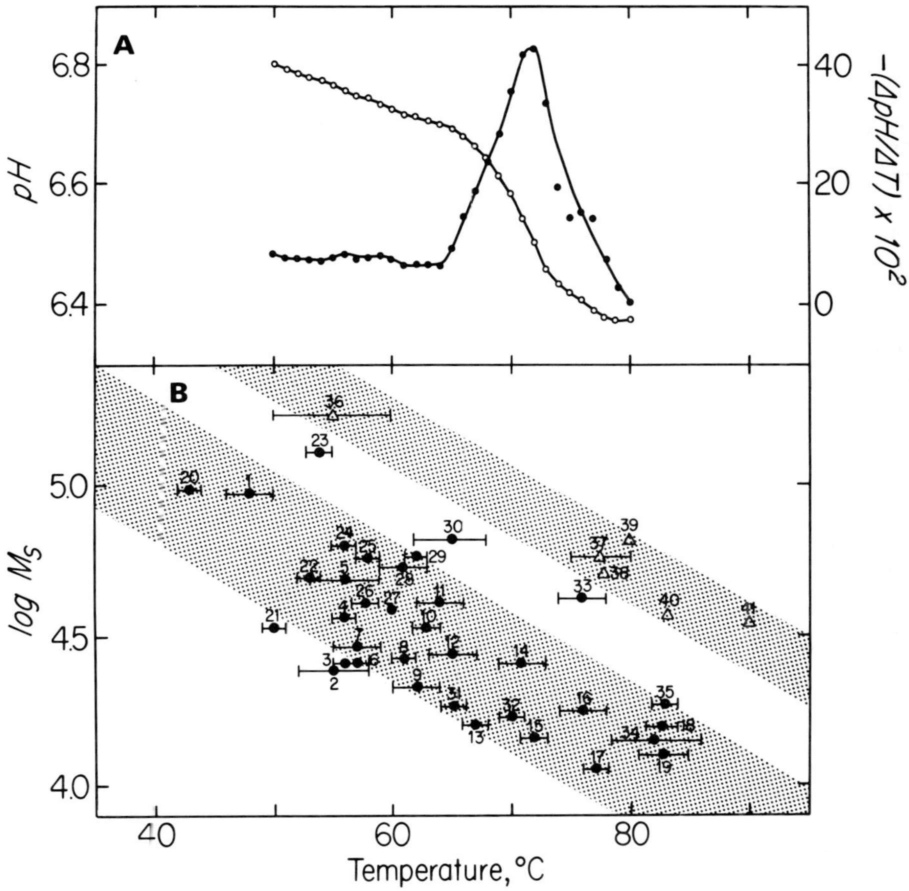

TABLE 1
Protein Parameters

Protein	*Source*	*Number*	T_M
Phosphorylase (14)	Rabbit Muscle	1	48±2
Trypsinogen	Bovine Pancreas	2	55±3
Chymotrypsinogen	Bovine Pancreas	3	56±1
Lactate Dehydrogenase	Rabbit Muscle	4	56±1
Phosphoglycerate Kinase	Yeast	5	56±3
Elastase	Porcine Pancreas	6	57±1
Carbonic Anhydrase	Bovine Red Blood Cell	7	57±2
Triosephosphate Isomerase	Rabbit Muscle	8	61±1
Adenylate Kinase	Rabbit Muscle	9	62±2
Carboxypeptidase	Bovine Pancreas	10	63±1
Alcohol Dehydrogenase	Horse Liver	11	64±2
Subtilisin	*B. subtilis*	12	65±2
Hemoglobin	Bovine	13	67±1
Conconavalin A	Jack Bean	14	71±2
Lysozyme	Hens Egg	15	72±1
Myoglobin	Sperm Whale	16	76±2
Trypsin Inhibitor	Soy Bean	17	77±1
Superoxide Dismutase	Bovine Red Blood Cell	18	83±1
Cytochrome *c*	Horse Heart	19	83±2
Myosin S-1 Fragment	Rabbit Muscle	20	43±1
Cytochrome c Peroxidase	Yeast	21	50±1
Fumarase	Porcine Heart	22	53±1
β-galatosidase	*E. coli*	23	54±1
Leucine Aminopeptidase	Porcine Kidney	24	56±1
Glutamate Dehydrogenase	Bovine Liver	25	58±1
Aldolase	Rabbit Muscle	26	58±1
Pepsinogen	Porcine	27	60±2
Catalase	Bovine Liver	28	61±2
Pyruvate Kinase	Rabbit Muscle	29	62±1
Serum Albumin	Bovine	30	65±3
Ferritin	Horse Spleen	31	65±1
Aspartate Transcarbamylase	*E. coli*	32	70±1
Ovalbumin	Hens Egg	33	76±2
Lactalbumin	Bovine Milk	34	82±3
β-Lactoglobin	Bovine Milk	35	83±1
RNA Polymerase (15)	*T. aquaticus*	36	55±5
ATPase (16)	*T. bacterium* PS 3	37	75

pH	Log M_S	n	V_R	$T_{M,P}$	Φ	%α	%β
8.06	4.97	4	88.9	85	886	38	25
5.80	4.39	1	79.7	55	770		
4.90	4.41	1	80.2	57	792	9	34
6.86	4.56	4	85.7	75	908	45	22
7.01	4.69	1	84.4	70	832	33	18
6.70	4.41	1	82.2	64	799	7	52
6.59	4.47	1	86.2	77	829	17	40
7.39	4.43	2	82.3	64	737	55	22
6.69	4.33	1	86.0	76	793	55	13
6.69	4.53	1	86.1	76	657	37	15
7.12	4.61	4	82.7	65	830	27	31
6.22	4.44	1	76.9	46	739	31	10
6.10	4.20	4	84.3	70	873	78	0
6.56	4.41	4	82.8	65	846	2	36
6.50	4.16	1	83.0	72	706	42	16
6.40	4.25	1	87.8	82	850	79	0
7.61	4.05	1	85.2	73	878	26	33
7.15	4.20	2	79.0	53	675	5	45
7.80	4.10	1	87.0	79	742	39	0
6.83	4.97	2	87.6	76	788		
6.49	4.53	1	87.1	79	831		
6.91	4.69	4	87.1	80	839		
7.48	5.13	4	86.1	76	826		
6.41	4.18	4	82.4	64	795		
6.59	4.76	6	85.2	73	835		
6.51	4.60	4	84.3	70	792		
6.40	4.59	1	82.1	63	924		
6.92	4.73	4	86.3	77	805		
6.97	4.76	4	84.5	71	793		
6.40	4.82	1	86.9	79	838		
5.69	4.26	24	87.4	80	858		
7.49	4.23	12	86.0	76	851		
6.47	4.64	1	86.1	72	866		
4.50	4:15	1	88.1	83	869		
6.10	4.27	2	88.2	83	939		
	5.23						
	4.75						

TABLE 1 (Continued)
Protein Parameters

Protein	Source	Number	T_M
Alkaline Phosphatase	*T. aquaticus*	38	78±2
DNA Polymerase (18)	*T. aquaticus*	39	80
Fumarase (19)	*T. aquaticus*	40	83
Enolase (3)	*T. aquaticus*	41	90

pH	Log M_S	n	V_R	$T_{M,P}$	Φ	%α	%β
	4.70						
	4.81						
	4.65						
	4.64						

The columns headed by the terms number, T_M, pH, Log M_S, n, V_R, $T_{M,P}$, Φ, %α and %β indicate the code number for the protein, the measured conformational melting temperature, the pH of the protein solution at that temperature, the $\log_{10}$ of the molecular weight of the largest polypeptide chain in the native protein, the number of polypeptide chains in the native protein, the average residue volume, the conformational melting temperature predicted from the average residue volume, the hydrophobic index, the percentage residues involved in helical structures, and the percentage residues involved in β-strands. Proteins 1, 5, 9, 23, 25, 26 were purchased from Boehringer; proteins 2, 4, 6-8, 13, 14, 24, 27-29 and 31 were purchased from Sigma and the remaining proteins were gifts. All protein samples used exhibited a single Coomassie blue staining band after polyacrylamide gel electrophoresis. The molecular weights, amino acid composition, and secondary structural content of the mesophilic proteins were obtained from The Enzymes, Atlas of Protein Sequence and Structure, *and references therein.*

inverse relationship for the thermophilic proteins suggests that the $\Delta H^{\ddagger}$ of each thermophilic protein is increased by a similar amount to increase thermostability. This would be most simply achieved by addition of a small number of new bonding interactions to the conformation of each mesophilic conformation by amino acid replacement without increasing chain length. Substitution (4) of thermophilic ferridoxin sequences in the mesophilic ferridoxin crystallographic structure and comparisons (7) of the crystallographic structures of mesophilic and thermophilic glyceraldehydephosphate dehydrogenases indicate the presence of small numbers of additional surface ionic interactions in the thermophilic structures. The results shown in Fig. 1B suggest this may be a common method used in nature to increase thermostability without reengineering protein structure. Presumably such additional interactions do not appear in mesophilic proteins since no thermal survival pressure exist for their selection.

SUMMARY

Midpoint temperatures for the thermal denaturation of 35 well characterized globular proteins obtained from mesophilic organisms measured in the neutral pH range by the $\Delta pH/\Delta T$ procedure were found to vary from 43° to 83°C. The midpoint temperatures of these proteins exhibited no correlation with their content of hydrophobic residues, average residue volume, or kind or amounts of secondary structure. However, the midpoint temperatures displayed an inverse dependence on the size of the polypeptide chains of the proteins examined indicating that activation entropy for thermal denaturation makes a major contribution to the thermal stability of native protein conformations. A parallel dependency between the optimal catalytic temperature and polypeptide chain size for enzymes purified from extreme thermophiles suggests that the enhanced thermostability of each of these enzymes is achieved by a small number of amino acid replacements which form supplemental noncovalent interaction without perturbing protein conformation.

REFERENCES

1. Singleton, R., Jr. and Amelunxen, R. E. (1973). Bacteriol. Revs. *37*, 320; Williams, R.A.D. (1975). Sci. Prog. Oxf. *62*, 373; Ljungdahl, L.G. and Sherod D. (1976). In *Extreme Environments* (M.R. Heinrich, ed.), p. 147, Academic Press, New York; Singleton, R., Jr. *ibid*, p. 189; *Enzymes and Proteins from Thermophilic Microorganisms* (H. Zuber, ed.), Birkhauser Verlag, Basel, 1976.
2. Ohta, Y., Ogura, Y. and Wada, A. (1966). J. Biol. Chem. *241*, 5919.
3. Barnes, L. D. and Stellwagen, E. (1973). Biochemistry *12*, 1559.
4. Perutz, M. F. and Raidt, H. (1975). Nature *255*, 256.
5. Bull, H. B. and Breese, K. (1973). Arch. Biochem. Biophys. *159*, 681.
6. Stellwagen, E. and Barnes, L. D. (1976). In *Enzymes and Proteins from Thermophilic Microorganisms* (H. Zuber, ed.), p. 223, Berkhauser Verlag, Basel.
7. Biesecker, G., Harris, J. I., Thierry, J. C., Walker, J.E. and Wonacott, A. J. (1977). Nature *266*, 328.
8. Bull, H. B. and Breese, K. (1973). Arch. Biochem. Biophys. *156*, 604.
9. Bull, H. B. and Breese, K. (1974). Arch. Biochem. Biophys. *161*, 665.
10. See R. Singleton, Jr. reference 1.
11. Joly, M. (1965). *A Physico-chemical Approach to the Denaturation of Proteins*, pp. 209-10, Academic Press New York.
12. Teller, D. C. (1976). Nature *260*, 729.
13. Goodno, C.C., Swenson, C. A. and Bull, H. B. (1975). Anal. Biochem. *67*, 220.
14. Goodno, C. C., Harris, T. A. and Swenson, C. A. (1976). Biochemistry *15*, 5157.
15. Fabray, M., Sumegi, J. and Venetianer, P. (1976). Biochim. Biophys. Acta *435*, 228.
16. Yoshida, M., Sone, N., Hirata, H. and Kagawa, Y. (1975). J. Biol. Chem. *250*, 7910.
17. Yeh, M.F. and Trela, J.M. (1976). J. Biol. Chem. *251*, 3134.
18. Chien, A., Edger, D. B., Trela, J.M. (1976). J. Bact. *127*, 1550.
19. Cook, W. R., and Ramaley, R. F. (1976). In *Enzymes and Proteins from Thermophilic Microorganisms* (H. Zuber, ed.), p. 207, Birkhauser Verlag, Basel.
20. This investigation was supported by Public Health Service grant GM-13215 from the Institute for General Medical Sciences.

EFFECT OF A SINGLE AMINO ACID SUBSTITUTION ON CONFORMATIONAL STABILITY OF THE TRYPTOPHAN SYNTHETASE α-SUBUNIT OF *ESCHERICHIA COLI*

Katsuhide Yutani
Kyoko Ogasahara
Minoru Suzuki
Yoshinobu Sugino

Institute for Protein Research
Faculty of Science
Osaka University
Suita, Osaka

Aizo Matsushiro

Research Institute for Microbial Diseases
Osaka University
Suita, Osaka

To understand the mechanism of thermostability of thermophilic enzymes, we have studied the role of calcium ions bound to protein on the thermostability of α-amylases from *B. stearothermophilus* and *B. subtilis* (Yutani, 1976). It has been found that the stabilizing effect of calcium ions on both enzymes is about the same. Stability of the thermophilic α-amylase in solution containing a denaturant such as urea is greater than that of the mesophilic enzyme, even under calcium-free conditions. This suggests that the difference in thermostability between thermophilic and mesophilic α-amylases is not directly correlated with binding of calcium ions, although calcium ions play a role in the stability of both enzymes.

Tajima *et al*. (1976) have also studied the role of calcium ions in the thermostability of thermolysin and neutral protease from *B. subtilis var. amylosacchalyticus*. It has been concluded that the stabilizing effect of calcium ions on

ISBN 0-12-268250-5

these enzymes is almost equal, and the extra thermostability of thermolysin is likely to derive from the amino acid sequence per se.

Levy *et al.* (1975) have compared the amino acid sequence of neutral protease A from *B. subtilis* with that of thermolysin, whose three-dimensional structure has also been elucidated. On the basis of sequence comparison, they have suggested that the two enzymes have similar three-dimensional structures and the difference in their thermostability may be caused by a few changes in amino acid residues.

The amino acid sequence of the ferredoxin of *C. tartarivorum* differs from that of the ferredoxin of *C. thermosaccharolyticum* at only two positions (Tanaka *et al.*, 1973). However, the ferredoxin of *C. tartarivorum* is inactivated 50 per cent after two hours at 70°C, whereas that of *C. thermosaccharolyticum* retains 90 per cent of its activity after similar treatment (Devanathan *et al.*, 1969). This shows that a few replacements of amino acid residues greatly affect the stability of ferredoxins.

We have been led by these findings to investigate the role of individual amino acid residues in stabilizing protein molecules. This problem is also interesting in connection with the physicochemical basis of temperature sensitive mutants used widely in molecular biology studies. So far, this problem has been approached mainly with the use of synthetic polypeptides (Ramachandran and Sasisekharan, 1968), chemical modification of natural proteins (Spande *et al.*, 1970), or homologous proteins from different species (Yutani, 1976; Tajima *et al.*, 1976; Levy *et al.*, 1975; Alexander and Pace, 1971; Puett, 1973; Devanathan *et al.*, 1969). However, many mutant microorganisms are available from genetic studies whose mutant proteins differ from those of the wild type only by a single amino acid substitution. Such proteins often exhibit marked differences in stability. Therefore, it was of interest for us to examine the effect of a single amino acid substitution on the stability of a protein. In this paper, we report our studies on the α-subunit of *E. coli* tryptophan synthetase (Yutani *et al.*, 1977 and 1978). We have found this protein to be excellent material for this type of investigation because many single substitution mutants are available, whose sequences (as well as that of the wild type) have been completely elucidated through the work of Yanofsky's group (Yanofsky and Horn, 1972).

TRYPTOPHAN SYNTHETASE α-SUBUNIT OF *E. COLI*

The tryptophan synthetase of *E. coli* is composed of α and β subunits in the form of an $\alpha_2\beta_2$ complex. The $\alpha_2\beta_2$ complex catalyzes the formation of tryptophan from indole-3-glycerol phosphate or indole. The α component of tryptophan synthetase alone catalyzes only the formation of indole from indole-3-glycerol phosphate, and the β_2 component alone catalyzes only the formation of tryptophan from indole, as shown below.

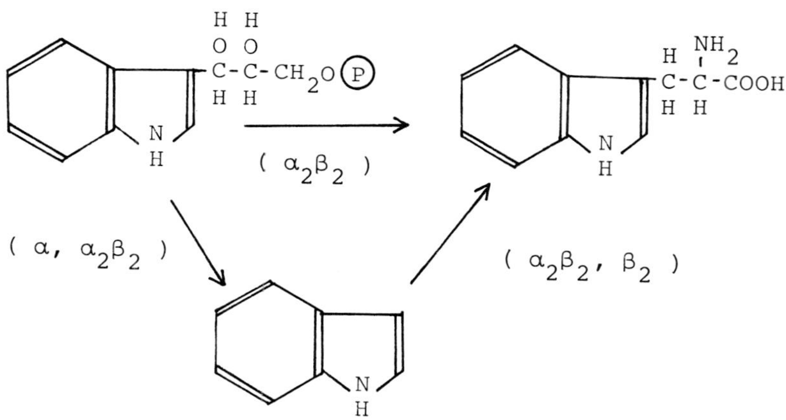

Some amino acid substitution mutations in the α chain result in complete or near-complete inactivity in reactions with indole-3-glycerol phosphate as substrate. However, all of these mutant α chains can form a complex with the β_2 subunit and activate it maximally in reactions with indole as substrate. The α subunit of *E. coli* tryptophan synthetase is a single polypeptide chain composed of 268 residues with a molecular weight of 29,000. This protein does not contain tryptophan and disulphide bonds (Yanofsky and Crawford, 1972).

HEAT INACTIVATION OF THE α-SUBUNIT OF TRYPTOPHAN SYNTHETASE

Heat Inactivation of Mutant Proteins

After heating tryptophan synthetase α-subunits in crude extracts from twelve mutant and wild type strains for 20 minutes at 58°C, the remaining activity was measured (Table 1). The wild type α-subunit retained approximately 50 per-

TABLE 1
Heat Inactivation of the α-subunit of Tryptophan Synthetase[a]

trp *Strain*	*Position of amino acid substitution*[b]	*Remaining activity*[c]
A218	22 Phe → Leu	10 ± 1 (%)
A11	49 Glu → Gln	22 ± 3
A33	49 Glu → Met	88 ± 13
A446	175 Tyr → Cys	60 ± 1
A487	177 Leu → Arg	16 ± 7
A223	183 Thr → Ile	35 ± 2
A23	211 Gly → Arg	13 ± 2
A46	211 Gly → Glu	76 ± 13
A187	213 Gly → Val	12 ± 4
A78	234 Gly → Cys	40 ± 8
A58	234 Gly → Asp	85 ± 10
A169	235 Ser → Leu	30 ± 4
B8	(Wild-type)	49 ± 4

[a]*Taken from Yutani* et al. *(1977) and reproduced with permission from Macmillan Journals Ltd.*

[b]*Yanofsky and Horn, 1972.*

[c]*After sonic treatment of cells, the supernatant solutions were obtained by centrifugation for 2.5 hours at 80,000 g. The supernatant solutions were used for the heat inactivation study. Heat treatment was carried out for 20 minutes at 58°C and pH 8.0. The activity of the α-subunit was measured by the disappearance of indole in the reaction of indole and serine to form tryptophan in the presence of pyridoxal phosphate and β_2-subunit at pH 8.0 and 37°C for 20 minutes. The β_2-subunit was prepared from mutant* trpA2, *an ochre mutant (Yanofsky and Crawford, 1972). Mutant* trpB8 *strain was used as the source of wild type α-subunit.* TrpB8 *strain is deleted in the β-component gene, but the α-component remains as in the wild type.*

cent activity after heating for 20 minutes at 58°C. The mutant proteins *trpA33*, *trpA46*, and *trpA58* were more stable than the wild type protein. On the other hand, seven of the mutant proteins examined, *trpA218*, *trpA11*, *trpA487*, *trpA223*, *trpA23*, *trpA187*, and *trpA169*, were more labile than the wild type protein. The mutant proteins *trpA446* and *trpA78* were inactivated approximately to the same degree as the wild type. These results confirm those reported by Maling and Yanofsky (1961). Though *trpA3* strain is not included in Table 1, Maling and Yanofsky (1961) have reported that the *trpA3* α-subunit is more stable than the wild type protein. In *trpA3*, the glutamic acid residue at position 49 of the wild type has been replaced by valine.

As shown in the following section, heat inactivation characteristics of purified proteins were identical to those of crude extracts. Therefore, we can assume that the difference in thermostability of crude enzymes reflects that of the α-subunit molecule itself. From these results, we should like to point out that four of the 13 mutant proteins examined were more stable than the wild type protein. This suggests the possibility that proteins may easily become more heat stable by a single amino acid substitution.

Correlation between Heat Inactivation and Amino Acid Residues Substituted

Kauzmann (1959) has proposed that hydrophobic interactions are important in stabilizing native protein molecules. The magnitude of "hydrophobicity" of amino acid residues has been estimated from the free energy change for transfer from ethanol to water (Tanford, 1962). The right hand column of Table 2 shows the difference of hydrophobicity between the substituted and the original amino acid residues in terms of difference of transfer free energy between two residues.

Mutant *trpA3*, *trpA33*, *trpA46*, and *trpA58* proteins were more stable than the wild type protein as shown in Table 1. For all of these, the difference of hydrophobicity is positive, indicating that the original residue is replaced by a more hydrophobic residue.

TrpA3, *trpA33*, and *trpA11* proteins are all substituted at position 49. The *trpA11* protein has glutamine in place of glutamic acid at position 49 and it is more labile than the wild type protein. Here the difference of hydrophobicity is negative, indicating that the original residue is replaced by a less hydrophobic residue. These data suggest that the increase or decrease in stability of mutant *trpA3*, *trpA33*, and *trpA11* proteins compared with the wild type is correlated with the difference of hydrophobicity between amino acid residues substituted. As discussed later, we tentatively assume that position 49 is buried in the hydrophobic interior of the molecule and it becomes exposed to water upon denaturation.

TABLE 2
Correlation between Heat Inactivation and Amino Acid Substitution

Stability[a]	trp *strain*	*Position of amino acid replacement*	*Difference of*[b] *hydrophobicity*
Stable	*A3*	49 Glu → Val	+ 1140 (cal/mol)
	A33	49 Glu → Met	+ 750
	A46	211 Gly → Glu	+ 550
	A58	234 Gly → Asp	+ 540
Labile	*A218*	22 Phe → Leu	- 230
	A11	49 Glu → Gln	- 650
	A487	177 Leu → Arg	- 1690
	A223	183 Thr → Ile	+ 2530
	A23	211 Gly → Arg	+ 730
	A187	213 Gly → Val	+ 1690
	A169	235 Ser → Leu	+ 2380

[a]*"Stability" was based upon the results of Table 1.*
[b]*Tanford, 1962.*

In the case of *trpA223*, *trpA23*, *trpA169*, and *trpA187* proteins, the difference in hydrophobicity is positive, but they are more labile than the wild type. This suggests that not only the nature of residue substituted, but also the position substituted greatly affects the stability of the protein.

COMPARISON OF DENATURATION BY UREA OR GUANIDINE HYDROCHLORIDE ON PURIFIED α-SUBUNITS OF *TRPA33*, *TRPA11* AND WILD TYPE

Purification of α-Subunit

To understand the effect of a single amino acid substitution on conformational stability in more detail, the denaturation of mutant proteins was studied by means of circular dichroic (CD) measurement. For this purpose it was necessary to purify the proteins. We selected mutants *trpA33* and *trpA11* for this study. The *trpA33* protein has methionine in place of glutamic acid at position 49 and is more stable than the wild type protein. The *trpA11* protein has glutamine in place of glutamic acid at the same position and is more labile than the wild type protein. The purification was performed according to the method described by Creighton and Yanofsky (1970) with slight modifications. The purified proteins gave a single band on polyacrylamide gel electrophoresis in each case. Heat inactivation characteristics of the purified samples (Fig. 1) corresponded with those of crude extracts (Table 1).

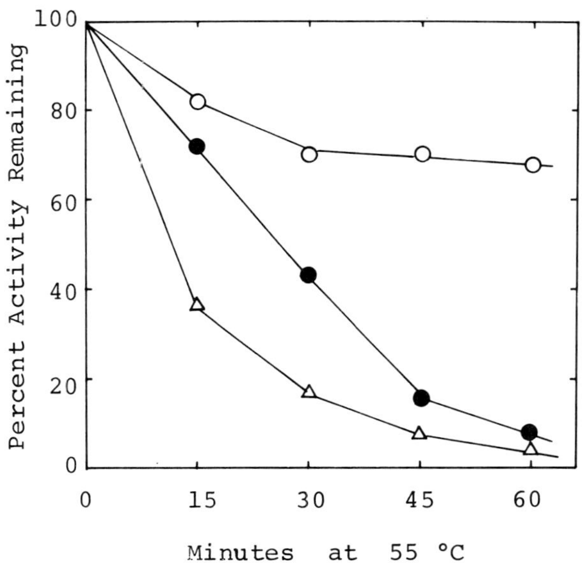

FIGURE 1. Heat inactivation of purified wild type (●), trpA33 *(O), and* trpA11 *(Δ) proteins at pH 8.0.*

Circular Dichroism Spectra of the α-Subunit

CD measurements were carried out with a JASCO model J-20 recording spectropolarimeter. The CD spectra were obtained at protein concentrations of 0.3 to 0.02 mg/ml using a 1 mm or 10 mm light path length. Fig. 2 shows the far ultraviolet CD spectra of the three proteins. The CD spectra of *trpA33* and *trpA11* proteins were not distinguishable from that of the wild type. This suggests that the conformation of the three proteins is similar. These spectra resembled those obtained by Heyn and Weischet (1975) for the wild type protein. The CD spectrum of this protein has a large negative Cotton effect at 222 nm, and thus conformational changes can be followed easily by measuring CD values at 222 nm. The helical contents of these proteins were estimated to be 55 per cent by the method of Chen *et al.* (1972).

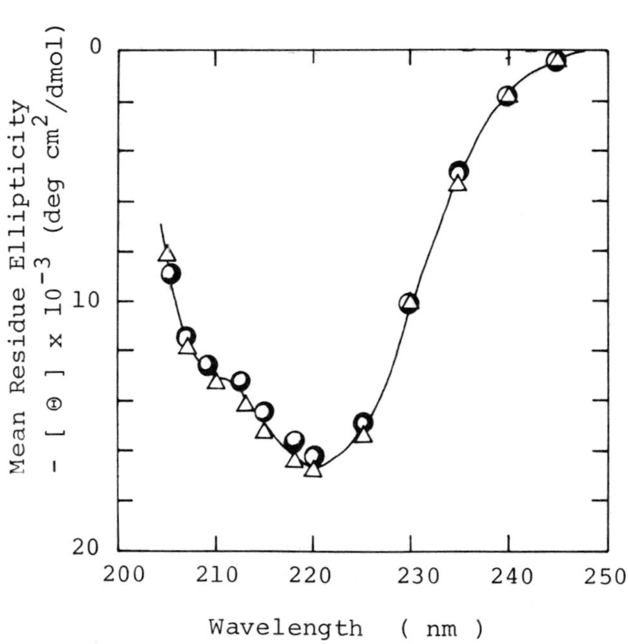

FIGURE 2. CD spectra of α-subunits of E. coli tryptophan synthetases at pH 7.0 and 25°C: wild type (●), trpA33 *(O), and* trpA11 *(Δ) proteins.*

Urea Denaturation of Wild Type and trpA33 *Proteins*

We examined denaturation of the proteins by urea or guanidine hydrochloride (GuHCl) in order to compare the conformational stability. Fig. 3 shows a comparison of denaturation by urea of purified wild type and mutant *trpA33* proteins measured by the change in CD at 222 nm at 25°C, 0.025 M Tris-HCl buffer, pH 8.0. The mutant *trpA33* protein was more resistant to urea than the wild type, corresponding to the relative stability to heat inactivation (Table 1). When the samples were diluted with buffer containing 0.1 mM EDTA and 0.2 mM dithioerythritol, the enzymatic activity and CD values at 222 nm recovered completely. Thus denaturation by urea is reversible under these conditions. The midpoints of urea denaturation curves at 25°C were 3.1 and 3.9 M urea for wild type and mutant *trpA33* proteins, respectively.

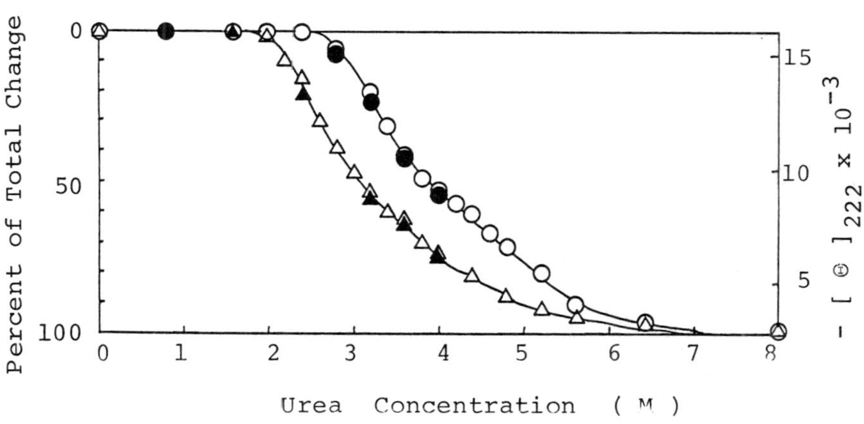

FIGURE 3. Urea denaturation of the α-subunit of wild type (△,▲) and trpA33 *(○,●). Temperature 25.0°C, pH 8.0, 0.025 M Tris-HCl buffer. The solution contained 0.1 mM EDTA and 0.2 mM dithioerythritol. Open symbols represent the ellipticity at 222 nm after incubation for 60 min in urea solutions at various concentrations. Filled symbols represent measurements of samples diluted after exposure to 8 M urea solution to demonstrate the reversibility of the denaturation. Taken from Yutani* et al. *(1977) and reproduced with permission from MacMillan Journals Ltd.*

Guanidine Hydrochloride Denaturation of Wild Type, TrpA33, *and* TrpA11 *Proteins*

Fig. 4 shows denaturation by GuHCl of the purified wild type, *trpA33* and *trpA11* proteins measured by the change in CD at 222 nm, 0.1 M potassium phosphate buffer containing 0.1 mM EDTA and 0.1 mM dithioerythritol, pH 7.0. Symbols in Fig. 4 represent the ellipticity at 222 nm after incubation for 60 minutes at 25.8°C in GuHCl solutions at various concentrations. The denatured proteins were completely renatured by lowering the GuHCl concentration by dilution. This shows that mutant *trpA33* protein was more resistant and *trpA11* protein was less resistant to GuHCl than the wild type protein, corresponding to their relative stability to heat inactivation. The curves for the three proteins are diphasic, suggesting the existence of a stable intermediate state.

Estimation of the Free Energy Change of Unfolding

In order to obtain some quantitative measure of the relative stability of these proteins, we wanted to estimate the free energy change of unfolding. This value could be calculated if we could obtain values for the equilibrium constants between the native and denatured forms. Since denaturation of a protein by GuHCl has been studied more extensively than denaturation by urea (Tanford, 1970), we analyzed the GuHCl denaturation curves for this purpose. We assume that there is one intermediate state and the denaturation by GuHCl can be represented by an equilibrium among a unique native form, an intermediate form, and a completely unfolded form as shown below:

$$N \rightleftharpoons I \rightleftharpoons D \qquad (1)$$

where N, I, and D are the native, intermediate, and completely unfolded forms, respectively. If the fractions of each state are expressed as f_n, f_i, and f_d, the equilibrium constants, K_{ni}, K_{id}, and K_{nd}, can be defined by the following equations:

$$K_{ni} = \frac{f_i}{f_n}, \quad K_{id} = \frac{f_d}{f_i}, \quad K_{nd} = \frac{f_d}{f_n} \qquad (2)$$

For a three state mechanism, an apparent equilibrium constant, K_{app}, is given by the following equation (Tanford, 1968):

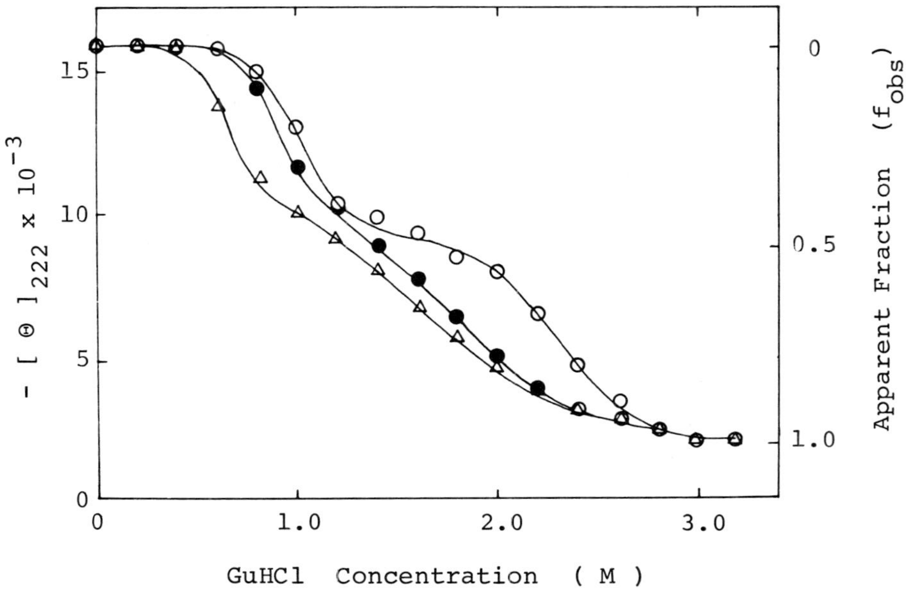

FIGURE 4. GuHCl denaturation of the α-subunit of tryptophan synthetase at 25.8°C and pH 7.0. Open circles, closed circles, and triangles are experimental data for trpA33, *wild type, and* trpA11 *proteins, respectively. Solid lines are calculated according to equation (8). The activity of GuHCl is calculated according to Hade's equation (equation (2) in Aune and Tanford, 1969). The apparent fractions of unfolding, f_{obs}, is given by the following equation:*

$$f_{obs} = \frac{[\theta]_n - [\theta]}{[\theta]_n - [\theta]_d}$$

where $[\theta]$ represents the values of $[\theta]_{222}$ at a given GuHCl concentration, and $[\theta]_n$ and $[\theta]_d$ represent the values of $[\theta]_{222}$ in the absence of GuHCl and in the presence of 3.2 M GuHCl, respectively.

$$K_{app} = \frac{f_{obs}}{1 - f_{obs}} = \frac{K_{nd} + \alpha K_{ni}}{1 + (1 - \alpha)K_{ni}} \qquad (3)$$

where α is the fraction of the total change in ellipticity at 222 nm, $[\theta]_{222}$, that occurs in going from N to I, i.e.,

$$\alpha = \frac{[\theta]_i - [\theta]_n}{[\theta]_d - [\theta]_n} \tag{4}$$

Rewriting equation (3), we have equation (5)

$$f_{obs} = \frac{K_{nd} + \alpha K_{ni}}{1 + K_{nd} + K_{ni}} \tag{5}$$

Aune and Tanford (1969) proposed equations (6) or (7) depending upon whether a protein molecule is assumed to be denatured by binding of the neutral or ionic form of GuHCl molecules, respectively.

$$K = K_0\ (1 + ka)^{\Delta n} \tag{6}$$

$$K = K_0\ (1 + ka_{\pm})^{\Delta n} \tag{7}$$

K_0 represents the equilibrium constant in the absence of denaturant, k the binding constant of GuHCl to a protein molecule, and Δn the difference between the two states in the number of GuHCl molecule bound per protein molecule, a and $a_{\pm}$ the activity and the mean ion activity of GuHCl, respectively. When equation (6) is substituted into equation (5), we have equation (8).

$$f_{obs} = \frac{K_{0,nd}(1 + ka)^{\Delta n_{nd}} + K_{o,ni}(1 + ka)^{\Delta n_{ni}}}{1 + K_{0,nd}(1 + ka)^{\Delta n_{nd}} + K_{0,ni}(1 + ka)^{\Delta n_{ni}}} \tag{8}$$

and f_{obs} in equation (8) is represented as a function of activity or GuHCl.

We used a computer program to give a least-square fit of the experimental data in Fig. 4 to equation (8), to obtain $K_{0,nd}$ and $K_{0,ni}$. Table 3 shows the five parameters at 25.8°C estimated from equation (8) when we put 3.0 for k, as was used for lysozyme by Aune and Tanford (1969). The value of the equilibrium constant between the native and the denatured states, $K_{0,nd}$, of the *trpA33* protein is smaller than that of the wild type protein. The corresponding value for the *trpA11* protein is larger than that of the wild type protein. The values of free energy change of unfolding in water, $\Delta G_{0,nd}$, for wild type, *trpA33*, and *trpA11* proteins at 25.8°C are calculated to be 8.8, 13.4, and 6.3 Kcal/mol. The order of the values of $\Delta G_{0,nd}$ obtained here corresponds very well with the order of stability of the three proteins, as seen from the denaturation profiles by GuHCl (Fig. 4). The order of the values of remaining activity among wild type, *trpA33*, and *trpA11* proteins after heating also agree with the order

TABLE 3
Thermodynamic Parameters of Wild Type, trpA33, *and* trpA11 *Proteins*

Parameter[a]	trpA33	*Wild type*	trpA11
$K_{0,nd}$	1.74×10^{-10}	3.92×10^{-7}	2.61×10^{-5}
$K_{0,ni}$	3.85×10^{-4}	4.03×10^{-4}	5.26×10^{-3}
$\Delta G_{0,nd}$*	13.35	8.77	6.27
$\Delta G_{0,ni}$*	4.67	4.64	3.12
Δn_{nd}	22.3	19.9	19.1
Δn_{ni}	11.9	13.8	14.1
α	0.48	0.37	0.36

[a]*These parameters were calculated according to equation (8) using the data in Fig. 4 and by assuming that GuHCl binds to protein in the neutral salt form (equation (6)) and k = 3.0.*

**Kcal/mol.*

of the values of the free energy change among them.

Values of Free Energy Change Obtained with Various Assumptions

Assuming euqation (6) for binding of the neutral GuHCl molecule with k = 3.0 or 0.1 and equation (7) for binding of the ionic GuHCl molecule with k = 1.2 or 0.1, respectively, values of free energy change of unfolding in water were obtained by curve-fitting (Table 4). The calculated values of the free energy change differed considerably depending on the different assumptions and the different values of the binding constant (k). Moreover, these values of $\Delta G_{0,nd}$ obtained here differed very much from those obtained with urea denaturation assuming a two-state transition (Yutani *et al.*, 1977), though the relative order of the values was the same in every case. At present, we can not judge which assumption is correct.

TABLE 4
Variation in the Estimation of Free Energy Change, $\Delta G_{0,nd}$, calculated by Assuming Different Values of Binding Constant and Binding of Neutral (equation (6)) or Ionic (equation (7)) Forms of GuHCl at 17.2°C and pH 7.0 for the trpA33 Protein

Binding form	Binding constant (k)	$\Delta G_{0,nd}$ (Kcal/mol)	$K_{0,nd}$
Equation (6)	3.0	14.03	2.79×10^{-9}
Equation (6)	0.1	7.12	4.42×10^{-4}
Equation (7)	1.2	21.84	3.65×10^{-15}
Equation (7)	0.1	14.49	1.26×10^{-9}

Fractions of the Intermediate and the Denatured States

The fractions of I and D states are expressed by the following equations,

$$f_i = \frac{K_{0,ni}}{1 + K_{0,ni} + K_{0,nd}} \tag{9}$$

$$f_d = \frac{K_{0,nd}}{1 + K_{0,ni} + K_{0,nd}} \tag{10}$$

Fig. 5 shows the fractions of the intermediate and the denatured states as a function of GuHCl concentration at 25.8°C calculated on the basis of two different assumptions. The curves of fractions calculated by assuming different values for the binding constant (k) overlapped with those shown in Fig. 5 except at very low GuHCl concentrations. This shows that the calculated equilibrium constant of unfolding in the presence of GuHCl is not affected by changing the assumption of binding or the binding constant.

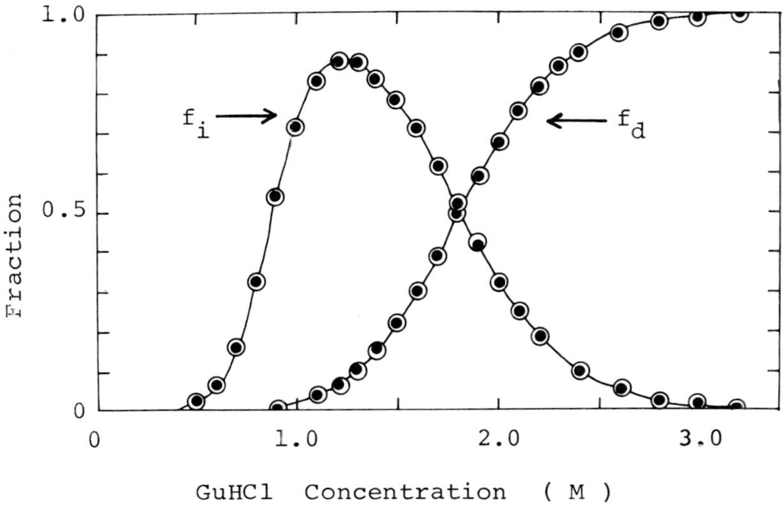

FIGURE 5. The fractions of the intermediate and the completely denatured forms of the wild type protein as a function of GuHCl concentration. The fractions of the intermediate (f_i) and the completely denatured (f_d) forms were calculated from equations (9) and (10), respectively. Closed circles were calculated assuming equation (6) and 3.0 for k; open circles were calculated assuming equation (7) and 1.2 for k. The calculations are based on the data of Fig. 4.

Mechanism of Stabilization

The order of the values for free energy change of unfolding calculated on various sets of assumptions, e.g., Table 3, correlates very well with the order of stability of these proteins as seen from the denaturation profiles by GuHCl (Fig. 4). The order of remaining activity after heating wild type, *trpA33*, and *trpA11* proteins also agrees with the order of free energy change among them.

X-ray diffraction studies have been performed by Schulz and Creighton (1969) on wild type and *trpA33* proteins. Their results show that the replacement of glutamic acid at position 49 by methionine caused little alteration of the conformation of the α-subunit. Our results, taken together with the X-ray data, show that a single amino acid substitution increased the stability of the molecule without causing a gross change in conformation.

The side chain contribution of an amino acid to free energy change for transfer from ethanol to water has been estimated to be 1,300, 550, and -100 cal per mol for methionine, glutamic acid, and glutamine, respectively (Tanford, 1962). The order of the values of transfer free energy change among methionine, glutamic acid, and glutamine coincides with the order of the free energy change of unfolding of *trpA33*, wild type, and *trpA11* proteins obtained here. This suggests that the increase or decrease in stability of *trpA33* or *trpA11* proteins compared with the wild type protein might be correlated with the difference of hydrophobicity of methionine or glutamine relative to glutamic acid.

Although hydrophilic residues such as glutamic acid are usually located on the surface of protein molecules, there are exceptions, especially when the residues are in the active sites of enzymes. In the case of glutamic acid 49 of the α-subunit of tryptophan synthetase, its substitution by methionine or glutamine results in loss of activity with indole glycerolphosphate and serine as substrates. We may therefore assume that glutamic acid 49 is at or near the active site of the enzyme and therefore positioned in a predominately hydrophobic interior environment.

Our results suggest that the stability of an enzyme may be greatly enhanced by an increase in hydrophobicity resulting from the substitution of a few suitably located amino acid residues. Stability of enzymes from thermophiles may be explained by a similar mechanism. We hope that further investigations of this type may clarify the role of individual amino acids in stabilizing proteins.

ACKNOWLEDGEMENTS

We thank Drs. T. Takagi and H. Iwasaki for their valuable comments. Mutant strains were kindly supplied by Drs. C. Yanofsky, I. P. Crawford, and S. Hiraga.

REFERENCES

Alexander, S. S. and Pace, C. N. (1971). Biochemistry *10*, 2738.

Aune, K. C. and Tanford, C. (1969). Biochemistry *8*, 4586.

Chen, Y. H., Yang, J. T., and Martinez, H. M. (1972). Biochemistry *11*, 4120.

Creighton, T. E. and Yanofsky, C. (1970). Methods Enzymol. *17*, Part A, 365.

Devanathan, T., Akagi, J. M., Hersh, R. T. and Himes, R. H. (1969). J. Biol. Chem. *244*, 2846.

Heyn, M. P., and Weischet, W. O. (1975). Biochemistry *14*, 2962.

Kauzmann, W. (1959). Adv. Protein Chem. *14*, 1.

Levy, P. L., Pangburn, M. K., Burstein, Y., Ericsson, L. H., Neurath, H. and Walsh, K. A. (1975). Proc. Natl. Acad. Sci. USA. *72*, 4341.

Maling, B. D. and Yanofsky, C. (1961). Proc. Natl. Acad. Sci. USA. *47*, 551.

Puett, D. (1973). J. Biol. Chem. *248*, 4623.

Ramachandran, G. N. and Sasisekharan, V. (1968). Adv. Protein Chem. *23*, 283.

Schulz, G. E. and Creighton, T. E. (1969). Eur. J. Biochem. *10*, 195.

Spande, T. F., Witkop, B., Degani, Y. and Patchornik, A. (1970). Adv. Protein Chem. *24*, 97.

Tajima, M., Urabe, I., Yutani, K. and Okada, H. (1976). Eur. J. Biochem. *64*, 243.

Tanaka, M., Haniu, M., Yasnobu, K. T., Himes, R. H. and Akagi, J. M. (1973). J. Biol. Chem. *248*, 5215.

Tanford, C. (1962). J. Am. Chem. Soc. *84*, 4240.

Tanford, C. (1968). Adv. Protein Chem. *23*, 218.

Tanford, C. (1970). Adv. Protein Chem. *24*, 1.

Yanofsky, C. and Crawford, I. P. (1972). In *The Enzymes*, 3rd ed. (P. D. Boyer, ed.) Vol. 7, p. 1, Academic Press, New York.

Yanofsky, C. and Horn, V. (1972). J. Biol. Chem. *247*, 4494.

Yutani, K. (1976). In *Enzymes and Proteins from Thermophilic Microorganisms* (H. Zuber, ed.), p. 91, Birkhauser Verlag, Basel.

Yutani, K., Ogasahara, K., Sugino, Y. and Matsushimo, A. (1977). Nature *267*, 274.

Yutani, K., Ogasahara, K., Suzuki, M. and Sugino, Y. (1978). in preparation.

STRUCTURAL FLUCTUATION OF PROTEIN AND THERMOPHILY

Masamichi Tsuboi
Shigeo Ohta
Mamoru Nakanishi

Faculty of Pharmaceutical Sciences
University of Tokyo
Bunkyoku, Tokyo

INTRODUCTION

In this paper, we compare a globular protein of a thermophilic bacterium with the corresponding protein of a mesophilic bacterium in a particular aspect of the motility of the protein molecules. The discussion is based upon our experimental result on the kinetics of the hydrogen-deuterium exchange reaction between the protein molecule and the solvent ^2H_2O. The samples chosen are the polypeptide chain elongation factor EF-Tu from *Thermus thermophilus* HB8 and EF-Tu from *Escherichia coli*. The preparation, purification, and some of the characterizations of these protein samples were made in collaboration with Professor Y. Kaziro's group, and our present paper may be considered jointly with that of Kaziro's group.

GENERAL SCHEME OF A STRUCTURAL FLUCTUATION IN A PROTEIN MOLECULE

Hydrogen Exchange Kinetics and Structural Fluctuation

Let us direct our attention to a peptide NH group in a protein molecule. Let us suppose that at time zero the protein molecule is suddenly brought into ^2H_2O. Then the exchange reaction,

ISBN 0-12-268250-5

$$\text{Protein}(N^1H) + {}^2H_2O = \text{Protein}(N^2H) + {}^1HO^2H \tag{1}$$

is considered to proceed as a simple first order reaction,

$$H(t) = \exp(-kt), \tag{2}$$

if the rate of diffusion is sufficiently high, so that the ^2H_2O concentration remains practically constant. Here, H(t) is the fraction of the hydrogen atom in the protein molecule remaining undeuterated at time t, and k is the first-order rate constant. The k value is often lower by several order of magnitude in a protein molecule than the "intrinsic" rate constant k_e of the peptide NH. Here, the "intrinsic" rate constant k_e means the rate constant of the exchange reaction of the hydrogen atom completely exposed to the solvent(^2H_2O). For example, a peptide NH group in "class 4" in *E. coli* EF-Tu·GDP has a k value of $1.7 \times 10^{-5} \text{sec}^{-1}$ at pH 7.5 and 6°C, whereas the k_e value is 3.2 sec^{-1} at this pH and temperature. This great lowering of the rate constant is attributable to a lower chance of the NH group of the protein now in question to be exposed to the solvent ^2H_2O. Thus, let us postulate that the hydrogen-deuterium exchange reaction takes place according to the following scheme:

$$N \underset{k_{DN}}{\overset{k_{ND}}{\rightleftarrows}} D \xrightarrow{k_e} D^* \ (\rightleftharpoons N^*), \tag{3}$$

where N means a conformation in which the NH group in question cannot be deuterated at all, and D means a conformation in which the NH group is deuterated with the rate constant k_e, i.e., with the rate constant equal to that of the exchange reaction of the NH group completely exposed to the solvent ^2H_2O. One may suppose two extreme cases (Hvidt and Nielsen, 1966):

(i) The case of the "EX_1 mechanism," where

$$k_{ND}, k_{DN} \ll k_e. \tag{4}$$

In this case the overall rate constant, k, of the hydrogen exchange reaction of the peptide group is given by

$$k = k_{ND}. \tag{5}$$

(ii) The case of the "EX_2 mechanism," where

$$k_{ND} + k_{DN} \gg k_e. \tag{6}$$

In this case,

$$k = k_e \gamma, \quad (7)$$

where γ is the D-form content

$$\gamma = [D] / ([N] + [D]). \quad (8)$$

The conformational change, $N \rightleftharpoons D$, which is assumed to exist (Eq. 3) in a solution at equilibrium is called "structural fluctuation." For the case of (i) "EX_1 mechanism," we can estimate k_{ND} from the observed k (see Eq. 5), and this is called "fluctuation rate." For the case of (ii) "EX_2 mechanism," on the other hand, we can estimate γ (see equations 7 and 8). This is called "fluctuation amplitude."

Temperature and pH Dependence of the Hydrogen Exchange Rate

The intrinsic rate constant, k_e, of the peptide NH group was examined by Hvidt and Nielsen (1966), and it has recently been re-examined in detail by Takahashi *et al.* (1978). In the pH = 6-9 and temperature = 6-50°C range, with which we deal in this paper, k_e is now known to be given by:

$$k_e = 6.4 \times 10^{(6 + pH - 3850/T)} \text{ sec}^{-1}, \quad (9)$$

where T is absolute temperature. Thus, k_e is proportional to the OH^- ion concentration, and it increases also with the temperature of the solution.

In the case of the "EX_2 mechanism," where the hydrogen exchange rate constant k is given by Eq. (7), and where γ is not greatly dependent upon pH and temperature, k should be found to be nearly proportional to k_e. In other words, hydrogen exchange rate constant k is nearly proportional to $[OH^-]$ and increases with the temperature of the solution.

In the case of the "EX_1 mechanism," on the other hand, the hydrogen exchange rate is equal to the fluctuation rate (Eq. (5)), and it does not depend upon k_e. Therefore, k may be independent of the pH of the solution.

Two Types of Structural Fluctuation Generally Supposed in a Protein Molecule

The results of a series of our experiments on hen egg-white lysozyme (Nakanishi *et al.*, 1972, 1973) are well explained by postulating two types of fluctuation of the molecular structure of this protein (Fig. 1). Around the transition point in the thermal denaturation, there takes

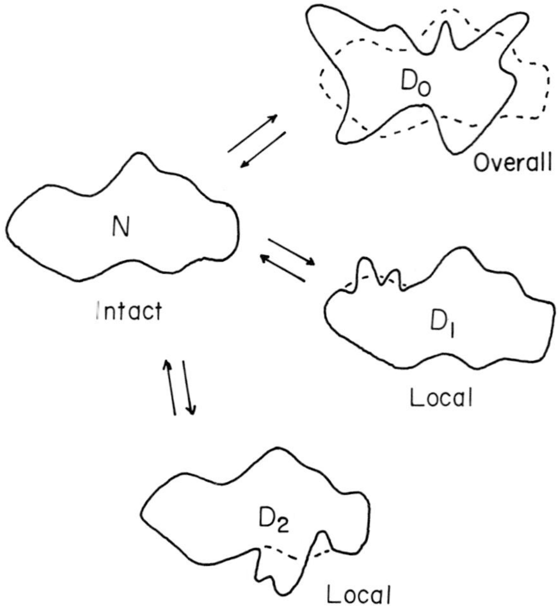

FIGURE 1. Overall and local fluctuation of the structure of a protein molecule.

place an "overall-type" structural fluctuation. That is a dynamic equilibrium between a native form, N, and a denatured form, D_o, in which almost all the intra-molecular peptide-peptide hydrogen bonds are broken. Such an overall-type fluctuation remains predominant on going from the thermal transition point down to the temperature (turning point), which is lower than the transition point by about 15°C. In the temperature range lower than the "turning point," the free energy difference, ΔF, between the N form and such a D_o form becomes so great that the $N \rightleftharpoons D_o$ fluctuation is not prevailing any more (see Fig. 2). Instead, localized fluctuation at various portions of the protein molecule would become predominant.

In a similar manner, two types of fluctuation are apparent in examining an effect of a denaturant. For hen egg-white lysozyme, a reversible denaturation is caused by LiCl. By means of circular dichroism measurement at pH 5.0 and at 20°C, a sharp transition in the conformation is found at 7.2 M LiCl. An exactly the same transition profile is observed when k/k_e is plotted against the LiCl concentration, where k is the hydrogen-deuterium exchange rate

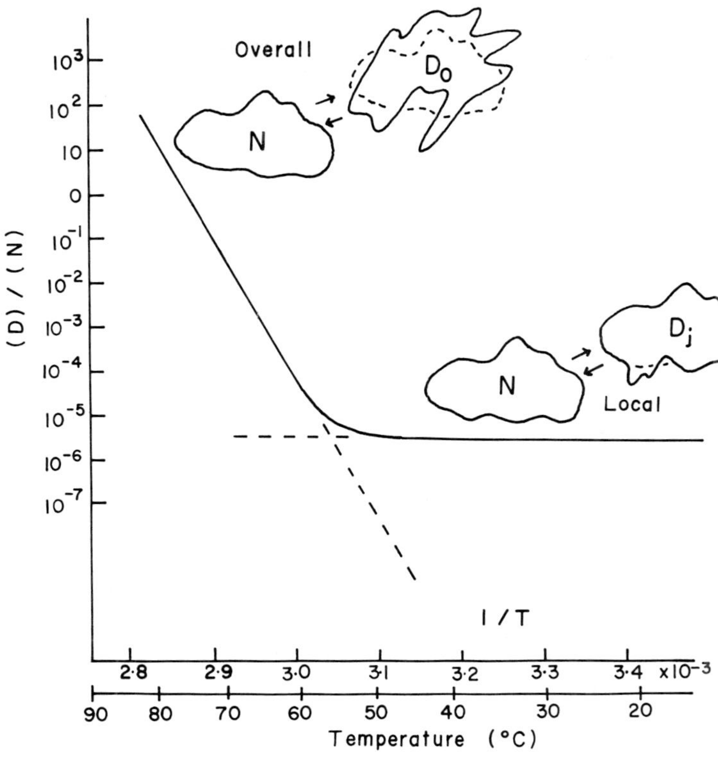

FIGURE 2. Van't Hoff plot for the equilibrium between the native form, N, and denatured form, D, of a protein.

constant of a particular class of the peptide NH groups in the lysozyme molecule. As far as the N⇌D equilibrium is examined by means of the circular dichroism measurement, the mole fraction, γ, of the D form should be taken as "zero" on lowering the LiCl concentration to 6.5 M; no difference can be detected between the solution with 6.5 M LiCl and the solution with 0 M LiCl. The hydrogen-exchange rate constant, k, on the other hand, is found to be greatly different in the solution with 6.5 M LiCl and that with 0 M LiCl. By examining such a k/k = γ value as a function of LiCl concentration, it is suggested that there is a "turning point" at about 6.0 M LiCl. In the denaturant-concentration range lower than this turning point, localized fluctuations rather than the overall-type fluctuation are considered to be predominant.

We would like to suggest that the appearance of these

two types of structural fluctuation is not a special situation for hen egg-white lysozyme, but is a general trend of a protein molecule. Evidence that supports this suggestion has been obtained for bovine α-lactalbumin (Takasada *et al.*, 1973) and for whale myoglobin (Nakanishi and Tsuboi, 1974). In Fig. 3 a phase diagram is shown, which gives an idea of such a general trend of a protein molecule.

Optimum Amplitude and Optimum Rate of Structural Fluctuation

It is speculated that for the proper enzyme action of a protein the molecule should have a suitable amount of flexibility and at the same time a suitable rigidity. In aqueous solution of hen egg-white lysozyme, for example, an effective amount of the D form is nearly constant in the 50-20°C region at various portions of the molecule; [D]/([N] + [D]) is estimated to be about 10^{-6}. This is probably the optimum fluctuation amplitude for the lysozyme molecule. Here, the fluctuation means a local fluctuation but not an overall-type fluctuation. The suitable flexibility and rigidity probably involve not only a proper fluctuation amplitude but also a proper rate of fluctuation; i.e., a proper rate of $N \rightleftharpoons D$ conformational change in the dynamic equilibrium. It is further speculated here that each protein in a thermo-

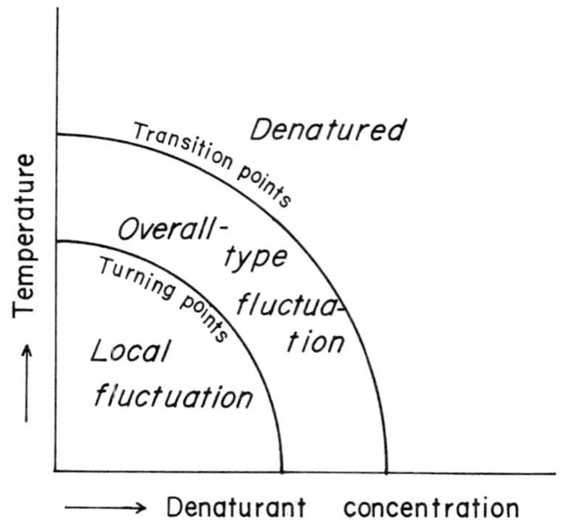

FIGURE 3. A phase diagram for three states of a protein molecule with different dynamic properties.

philic bacterium has its optimum amplitude and rate of fluctuation at a higher temperature than that in a mesophilic bacterium.

A COMPARATIVE STUDY OF EF-Tu FROM *THERMUS THERMOPHILUS* HB8 AND EF-Tu FROM *ESCHERICHIA COLI*

A discussion is now given on the structural fluctuation of the polypeptide chain elongation factor EF-Tu molecule from *Thermus thermophilus* HB8 in comparison with that from *Escherichia coli*. Available data suggest that the properties of EF-Tu from this thermophilic bacterium resemble those for *E. coli* EF-Tu. For example, the molecular weights of *T. thermophilus* EF-Tu and *E. coli* EF-Tu are 49,000 and 47,000, respectively, and their amino acid compositions are similar (Arai, Oshima and Kaziro, unpublished results). Their circular dichroic spectra are almost identical and the activation energy in the dissociation reaction of EF-Tu·GDP or EF-Tu·GTP is again nearly equal. On the other hand, the thermal stability of EF-Tu differs very much between these two microorganisms (Fig. 4). *T. thermophilus* EF-Tu is also stable to acid, alkali, and other protein denaturants; no irreversible denaturation was detected even at pH 1, pH 12, or in 8 M urea.

Hydrogen Exchange Kinetics: The Kinetic Classes

The hydrogen exchange study (Ohta *et al.*, 1977) indicated that each of these proteins now in question contains peptide NH groups in a great variety of the intramolecular environments. Instead of Eq. (2), therefore, the kinetics must in general be represented by the relationship:

$$H(t) = \sum_{j=1}^{n} c_j \exp(-k_j t), \tag{10}$$

where c_j is the number of the peptide groups in the jth kinetic class, k_j is the rate constant of the jth kinetic class and n is the total number of kinetic classes. In an attempt to facilitate the comparison, we classified all the peptide groups (412 in *E. coli* EF-Tu) into seven classes, and assumed that the number(n) of the kinetic classes as well as the size(c_j) of each class(j) is independent of the experimental conditions. It was also assumed that the same classification is valid for both *E. coli* EF-Tu and *T.*

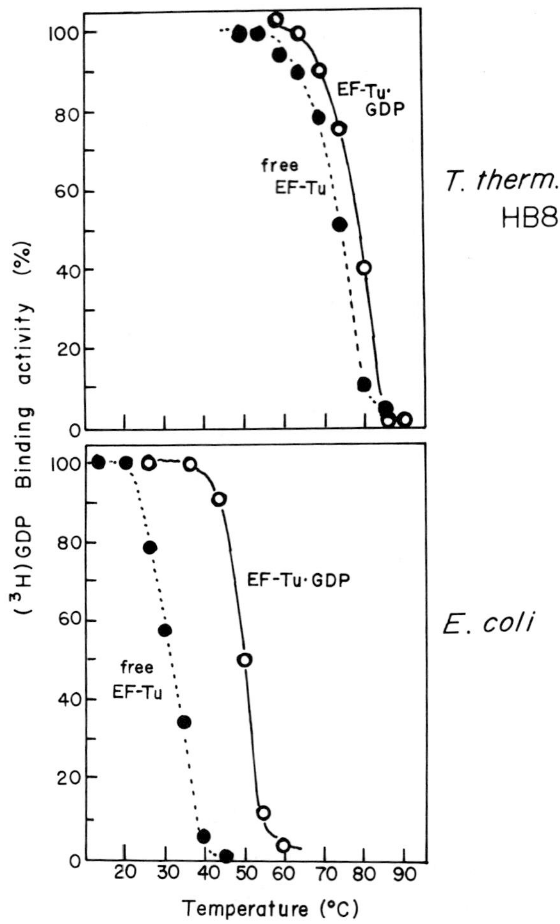

FIGURE 4. [^{3}H]GDP-binding activity remaining after incubation for 5 min at a given temperature is plotted against the temperature of incubation. The mixtures contained 50 mM Tris-HCl (pH 7.9 at 25°C), 5 mM 2-mercaptoethanol and 1 mM EDTA with (for EF-Tu·GDP) or without (for free EF-Tu) 10 mM $MgCl_2$.

thermophilus EF-Tu. Thus, we assumed that for every hydrogen-deuterium exchange curve observed, the values given in the first three columns of Table 1 hold. This assumption is an arbitrary one, but this does not essentially affect the conclusion given below. On the basis of this assumption, the rate constants, k_j, were determined for free EF-Tu,

EF-Tu·GDP, and EF-Tu·GTP of *E. coli* as well as of *T. thermophilus*. Some of the k_j values for EF-Tu·GDP are given in Table 1.

EF-Tu·GTP versus EF-Tu·GDP

For the peptide groups in the kinetic classes 3 and 4, the rates of exchange were found to be slightly greater in EF-Tu·GTP than in EF-Tu·GDP. This difference was observed in a similar manner for the factor from *E. coli* to that from *T. thermophilus*. On the basis of the observed pH-dependence of k_3 and k_4, the deuteration of the peptide groups of these classes is considered to proceed along the "EX_2 mechanism." Therefore, the difference between EF-Tu·GTP and EF-Tu·GDP just mentioned is attributable to the difference in the γ_3 and γ_4 values. Such a difference in γ values may in turn be associated with a difference in the amount of free energy stored in the molecular conformation of EF-Tu·GTP and EF-Tu·GDP.

It is suggested that kinetic classes 3 and 4 are related to the function of EF-Tu. It is interesting that the structural fluctuation of these classes is similar in both *T. thermophilus* EF-Tu and in *E. coli* EF-Tu.

Free EF-Tu versus EF-Tu·GDP

The hydrogen-deuterium exchange reaction of free EF-Tu from *E. coli* is extremely fast at 38°C (for all classes). At this temperature, *E. coli* free EF-Tu is rapidly inactivated, but EF-Tu·GDP is not (Fig. 4). The hydrogen-deuterium exchange kinetics of EF-Tu·GDP are found to depend considerably upon the concentration of GDP in the solvent (if it is lower than 0.6 mM). This fact may indicate that the hydrogen-deuterium exchange reaction of EF-Tu·GDP takes place at higher temperatures through its dissociation into free EF-Tu and GDP. The structural fluctuation of the free EF-Tu molecule would be the "overall-type" fluctuation.

In contrast to the free EF-Tu from *E. coli*, the free EF-Tu from *T. thermophilus* is found to be very rigid even at 38°C; its structural fluctuation is almost as small as that of EF-Tu·GDP. As shown in Fig. 5, the kinetic curve with free EF-Tu is quite similar to that of EF-Tu·GDP at 38°C and 55°C. It is interesting to note that, as far as EF-TU·GDP is concerned, the preparations from *E. coli* and *T. thermophilus* are very similar at 38°C and pH 7.5 (see Fig. 5). In *E. coli*, however, the ligand GDP (or GTP) is essential to reduce the flexibility, whereas in *T. thermophilus*, as far

TABLE 1
Rate Constants, k_j (sec^{-1}), of Hydrogen-deuterium Exchange Reactions of the Peptide Groups

E. coli *EF-Tu·GDP*

Class	c_j[a]	A(amide II)/[b]	6°C				38°C			
			pH 7.5		8.9		7.4		8.5	
j		A(amide I)	k_j	γ_j	k_j	γ_j	k_j	γ_j	k_j	γ_j
			(k_e = 3.2)		(81)		(67)		(8.4 x 10^3)	
1	119	0.45 - 0.32								
2	55	0.32 - 0.26	1.7 x 10^{-3}	5.3 x 10^{-4}						
3	55	0.26 - 0.20	8.3 x 10^{-5}	2.6 x 10^{-5}			2.5 x 10^{-3}	3.7 x 10^{-5}		
4	46	0.20 - 0.15	1.7 x 10^{-5}	5.3 x 10^{-6}	1.2 x 10^{-3}	1.5 x 10^{-5}	6.7 x 10^{-4}	1.0 x 10^{-5}	3.6 x 10^{-3}	4.3 x 10^{-6}
5	37	0.15 - 0.11			2.5 x 10^{-4}	3.1 x 10^{-6}	1.7 x 10^{-4}	2.5 x 10^{-6}	8.8 x 10^{-4}	1.0 x 10^{-6}
6	73	0.11 - 0.03			4.2 x 10^{-6}	5.2 x 10^{-8}	5.0 x 10^{-6}	7.5 x 10^{-8}	6.9 x 10^{-5}	8.2 x 10^{-8}
7	27	0.03 - 0								

T. thermophilus *EF-Tu·GDP*

Class[c]	6°C		38°C			45°C		55°C	66°C		
j	*pH 6.8*	*8.9*	*6.5*	*7.4*	*8.9*	*7.3*	*8.5*	*7.2*	*6.4*	*7.6*	*8.3*
1											
2	5.0×10^{-4}										
3	8.3×10^{-5}										
4	1.7×10^{-6}		1.0×10^{-3}								
5		2.5×10^{-4}	1.0×10^{-4}	1.7×10^{-4}	3.3×10^{-4}	1.2×10^{-5}	1.4×10^{-3}	1.0×10^{-3}			
6		2.5×10^{-6}	6.7×10^{-6}	8.3×10^{-6}	1.5×10^{-5}	2.3×10^{-5}	2.7×10^{-5}	1.7×10^{-4}	1.8×10^{-4}	2.2×10^{-4}	7.5×10^{-4}
7								1.7×10^{-6}	1.7×10^{-6}	1.7×10^{-5}	4.2×10^{-5}

[a] *Number of the peptide groups per molecule.*

[b] *Intensity ratio of the infrared absorption band amide II at 1540* cm^{-1} *versus amide I at 1650* cm^{-1}.

[c] *The classification was made in the same way as that for* E. coli *EF-Tu·GDP by the use of the A(amide II)/A(amide I) scale.*

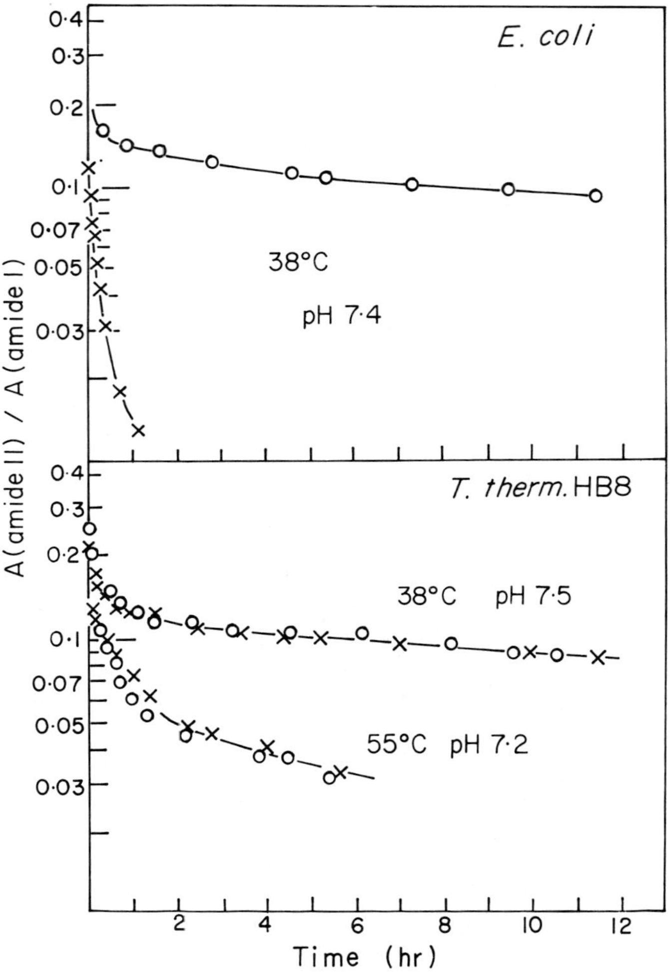

FIGURE 5. Semilogarithmic plots of hydrogen-deuterium exchange of the peptide groups in E. coli *EF-Tu and* T. thermophilus *EF-Tu. (o) EF-TU·GDP, (x) free EF-Tu.*

as the low flexibility is concerned, the ligand is not required at all even at 55°C. At 66°C, however, free EF-Tu has a greater fluctuation amplitude than EF-Tu·GDP. This is the temperature at which the inactivation of free EF-Tu from *T. thermophilus* starts (Fig. 4).

Kinetic Classes 5 and 6

In the pH range 6-9, it is known that $\log_{10}k_e \propto$ pH. Therefore, if $k_j = k_e\gamma_j$ (Eq. (7)), and if γ_j is independent of pH, then $\log_{10}k_j \propto$ pH. This seems to be the case for k_3 and k_4 of *T. thermophilus* EF-Tu·GDP. For k_5 and k_6, however, this is not the case. As shown in Fig. 6, the kinetics depend only slightly upon pH at 38°C and 45°C. In contrast

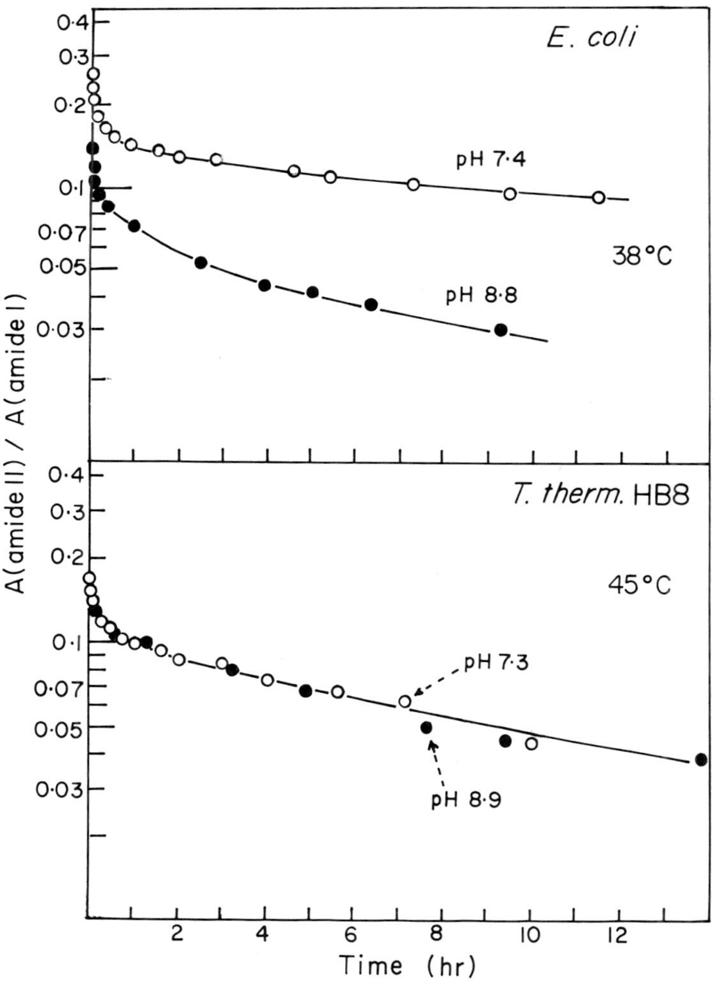

FIGURE 6. Semilogarithmic plots of hydroden-deuterium exchange of the peptide groups in E. coli *EF-Tu·GDP and* T. thermophilus *EF-Tu·GDP.*

with this, in the case of EF-Tu·GDP from *E. coli*, the kinetics of the hydrogen-deuterium exchange depends upon pH as is expected (see Fig. 6). This fact suggests that the hydrogen exchange reaction proceeds along "EX_1-mechanism" rather than "EX_2-mechanism" for the kinetic classes 5 and 6 of the peptide groups of *T. thermophilus* EF-Tu·GDP. In other words, it has been suggested that

$$k_{ND,5}, k_{ND,6} \ll k_e$$

for *T. thermophilus* EF-Tu·GDP, whereas

$$k_{ND,5}, k_{ND,6} \gg k_e$$

for *E. coli* EF-Tu·GDP.

Possible Significance of a Slow Fluctuation

The peptide groups of classes 5 and 6, just mentioned, are assignable to those involved in intra-molecular hydrogen bonds which form a core structure of this protein molecule. The slow fluctuation of such a structure may be associated with a slow denaturation of this protein.

We have recently examined the rate of denaturation of *T. thermophilus* EF-Tu·GDP caused by urea at 38°C and 45°C by means of circular dichroism measurements. In the urea range of 4-7.5 M at pH 8.0, the logarithm of the initial rate was found to be nearly proportional to the urea concentration. By extrapolation, the denaturation rate at zero concentration was found to be 4.9×10^{-5} sec^{-1} at 38°C, and 5.5×10^{-5} sec^{-1} at 45°C. A similar examination was made for a guanidium chloride denaturation by Mr. H. Nojima, Department of Biochemistry and Biophysics, University of Tokyo (personal communication). The rate of denaturation (extrapolated to zero concentration) was found to be 2.4×10^{-5} sec^{-1} at 25°C. These rate constant values of the denaturation reaction are all unusually low. As may be seen in Fig. 7, each of them is nearly equal in order of magnitude to the fluctuation rate estimated at the same temperature from the hydrogen exchange kinetics of the class 6 peptide groups. We should be reminded here, however, that the latter is the rate constant of a local fluctuation whereas with urea or with guanidium chloride we are observing an overall-type conformation change.

Lastly, an additional speculation is presented that the slow fluctuation in the 38-45°C region may also be associated with a proper fluctuation rate at a higher temperature. An extrapolation of the Arrhenius plot given in Fig. 7

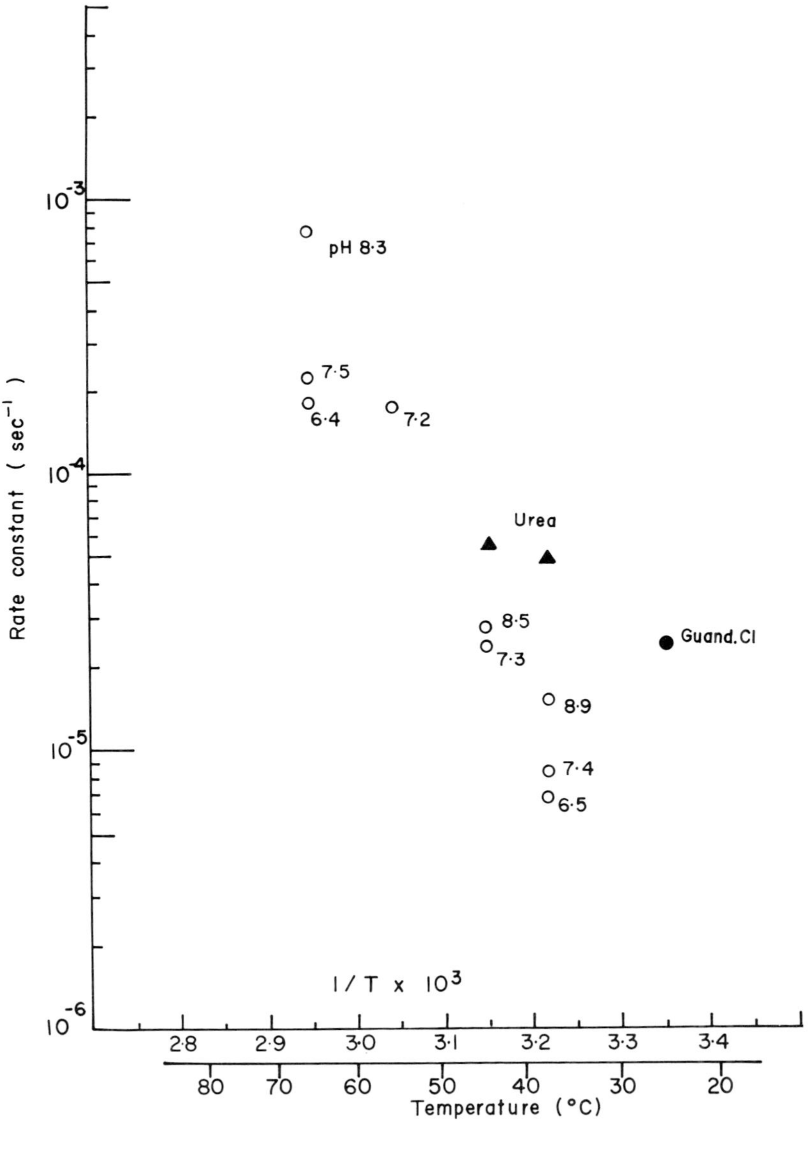

FIGURE 7. Rate constants for structural fluctuation (o), urea denaturation (▲), and guanidium chloride denaturation (●) of T. thermophilus *EF-Tu·GDP plotted on a logarithmic scale against reciprocal absolute temperature.*

indicates that the rate of the structural fluctuation now in question is about 2 x 10^{-3} sec^{-1} at 80°C, for example. Even when the fluctuation amplitude is intolerably great (at 80°C), such a low fluctuation rate may serve as an effective thermostabilizer.

REFERENCES

Hvidt, A. and Nielsen, S. O. (1966). Advan. Protein Chem. *21*, 287.

Nakanishi, M. and Tsuboi, M. (1974). Bull. Chem. Soc. Japan *47*, 293.

Nakanishi, M., Tsuboi, M. and Ikegami, A. (1972). J. Mol. Biol. *70*, 351.

Nakanishi, M., Tsuboi, M. and Ikegami, A. (1973). J. Mol. Biol. *75*, 673.

Ohta, S., Nakanishi, M., Tsuboi, M., Arai, K. and Kaziro, Y. (1977). Eur. J. Biochem. *78*, 599.

Takahashi, T., Nakanishi, M. and Tsuboi, M. (1978). Bull. Chem. Soc. Japan, to be published.

Takesada, H., Nakanishi, M. and Tsuboi, M. (1973). J. Mol. Biol. *77*, 605.

COMPARATIVE STUDIES OF THERMOPHILIC AND MESOPHILIC ENZYMES: OBJECTIVES, PROBLEMS, RESULTS

H. Zuber

Institut für Molekularbiologie und Biophysik
Eidgenössische Technische Hochschule
Zürich, Switzerland

INTRODUCTION

Comprehension of thermophilic microorganisms and their special ability to live at high temperature is based on an understanding of both their metabolism and the structure and function of their cellular components. There was much early interest in the enzymes of thermophiles but this interest was focused primarily on the increased thermostability of these enzymes and on the structural basis of this stability. Much less effort was directed towards an understanding of the biological function, i.e., substrate binding, catalytic mechanism, effector binding of these enzymes at higher temperatures. However, it must be emphasized that in terms of an optimal adaptation to higher temperature both thermostability and biological function must be considered in the definition of a thermophilic enzyme.

Studies of numerous thermophilic enzymes have shown that because of their structure these enzymes possess higher temperature optima, increased thermostability and lower efficiency than enzymes from corresponding mesophilic species (1). Structural investigations, however, have shown that, contrary to original expectations, the special thermophilic structural elements are in no way pronounced or extensive (1, 2). Thermophilic enzymes differ only slightly from the corresponding mesophilic ones. For thermodynamic reasons this result is not unexpected since a difference of only a few non-covalent interactions has the potential to significantly alter the free energy of activation for the

ISBN 0-12-268250-5

denaturation process. However, based on phylogenetic considerations, i.e., adaptation of the enzyme structure over a long time period by mutation-selection, one would expect more significant structural differences. The true picture remains clouded since these differences are not easily discernible from among the changes in amino acid sequences due to species differentiation.

Because of this uncertainty the situation in the field of thermophilic enzymes presents a dilemma. One has the desire to know how thermophilic enzymes differ structurally from mesophilic enzymes but one wonders whether it is at all possible to perform significant and unequivocal experiments which will detect these typical and possibly very slight structural differences. And is it worth the effort?

I regard my task at this conference primarily as one of discussing this very situation and the problem of structural comparison of mesophilic and thermophilic enzymes. I shall try to answer the last two questions with the aid of the small amount of experimental data available today.

These structural data, however, are only meaningful when considered in relation to the special properties of thermophilic enzymes. Recent studies of many groups on the characteristics of thermophilic enzymes have been for understandable reasons concerned primarily with thermostability and the denaturation process. Such experiments should definitely be continued on typical examples and dealt with in greater detail. But "thermophilic enzyme catalysis" should also be given more consideration. We have performed some informative experiments in this regard with dehydrogenases and I will present these in the second part of my paper.

STRUCTURE OF THERMOPHILIC ENZYMES

It is reasonable to assume that the special structure of thermophilic enzymes can only be worked out in comparison with mesophilic enzymes. In such comparative studies involving determination of the primary and three-dimensional structures one should pay attention to one point in particular. The compared enzymes should come from organisms as closely related phylogenetically as possible. The typical structural differences between thermophilic and mesophilic enzymes are a part of the overall species differences as manifested in the amino acid sequences of these enzymes. They were developed through mutation-selection as part of a metabolism adapted to higher or lower temperatures, exactly as other species differences were selected in response to other metabolic and environmental factors. One would therefore expect that the thermophilic-mesophilic differences in amino acid sequence would be most visible when the other differences are small, e.g., of the same order of magnitude as the thermophilic-mesophilic ones. This should be the case, for example, with thermophilic and mesophilic enzymes from closely related bacteria whose metabolism is nearly the same. Considering this minimum requirement it is not surprising that extensive statistical comparisons of amino acid compositions of unrelated thermophilic and mesophilic enzymes have provided little information about the special structural features of thermophilic enzymes (3). However, if one restricts himself to the conditions discussed above there are indications that in some special cases a thermophilic enzyme can be recognized on the basis of its amino acid composition.

To show that comparative studies which take this consideration into account do in fact lead to the determination of some structural details of thermophilic enzymes one may examine the data on the structure and properties of the low molecular weight thermophilic and mesophilic ferredoxins. Comparison of the structures of these proteins has already provided concrete indications of structural elements which result in thermostabilization. It has been shown that in conformity with the slight differences in the denaturation free energy of activation between thermophilic and mesophilic ferredoxins, a change of only a few amino acid residues can cause large changes in thermostability (4, 5). The hope exists that at least a part of this knowledge can be extrapolated to the case of the larger enzymes.

Today the amino acid sequence of ferredoxin is known for a large number of organisms. As expected, the ferredoxins from organisms of a single genus, *Clostridium*, show a much higher degree of sequence homology when compared to each other (60-96%) than when compared to ferredoxins of other organisms (14-51%) (5, 6, 7). Since the clostridia ferredoxins are from both thermophilic and mesophilic species the question of whether or not thermophilic structural details can be better recognized when one compares closely related organisms may be clarified. In fact, if one concentrates on the basic and acidic amino acid residues differences between thermophilic and mesophilic ferredoxins are already discernible in the amino acid composition (Table 1). Devanathan, *et al.* (4) have found that thermophilic ferredoxins contain more basic amino acid residues. Ferredoxin models constructed by Perutz and Raidt (5) showed that this difference may be significant in that it allows additional ionic bonds to be formed in the thermophilic protein. Data presented in 1975 in Zürich (8, 9, 11), which showed that in the case of some thermophilic enzymes the Arg and Glu content is also increased, led to much discussion on the role of ionic bonds in thermostability.

TABLE 1

*Mesophilic and Thermophilic Ferredoxins: Differences in Basic and Acid Amino Acid Residues and Possible Salt Bridges***

	Arg	Lys	His	Total Arg,Lys,His	Glu	Gln	Asp	Asn	Total Glu,Asp	Total Gln,Asn	Salt Bridges (possible) I	II	III
M. aerogenes	0	1	0	1 ↓	2	2	5	3	7 ↓	5 ↑	1	0	0
C. pasteurianum	0	1	0	1	2	2	5	3	7	5	1	0	0 (8.5*)
C. acidiurici	1(1)	0	0	1(1)	2	2	6	1	8	3	1	1	(1)(6.6*)
P. elsdenii	0	2(1)	1(1)	3(2)	6(3)	0	3	0	9	0	3	1	1
C. tartarivorum	0	2(2)	2(2)	4(4)	5(2)	2	4	0	9	2	4	2	2 (4.6*)
C. thermosacharo-lyticum	0	2(2)	2(2)	4(4)	7(4)	0	4	0	11	0	4	4	4
B. stearothermo-philus a)	0	3	0	3	5	1	11	1	16	2	3	2	-
D. gigas	1(1)	1	0	2	3	6	6	5	9	11	2	1	-
Chlorobium lim I	0	0	0	0	5	2	4	2	9	4	0	-	-
" " II	1	1	1	3	8	0	5	1	13	1	3	-	-
Chromatium vin. b)	0	1	1	2	6	4	4	3	10	7	2	-	-
Spinach c)	1	2	0	3	3	0	7	1	10	1	3	-	-

I Amino acid composition

II Amino acid sequence

III Three-dimensional structure

) ΔΔG to C. thermosacharolyticum (70°C, KJ)

a) 67 amino acid residues from the N-Terminus

b) 63 amino acid residues from the N-Terminus

c) amino acid residues from position 35 to 100

Numbers in brackets: critical basic and acid amino acid residues

**) The data for the ferredoxins from Chlostridia, Micrococcus aerogenes and from Peptostreptococcus elsdenii have been taken from Perutz and Raidt [5], those from Bacillus stearothermophilus, Desulfovibrio gigus, Chlorobium limicula I and II and Chromatium vinosum from Hase et al. [6] and from Spinach from [7].

In the case of the clostridia ferredoxins an increase in thermostability parallels an increase in the total number of basic amino acid residues as well as an increase in Glu. Asn and Gln decrease while Asp remains effectively constant. Since the acidic residues greatly exceed the basic residues the increase leads to an increased probability of ionic bond formation. In the thermophilic ferredoxins a maximum of 4 ionic bonds would be possible. It is probable from this comparison that basic amino acid residues as well as Glu are important for increased thermostability. If, however, one compares the amino acid composition of the clostridia ferredoxins with those from less closely related species, such a relationship is no longer discernible. This comparison shows that it is justified to demand that only proteins of organisms as closely related as possible be compared and that in such a series typical thermophilic-mesophilic differences may already be visible in the amino acid composition. In making this comparison special attention must be paid to the basic amino acid residues and since these presumably become effective only via ionic bonds the Glu/Gln and Asp/Asn ratios must also be determined (Table 1, salt bridges I). Obviously, such a comparison can only provide a basis. The primary and three-dimensional structure provide the decisive details.

With the ferredoxins and perhaps with other proteins, one can try to localize the critical thermostabilizing amino acid residues in the primary structure. Sequence comparisons are shown in Table 2. With the most thermostable ferredoxin from *Clostridium thermosacharolyticum* (VI) possibly all 4 basic amino acid residues and 4 of the critical Glu residues are concerned in the formation of 4 ionic bonds. With the more thermolabile *C. tartarivorum* (V) the 4 basic amino acids have been conserved but 2 of the 4 critical Glu residues are now Gln and unable to form ionic bonds. The moderately thermostable ferredoxin from *C. acidiurici* (III) possesses 1 of the critical basic residues and one acidic amino acid residue in a critical region (position 6). In the most thermolabile ferredoxins from *C. pasteurianum* (II) and from the homologous *Micrococcus aerogenes* (I) all of the critical basic residues are missing. However, at position 15 in I and 3 in II, a Lys is now present. Therefore, for these two proteins one ionic bond, but not as critical, is possible. Naturally, confirmation of the ionic bonds requires knowledge of the three-dimensional structure. It is therefore interesting to note that good agreement exists for this perhaps simple case between the predictions based on amino acid composition and sequence and the results from the three-dimensional models (5) (Table 1 possible salt bridges).

Of course it would be wrong to consider only the thermostabilization effects of ionic bonds. These interactions

TABLE 2

*Amino Acid Sequences of Mesophilic and Thermophilic Ferredoxins**

		1	2	3	4	5	6	7	8	9	10	11	12	13	14	15	16	17	18	19	20	21	22	23	24	25	26	27	28
I	M. AEROGENES	A	Y	V	I	N	D	S	C	I	A	C	G	A	C	K	P	E	C	P	V	N	-	I	Q	Q	G	-	S
II	C. PASTEURIANUM	A	Y	K	I	A	D	S	C	V	S	C	G	A	C	A	S	E	C	P	V	N	A	I	S	Q	G	D	S
III	C. ACIDI URICI	A	Y	V	I	N	E	A	C	I	S	C	G	A	C	D	P	F	C	P	V	D	A	I	S	Q	G	D	S
IV	P. ELSDENII	M	H	V	I	S	D	E	C	V	K	C	G	A	C	A	S	T	C	P	T	G	A	I	E	E	G	E	T
V	C. TARTARIVORUM	A	H	I	I	T	D	E	C	I	S	C	G	A	C	A	A	E	C	P	V	E	A	I	H	E	G	T	G
VI	C. THERMOSACHARLYTICUM	A	H	I	I	T	D	E	C	I	S	C	G	A	C	A	A	E	C	P	V	E	A	I	H	E	G	T	G
**) VII	B. STEAROTHERMOPHILUS	P	K	T	I	K	E	T	C	I	A	C	G	A	C	G	A	A	A	P	-	D	I	Y	-	D	Y	D	E
			Y		(VD)																								

	29	30	31	32	33	34	35	36	37	38	39	40	41	42	43	44	45	46	47	48	49	50	51	52	53	54	55	56
I	I	Y	A	I	D	A	D	S	C	I	D	C	G	S	C	A	S	V	C	P	V	G	A	P	N	P	E	D
II	I	F	V	I	D	A	D	T	C	I	D	C	G	N	C	A	N	V	C	P	V	G	A	P	V	Q	E	-
III	R	Y	V	I	D	A	D	T	C	I	D	C	G	A	C	A	G	V	C	P	V	G	A	P	V	Q	A	-
IV	K	Y	V	V	T	-	D	S	C	I	D	C	G	A	C	E	A	V	C	P	T	G	A	I	S	A	E	-
V	K	Y	Q	V	D	A	D	T	C	I	D	C	G	A	C	Q	A	V	C	P	T	G	A	V	K	A	E	-
VI	K	Y	E	V	D	A	D	T	C	I	D	C	G	A	C	E	A	V	C	P	T	G	A	V	K	A	E	-
VII	D	G	I	A	D	D	N	Q	G	I	E	V	M	D	A	F	E	G	C	P	T	E	S	I	K	V	A(+14	
				(YVTL)						V		(PD-ILIDDM)																

*) DATA FOR I-IV IS FROM PERUTZ AND RAIDT [5] AND FOR VII IS FROM HASE ET AL. [6]

**) PRELIMINARY ALIGNMENT

only become effective in cooperation with other non-covalent interactions. There are obviously stabilization contributions to be expected from hydrophobic interactions and hydrogen bonds, especially in the case of oligomeric proteins. A remote, future objective is to differentiate the thermostabilizing contributions of the different non-covalent interactions.

The harmonious results concerning the role of the basic and acidic amino acid residues in the case of the ferredoxins encourage the assumption that these residues are also important for stabilization of other thermophilic proteins. Perhaps a similar comparative approach would be successful and provide a positive answer to the question of whether or not thermophilic structural features are recognizable. Still open to discussion is the amount of effort involved.

Examination of the amino acid composition of some thermophilic enzymes shows that, in comparison to mesophilic enzymes, the arginine content is significantly increased. This conclusion is based on data presented in Zürich by O'Brien *et al.* (11), Ljungdahl *et al.* (8) and myself (9), as well as on the statistical analysis of Singleton (3). In fact, as we first established (9) with thermophilic lactate dehydrogenase (LDH), the increased Arg content practically always is accompanied by a decreased Lys content (Tables 3 and 4). Apparently, for structural or functional reasons the sum of the basic amino acid residues in the enzymes remains approximately constant, in contrast to the ferredoxins. The molecule is, however, made more thermostable by increased Arg incorporation at specific positions at the expense of Lys residues at other positions. This means that these thermophilic and mesophilic enzymes probably differ in the arrangement of ionic bonds. The use of a different arrangement in the case of the larger enzymes (5-8 times as many residues as the ferredoxins) instead of an increase in the number of basic residues may be due to structural and functional constraints. Also, since the number of basic amino acids in mesophilic ferredoxins is so small in comparison to that found in larger mesophilic enzymes, it is necessary to increase this number in order to provide additional ionic bonds. For the enzymes, however, the basic residues only need be exchanged and have their position in the three-dimensional structure altered in order to provide additional thermostability. The same situation holds for the acidic residues.

With the ferredoxins therefore the critical thermostabilizing amino acids are much more easily identified than with the larger enzymes. One variable "mesophilic" basic amino acid residue will be in contrast to 4 critical rather fixed "thermophilic" ones. However, for the thermophilic and mesophilic enzymes the extent of the basic and acidic amino acid

LDH from	Arg	Lys	His	Total Arg,Lys	Total Arg,Lys,His	Difference Arg,Lys
B.stearothermophilus	16	12	8	28	36	+10 Arg −12 Lys; +10 Arg −11 Lys; +11 Arg − 9 Lys
B. caldotenax	17	15	7	32	39	
B. X_1 (37°C)	6	24	10	30	40	
B. subtilis(ATCC 6051)	6	23	9	29	38	
Dogfish	9	29	11	38	49	
Rabbit	8	24	7	32	39	

Ala DH from	Arg	Lys	His	Total Arg,Lys	Total Arg,Lys,His	Difference Arg, Lys
B. caldotenax	21	29	9	43	51	+8 Arg −3 Lys; +7 Arg +1 Lys
B. X_2 (37°C)	13	32	8	46	54	
B.subtilis(ATCC 6051)	14	28	9	42	51	

Neutral Proteinase from	Arg	Lys	His	Total Arg,Lys	Total Arg,Lys,His	Difference Arg,Lys
B. Thermoproteolyticus	10	11	8	21	29	+1 Arg −3 Lys; +2 Arg −5 Lys
B. X_2 (37°C)	9	14	8	23	31	
B.subtilis(NRRLB 3411)	8	16	5	24	29	

*) Amino acid compositions were determined in our laboratory.

TABLE 3
*Basic Amino Acid Residues in Thermophilic and Mesophilic LDH (303-329), Ala DH (450-490) and Neutral Proteinases (300-316)**

residue exchange is not easily detectable in the amino acid composition. Against a background at 30-60 basic amino acid residues only 1-10 residues are exchanged. An additional complication is that in most amino acid analyses of thermophilic and mesophilic enzymes details concerning the Glu/Gln and Asp/Asn ratios are missing. Thus, one is unable to detect potentially important differences in the number of acidic amino acid residues.

It is also interesting that in some cases the Arg-Lys differences exist not only between thermophilic and mesophilic enzymes from bacteria but also between bacterial thermophilic enzymes and mesophilic enzymes from higher organisms. This points to a potentially fundamental structural difference between thermophilic and mesophilic enzymes, visible even within large species differences. Thus,

Enzyme	Arg	Lys	His	Total Arg,Lys	Total Arg,Lys,His	Differences Arg-Lys
Triosephosphate Isomerase (474-496 residues [a])						
C.thermosacharolyticum	15	40	8	55	63	+5 Arg −9 Lys; +4 Arg −7 Lys
C.pasteurianum	10	49	9	59	68	
C.sp. strain 69	11	47	7	58	65	
Rabbit muscle	16	42	8	58	66	
Formyltetrahvdrofolate Synthetase (2209 residues [b])						
C.thermoaceticum	97	133	32	230	262	+25 Arg −53 Lys; +36 Arg −48 Lys; +31 Arg −53 Lys
C.formicoaceticum	72	186	35	258	293	
C.cylindrosporum	61	181	32	242	274	
C.acidi urici	66	186	45	252	297	
Glyceraldehyde-3-Phosphate Dehydrogenase (321-331 residues [c])						
Thermus thermophilus	17	19	8	36	44	+4 Arg −6 Lys
Thermus aquaticus	16	23	10	39	49	+3 Arg −2 Lys
B.stearothermophilus	15	22	9	37	46	+2 Arg −3 Lys
B.cereus	13	25	7	38	45	
E.coli	12	26	6	38	44	
Yeast	11	26	8	37	45	
Lobster	9	28	5	37	42	

a) Data from Shing et al.[10)]

b) Data from O'Brien et al.[11)]

c) Data from Fujita et al.[12)]

TABLE 4
Arg-, Lys- and His-Residues in Enzymes from Thermophilic and Mesophilic Organisms

the variable Arg-Lys ratio may represent within the overall species differences in the primary structure of these enzymes a typical thermophilic-mesophilic species differentiation developed during evolution. It is reasonable to assume that other protein-types, which have different properties and functions than the enzymes, will show other thermophilic-mesophilic differences in respect to the basic and acid amino acid residues. The reverse, e.g., an increased Lys content and decreased Arg content, has been reported for 10 of the ribosomal proteins from *B. stearothermophilus* when compared to the equivalent proteins from *E. coli* (13).

With regard to comparative studies on the primary structure of thermophilic and mesophilic enzymes the ferredoxin example has demonstrated that it is important to compare highly homologous enzymes and that at least 2-3 thermophilic proteins must be compared with 2-3 mesophilic proteins in order to establish critical thermostabilizing amino acid residues. For the ferredoxins 3 thermophilic and 3 mesophilic proteins were necessary to differentiate between the variable "mesophilic" basic amino acid residues and the fixed "thermophilic" basic residues. In the past, investigations were usually confined to a thermophilic protein alone or to bacterial thermophilic proteins and mesophilic proteins from not closely related microorganisms or even higher organisms. As with the ferredoxins, one might expect that the amino acid sequence data from mesophilic and thermophilic enzymes would allow identification of critical basic and acidic amino acid residues. The position of these residues in the three-dimensional structure should make it possible to recognize key ionic bonds. Coleman *et al.* (14) were able to localize possible salt bridges on the surface of the thermolysin molecule. Biesecker *et al.* (15) compared the three-dimensional structure of glyceraldehyde-3-phosphate dehydrogenase from phylogenetically different organisms (lobster muscle, *B. stearothermophilus* and *Thermus aquaticus*) and were able to postulate the presence of two "buried" thermostabilizing ionic bonds. These were intersubunit ionic bonds and it was not possible to identify any potential thermostabilizing surface ionic bonds. In light of the earlier considerations it is reasonable to assume that such ionic interactions at or near the outer enzyme surface (perhaps shielded partly by hydrophobic interactions from the bulk water) could only be identified by comparison of several closely related thermophilic and mesophilic enzymes.

Some years ago, we began isolation and characterization of suitable thermophilic and mesophilic enzymes from spore-forming bacilli. The enzymes were LDH, for which some primary and three-dimensional structural data were available, neutral proteinase, and alanine dehydrogenase (AlaDH). We compared: 1) Thermophilic LDH from *B. stearothermophilus* and *B. caldotenax* with mesophilic LDH from *Bacillus B.* X_1 and *B. subtilis*; 2) Thermophilic neutral proteinase from *B.* X_2, *B.* X_3, *B. cereus* and *B. subtilis*; 3) Thermophilic AlaDH from *B. caldotenax* with mesophilic AlaDH from *B.* X_2 and *B. subtilis* (Marburg). Sequence analysis of the N-terminal region (30-40 residues) showed that an 80-100% homology existed between the thermophilic enzymes compared to a 60-70% homology between the thermophilic and mesophilic enzymes. This is in agreement with the homology relationship of the ferredoxins within the Clostridia. Amino acid composition data revealed an Arg-Lys exchange in all cases (Table 3). On the basis of these results the three enzymes appeared suitable for comparative sequence analysis. Sequence analysis of LDH is currently in progress.

The mesophilic bacilli *B.* X_1, *B.* X_2 and *B.* X_3 were obtained by temperature adaptation of *B. stearothermophilus*, *B. caldotenax* and *B. caldolyticus*, respectively. These adaptation experiments and the positive identification of the mesophilic variants are unfortunately not yet concluded. We have recently obtained some rather contradictory results. Digestion of the DNA from the mesophilic variants by restriction enzymes yielded a fragmentation pattern which was often not identical with that of the corresponding thermophilic strain. Immunological experiments also suggest a difference between the thermophilic bacilli and their corresponding mesophilic variants. Antibodies produced against the spores of the mesophilic variant of *B. stearothermophilus* cross-react with the spores of *B. subtilis* and to some extent with the spores of *B. coagulans* and antibodies produced against the spores of the mesophilic variant of *B. caldotenax* cross-react with the spores of *B. cereus*. However, these antibodies show no cross-reaction with the spores of the thermophilic strains. Although the cultures were cloned before, during and after the temperature adaptation, one cannot rule out completely the possibility that under the growth conditions used a mesophilic variant of *B. stearothermophilus* with properties similar to *B. subtilis* and of *B. caldotenax* with properties similar to *B. cereus* was selected. Similar results involving the selection of thermophilic variants from apparently pure mesophilic cultures were reported several years ago (16).

PROPERTIES OF THERMOPHILIC ENZYMES

We have compared the thermostability and enzymatic activity of thermophilic and mesophilic bacterial LDH, AlaDH and neutral proteinases. The thermostability, as measured by the residual activity after 30 min. heating at various temperatures, reveals the usual differences between thermophilic and mesophilic enzymes. Whereas enzymes from thermophilic bacilli retain 100% of their activity after heating to 70-90°C, enzymes from mesophilic bacilli (cultivated at 37°C) are more thermolabile, retaining 100% activity after heating to 40-60°C. The thermostability is strongly dependent on the buffer used, e.g., LDH from *B. stearothermophilus*, *B. caldotenax* and *B.* X_1 is at pH 6 more stable in the presence of 0.1 M Triethanolamine-buffer than in 0.1 M phosphate-buffer. It is reasonable to assume that the buffer anions or cations interact differently with surface acidic or basic amino acid residues which are important for thermostability.

Thermostability measurements are based on residual activity measured at a temperature lower than the heating temperature. Under these conditions one only measures the poorly understood process of irreversible denaturation and therefore such measurements are not ideal for the study of denaturation kinetics. We are currently using other methods (fluorescence, UV and CD) to more accurately follow the complex denaturation kinetics. Before the denaturation kinetics on both thermophilic and mesophilic enzymes have been examined in more detail any consideration of the structural differences between these proteins based on the free energy of activation for denaturation should not be considered conclusive.

Comparative studies on the catalytic function of thermophilic and mesophilic enzymes are also important. The ascending part of a temperature optimum curve represents the temperature dependence of this function whereas the descending part is based on a combination of the thermostability of the enzyme-substrate complex and of the entire enzyme molecule. For LDH we have found that the temperature optimum for the forward reaction (lactate oxidation) differs from that for the reverse reaction (pyruvate reduction) as shown in Figure 1. The values for *B. stearothermophilus* are 65°, 55°, for *B. caldotenax* 70°, 60°, for $B.X_1$ 65°, 55°, and for *B. subtilis* 50°, 45°C. These data show that both the temperature dependence of the rate limiting step and the thermostability of the active enzyme are different for the two reactions. In part this may be due to the fact that the two reactions are studied at different pH values, 8.5 for the oxidation and 6.0 for the reduction. The different temperature optima also show that the temperature dependence of the

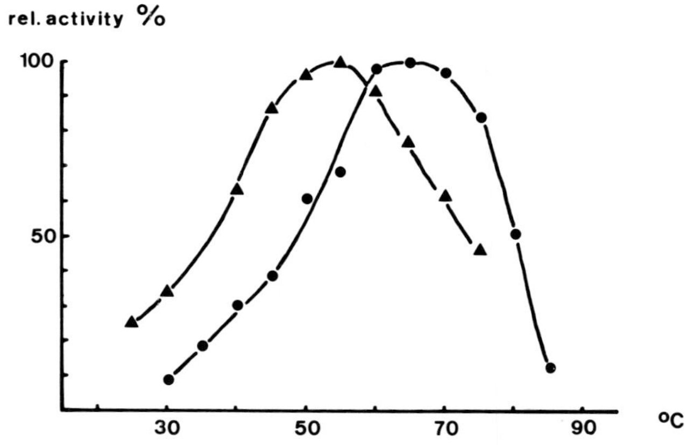

FIGURE 1. Temperature optima of LDH from B. stearothermophilus *grown at 55°C. Lactate oxidation (circles) carried out at pH 8.5; pyruvate reduction (triangles) carried out at pH 6.0.*

enzyme activity (enzyme efficiency) and the thermostability are related in an inverse manner: a high enzyme efficiency at low temperature occurs in enzymes which have a low thermostability at high temperatures and vice versa. For thermophilic LDH we have also found that the enzyme molecule itself is more thermostable than the enzyme-coenzyme-substrate complex, e.g., the LDH from *B. stearothermophilus* alone retains 100% activity after being heated for 30 min. at 85°C whereas the complex shows a maximum of 60°C for pyruvate reduction and 65°C for lactate oxidation. For mesophilic LDH the differences are less pronounced, 55°C for the enzyme and 45-50°C for the complex.

From analysis of the ascending part of the temperature dependence curves we determined the activation parameters ΔH*, ΔG* and ΔS* for reactions catalyzed by LDH and AlaDH from mesophilic and thermophilic bacilli. We have also compared these values with those obtained by Low *et al.* (17) for LDH from higher organisms, which included both cold-adapted (tuna, halibut) and warm-adapted (rabbit) species. In comparing the catalytic efficiency of the different enzymes one should pay special attention to the value of ΔG*. Since this parameter represents the height of the energy barrier over which the reactants must pass to become products a lower value of ΔG* may be equated with a more efficient enzyme. As expected, ΔG* is lowest for the LDH from tuna and halibut and highest for the LDH from thermophilic bacilli (Table 5). This appears to indicate that these enzymes are adapted in their efficiency to the organism's environmental temperature.

The free energy of activation is of course a combination of the enthalpy and entropy of activation. It was therefore of interest to determine the contributions made by these two latter activation parameters to ΔG* for the reaction catalyzed by the different lactate dehydrogenases. The ΔH* values, which represent heat exchange with the internal energy of the system, were calculated from the Arrhenius activation energy and the ΔS* values were calculated from absolute reaction rate theory (18). We found for both mesophilic and thermophilic enzymes that the values of ΔH* and ΔS* were lower than those for rabbit LDH but very similar to those for halibut and tune LDH (Table 5). Such a lowered value of ΔH* makes the enzyme-catalyzed reaction less temperature dependent, which may be of selective value to organisms which live over a relatively broad temperature range (poikilotherms). Somero and Low (19) also showed that the ΔH* and ΔS* values for LDH from tuna, halibut, rabbit and chicken ovary in such a way that a change in ΔH* of 342 calories/mole is associated with a change in ΔS* of 1 entropy unit. Our data for thermophilic bacterial LDH below 45°C and for mesophilic bacterial LDH are in agreement with this relationship. With

TABLE 5
LDH: ΔH^*-, ΔG^*-Values

	Organisms	Test Temp. (°C)	V_{max} a)	E_a (cal/mole)	ΔH^* (cal/mole)	ΔS^* (e.u.)	ΔG^* (cal/mole)	Break Arrh. Plot	Thermostab. (30 min)
Homoiotherms	Rabbit **)	5	$0,975.10^2$	13'100	12'550	- 2.3	13'200		
		35	$1,08.10^3$	13'100	12'500	- 2.5	13'250		
Poikilotherms	Tuna b)	5	$8,6.10^2$	9'350	8'800	-13.3	12'500		
		35	$4,5.10^3$	9'350	8'750	-13.5	12'900		
	Halibut b)	5	$4,1.10^2$	9'300	8'750	-13.5	12'500		
		35	$2,1.10^3$	9'300	8'650	-13.7	12'900		
	B.stearothermophilus (55°) c)	35	$6,75.10^2$	7'800	7'200	-20.6	13'500	45°	85°
		55	$1,50.10^3$	1'850	1'200	-38.6	13'850		
	B.caldotenax	35	$6,8.10^2$	9'500	8'900	-14.9	13'500	45°	70°
		55	$1,5.10^3$	2'600	1'950	-36.5	13'900		
	B. X_1 (37°) c)	35	$1.6.10^3$	10'100	9'500	-11.3	13'000	none	55°
	B.subtilis	35	$1.5.10^3$	7'300	6'700	-19.4	12'700	none	50°

a) V_{max} = μMol substrate (pyruvatereduced) per min. per mg enzyme
b) P.S. Low et al., Proc.Nat.Acad.Sci. USA, 70, 430 (1973)
c) cultivation temperature

this variation the different enzymes can adapt their activation parameters to different temperatures. These authors suggest that the observed differences in activation parameters among the LDH homologues result from a single type of activation process taking place to different extents in the various enzymes. The possible types of molecular processes which could produce such a covariance are weak non-covalent interactions occurring during the activation step within the enzyme, the enzyme-substrate complex or between the enzyme and the solvent (20).

The temperature dependence of V_{max} for the pyruvate reduction catalyzed by the thermophilic lactate dehydrogenases is different above and below 45°C. This results in a non-linear Arrhenius plot and gives much lower values of ΔH^* and more negative values of ΔS^* above 45°C than below (Table 5). When considering these data it is important to remember that the thermophilic LDH normally functions above 45°C in the bacillus and that the enzyme-coenzyme-substrate complex is stable up to 60°C.

The altered activation parameters indicate the existence of special catalytic conditions at high temperatures. The lower heat uptake and greater entropy change during the activation step could indicate that during the formation of the enzyme-coenzyme (binary) complex the LDH molecule has taken up more heat energy and become more ordered. The temperature dependence of the fluorescence of the binary complex shows a break at 40°C demonstrating that a conformational change occurs in this complex above 40°C. This conformational change may be the basis not only for changes in the binding parameters of the binary complex but also for changes in the activation parameters. Determination of these binding parameters, important for the understanding of the enzyme-coenzyme binding energetics and of the activation energetics, is currently in progress in our laboratory.

The activation parameters of bacterial LDH are taken from the overall enzyme-catalyzed reaction and represent the parameters of the rate-limiting step. As for higher organisms (21), this is most likely the isomerization step for pyruvate reduction at pH 6 and the NADH dissociation step for lactate oxidation at pH 8.5. In this context, it is not surprising that a break in the Arrhenius plot is only observed for the pyruvate reduction and not for the lactate oxidation. The involvement of a conformational change in the rate-determining isomerization step should make this reaction more temperature sensitive. This may also explain the lower temperature optimum of the pyruvate reduction as compared to the temperature optimum of the lactate oxidation (Fig. 1). Mesophilic bacterial lactate dehydrogenases do not show these conformational changes at 45°C, since at this temperature denaturation of the active site has already occurred. No break in the Arrhenius plot is observed between 20 and 40°C.

Similar results were obtained with thermophilic AlaDH. Conformational changes at higher temperatures seem to be a common property of thermophilic enzymes. Thus, these enzymes may be especially suitable for investigations of the energetics of enzyme catalysis, both at a general level and at the level of the fast rate-limiting steps.

The thermophilic enzymes are stabilized at higher temperatures by, for example, ionic bonds in cooperation with other non-covalent interactions on the enzyme surface. Under these conditions the enzyme-coenzyme-substrate complex in the active site of thermophilic LDH is kept active up to 60°C by the more rigid enzyme molecule. However, conformational changes of important parts of the active site which occur at higher temperatures and which change drastically the binding and activation parameters are discernible.

Activation parameters should also have significance for the denaturation process. An enzyme molecule is a delicately balanced macromolecule possessing a relatively small negative free energy of stabilization, in the case of mesophilic proteins of the order of -5 to -10 kcal/mole. The molecule possesses the necessary flexibility for catalysis, but is also rather sensitive to thermal effects. The free energy of stabilization is a combination of the stabilizing enthalpy terms from non-covalent interactions, the stabilizing entropy caused by removal of hydrophobic groups from the water, and the destabilizing entropy term associated with the molecule's desire to become a random structure. At higher temperatures this last term dominates and denaturation occurs.

At a given temperature thermophilic enzymes should have a more negative free energy of stabilization than mesophilic enzymes. This of course must be due to the presence of additional stabilizing interactions. To date, only a few comparative studies of the activation parameters of the reversible and irreversible denaturation of thermophilic and mesophilic enzymes have been made (22). It is necessary to investigate more examples in order to better understand the denaturation kinetics and thermodynamics of thermophilic enzymes. This is especially important for considerations of the structural basis of thermostability and increased enzyme rigidity. Increased rigidity (lower flexibility) of the thermophilic enzyme molecule and especially of the active site is thought to result in a reduced enzyme efficiency.

This last point should be of special importance with regard to enzyme surface. Here there exists a delicately balance equilibrium between hydrophilic and hydrophobic amino acid side chains which may be significantly influenced by stabilizing or destabilizing ionic bonds. In the case of thermophilic enzymes a "clamping effect" of such special stabilizing bonds should keep the enzyme surface in order up to a higher temperature. However, this should also affect the active site region and lower the flexibility important for enzyme efficiency. This should be most evident at lower temperatures since at higher temperatures additional conformational changes could increase the local flexibility and thereby the catalytic efficiency. This picture is in agreement with our results.

Such "clamping effects," which influence the thermostability, flexibility, efficiency and state of the surface of the enzyme have been found for the aminopeptidase I (AP I) from *B. stearothermophilus* (23). In this case, however, the effect is not caused by ionic bonds between basic and acidic amino acid residues but by chelates between extrinsic metal cations (Zn^{2+}, Co^{2+}) and acidic amino acid residues. The AP I from *B. stearothermophilus* consists of 12 subunits and 2 metal cations can be bound by each subunit. One ion binds at the active site and the other probably serves to provide additional thermostabilization. The more flexible Co-aminopeptidase is 20 times more active than the more rigid Zn-enzyme, but is less thermostable. After heating for 1 hour at 80°C without metal ions in the incubation mixture, the Zn-enzyme retains full activity while the Co-enzyme loses 35% of its activity. We also tested the surface hydrophobicity of AP 1 with the fluorescent hydrophobic marker 8-anilinonaphtalene-1-sulfonic acid. The fluorescent yield, a measure of the hydrophobicity, was greatest for the inactive apoenzyme, with the Zn-enzyme next, and was least for the fully active Co-enzyme. Hachimori and Nosoh (24) have also used this marker to monitor the effect of different temperatures on the conformation of *B. stearothermophilus* ATPase. Such fluorescence measurements as well as modification reactions and limited proteolysis seem to be suitable as probes to compare the surface of thermophilic and mesophilic enzymes.

Finally, one of our future objectives is to correlate the data on the properties of thermophilic LDH, AlaDH and neutral proteinase with the structural data. This should provide a better understanding both of the structure-function relationships of these enzymes and of how enzymes adapt to high temperatures. Enzyme molecules are formed during evolution by optimization of both their protein structure and biological function not only on the basis of their cellular function but also of environmental factors such as temperature. Thermophilic and mesophilic enzyme structures are selected at high and low temperatures, respectively, from a physiological viewpoint. They should possess at each normal cellular temperature the optimal catalytic efficiency and structural stability. Hopefully, from amino acid sequence data as well as three-dimensional structures some of these particular structural details will become visible and shed some light on the evolution of thermophilic enzymes.

In conclusion, taking into account all the results and considerations on the structure and properties of thermophilic enzymes, I can, in spite of the many problems to be solved, give a positive answer to the major question asked at the beginning. Thermophilic enzymes are worth our further effort.

ACKNOWLEDGEMENT

These investigations were supported by the Swiss National Science Foundation, project No. 3.005-0.76. I wish to express my thanks to Dr. D. M. Pederson for valuable discussions.

REFERENCES

1. Singleton, R. and Amelunxen, R. (1973). Bacteriol. Rev. *37*, 320.
2. Zuber, H. (ed.) (1976). *Enzymes and Proteins from Thermophilic Microorganisms*, Experentia Suppl. 26, Birkhauser, Basel.
3. Singleton, R., Middaugh, C. R., and MacElroy, R. D. (1977). Int. J. Peptide Protein Res. *10*, 51.
4. Devanathan, T., Akagi, J. M., Hersh, R. T., and Himes, R. H. (1969). J. Biol. Chem. *244*, 2846.
5. Perutz, M. F. and Raidt, H. (1975). Nature *255*, 256.
6. Hase, T., Ohmiya, N., Matsubara, M., Mullinger, R. N., Rao, K. K., and Hall, D. O. (1976). Biochem. J. *159*, 55.
7. Dayhoff, M. O. (Ed.) (1976). *Atlas of Protein Sequence and Structure*, Vol. 5, Suppl. 2, Nat. Biomed. Res. Found., Washington.
8. Ljungdahl, L. G., Sherod, D. W., Moore, M. R., and Andreesen, J. R. (1976). In *Enzymes and Proteins from Thermophilic Microorganisms*, (H. Zuber, ed.), p. 237, Birkhäuser, Basel.
9. Frank, G., Haberstich, H. U., Schär, H. P., Tratschin, J. D., and Zuber, H. (1976). In *Enzymes and Proteins from Thermophilic Microorganisms*, (H. Zuber, ed.), p. 375, Birkhäuser, Basel.

10. Shing, Y. W., Akagi, J. M., and Himes, R. (1975). J. Bacteriol. *122*, 177.
11. O'Brien, W. E., Brewer, J. M., and Ljungdahl, L. G. (1976). In *Enzymes and Proteins from Thermophilic Microorganisms* (H. Zuber, ed.), p. 249, Birkhäuser, Basel.
12. Fujita, S. C., Oshima, T., and Imahori, K. (1976). Eur. J. Biochem. *64*, 57.
13. Ljungdahl, L. G. and Sherod, D. (1976). In *Extreme Environments: Mechanisms of Microbial Adaptation*, (M. R. Heinrich, ed.), p. 147, Academic Press, New York.
14. Coleman, P. M., Jansonius, J. N., and Matthews, B. W. (1972). J. Mol. Biol. *70*, 701.
15. Biesecker, G., Harris, J. I., Thierry, J. C., Walker, J. E., and Wonacott, A. J. (1977). Nature *266*, 328.
16. Allen, M. B. (1953). Bacteriol Rev. *17*, 125.
17. Low, P. S., Bada, J. L., and Somero, G. N. (1973). Proc. Natl. Acad. Sci. USA *70*, 430.
18. Johnson, F. H., Eyring, H., and Polissar, M. Jr. (1954). In *The Kinetic Basis of Molecular Biology*, John Wiley & Sons, Inc., New York.
19. Low, P. S. and Somero, G. (1974). Comp. Biochem. Physiol. *49B*, 307.
20. Somero, G. N. and Low, P. S. (1977). Nature *266*, 276.
21. Holbrook, J. J., Liljas, A., Steindel, St. J. and Rossmann, M. J. (1975). In *The Enzymes*, Vol. XI, p. 191, Academic Press, New York.
22. Fujita, S. C. and Imahori, K. (1974). In *Peptides, Polypeptides and Proteins*, (E. R. Blout, F. A. Bovey, M. Goodman and N. Lotan, eds.), p. 217, John Wiley & Sons, Inc., New York.
23. Roncari, G., Zuber, H., and Wyttenbach, A. (1972). Int. J. Peptide Protein Res. *4*, 267.
Deranleau, D. A. and Zuber, H. (1977). Int. J. Peptide Protein Res. *9*, 258.
24. Hachimori, A. and Nosoh, Y. (1973). Biochem. Biophys. Acta *315*, 481.

PURIFICATION AND PROPERTIES OF MALATE DEHYDROGENASE AND ISOCITRATE DEHYDROGENASE FROM AN EXTREME THERMOPHILE, *THERMUS FLAVUS* AT-62

Takashi Saiki
Shinji Iijima
Kazuo Tohda
Teruhiko Beppu
Kei Arima

Department of Agricultural Chemistry
University of Tokyo
Yayoi, Bunkyoku, Tokyo

We have been interested in the distribution and taxonomic characteristics of extremely thermophilic bacteria found in Japanese hot springs belonging to the genus *Thermus*. The properties of different enzymes from *Thermus flavus* AT-62, one of these thermophilic strains, have been investigated to elucidate the molecular mechanism of their thermostability (1-6). NAD^+-specific malate dehydrogenase (EC 1.1.1.37) (MDH) and $NADP^+$-linked isocitrate dehydrogenase (EC 1.1.1.42) (IDH) from *T. flavus* AT-62 are thermostable dimeric enzymes and simple continuous measurement of their enzyme activities are possible even at elevated temperatures. Therefore, among oligomeric enzymes, MDH and IDH are good choices as simple model enzymes for the study of kinetics of folding and unfolding, dissociation and reassociation of subunits, and inactivation and reactivation. A brief description of *T. flavus* AT-62, purification of the two enzymes by affinity chromatography, and thermal and some other properties of the enzymes will be presented here.

ISBN 0-12-268250-5

EXTREMELY THERMOPHILIC STRAIN *THERMUS FLAVUS* AT-62

Since Japan is a volcanic country, many hot springs are distributed throughout the land (Fig. 1a). The hot springs which we examined are indicated as solid and open circles in Fig. 1b. Solid circles in Fig. 1b are those hot springs from which we could isolate extremely thermophilic bacteria belonging to genus *Thermus*. One of the isolates, *Thermus flavus* AT-62, is a Gram-negative, nonsporulating, non-motile, yellow pigmented rod, with a temperature range for growth of 40-80°C (1). Oshima and Imahori (7, 8) isolated *T. thermophilus* (7, 8), whereas Brock and Freeze (9) isolated *T. aquaticus*. In describing the genus *Thermus* in *Bergey's Manual of Determinative Bacteriology* (8th ed.), Brock states that "detailed taxonomic studies of the genus have not yet been made, and more than one species may ultimately be recognized" (10). In a comparative study of *Thermus flavus* and *Thermus aquaticus*, we pointed out that the two strains differ in (G+C) content of DNA and in physiological characteristics, such as the ability to assimilate ammonium salts, liquefy gelatin and coagulate milk (6). *T. thermophilus* is distinguished from

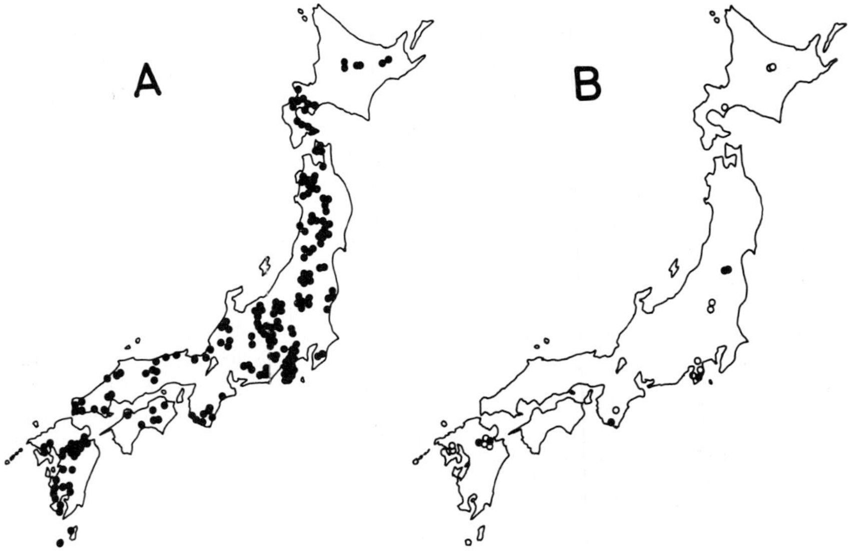

FIGURE 1. A. Distribution of hot springs in Japan. B. Hot springs from which extreme thermophiles were isolated (solid circles).

other *Thermus* species particularly in its higher temperature range for growth (7, 8). We also isolated a *Thermus* species which resembled *Thermus* X (11) in sharing the lack of yellow pigment (T. Saiki *et al.*, unpublished). Further taxonomic data on different types of *Thermus* strains must be accumulated for resolving the problems concerning the classification of species included in *Thermus* and the relationship of this genus to other genera.

PURIFICATION OF MALATE DEHYDROGENASE AND ISOCITRATE DEHYDROGENASE BY BLUE DEXTRAN-SEPHAROSE COLUMN

We already reported on the purification of IDH from *T. flavus* AT-62 by rather laborious procedures, including ion exchange chromatography and Ampholine electrofocusing (4). The introduction of blue dextran affinity chromatography enabled us to purify MDH and IDH from the organism simultaneously and to improve the purification procedures in general. The chromophore of blue dextran (Cibacron blue F3GA) is known to be a specific ligand for these enzymes possessing a dinucleotide fold structure and was demonstrated to be useful for the purification of a wide range of proteins including dehydrogenases (12).

MDH and IDH were partially purified from cell extracts of *T. flavus* AT-62 employing the conventional procedures of DEAE-cellulose and DEAE-Sephadex A-50 column chromatography (Table 1). Both enzymes were further purified with an affinity column of blue dextran-Sepharose which was prepared according to the procedure of Ryan and Vestling (13). The MDH activity was eluted with a linear gradient of NADH from 0 to 0.3 mM on first and second affinity column chromatography. On the third affinity column, MDH was eluted with a buffer solution containing 0.2 M KCl after washing the column with a 1 mM ATP-10mM $MgCl_2$ solution. The IDH, on the other hand, was purified through the affinity columns, first by ATP (0 to 0.3 mM) gradient elution and then by DL-isocitrate (0 to 2 mM) gradient elution. The MDH and IDH were finally purified 220 fold and 370 fold, respectively (Tables 2 and 3).

EFFECTS OF VARIOUS LIGANDS ON ELUTION OF MDH AND IDH FROM THE AFFINITY COLUMN

Various ligands of the dehydrogenases were tested for their ability to elute these enzyme activities from the affinity column. Aliquots (0.5 ml) of partially purified

TABLE 1
Partial Purification of Malate Dehydrogenase and Isocitrate Dehydrogenase from T. flavus *AT-62*

Fraction	Protein mg	Malate dehydrogenase			Isocitrate dehydrogenase		
		Total units	Units per mg	Yield %	Total units	Units per mg	Yield %
Crude extract	99,000	1,120 x 10^3	11.3	100	77,000	0.78	100
1st DEAE-cellulose	61,000	844 x 10^3	13.8	75	60,500	1.0	79
2nd DEAE-cellulose	7,200	651 x 10^3	90.4	58	33,600	4.7	44
DEAE-Sephadex A-50	1,020	410 x 10^3	406.0	37	21,600	21.4	28

TABLE 2
Purification of Malate Dehydrogenase by Blue Dextran-Sepharose Affinity Chromatography

Fraction	*Protein mg*	*Total units*	*Units per mg*	*Yield %*
DEAE-Sephadex A-50	168	68,000	406	37 (100)
1st blue dextran-Sepharose (elution with NADH)	32	57,000	1,810	31 (83)
2nd blue dextran-Sepharose (elution with NADH)	22	53,000	2,400	29 (78)
3rd blue dextran-Sepharose (elution with KCl)	18	46,000	2,520	25 (68)

TABLE 3
Purification of Isocitrate Dehydrogenase from T. flavus *by Blue Dextran-Sepharose Affinity Chromatography*

Fraction	*Protein mg*	*Total units*	*Units per mg*	*Yield %*
DEAE-Sephadex A-50	168.0	3,600	21.4	28 (100)
1st blue dextran-Sepharose	48.0	3,250	68.0	25 (90)
2nd blue dextran-Sepharose (elution with ATP)	10.7	2,770	259.0	22 (77)
3rd blue dextran-Sepharose (elution with isocitrate)	3.4	980	288.0	7.6 (27)

preparations having both enzyme activities were applied to the affinity column (1 x 5 cm) and 5 ml of ligand solutions were used for the elution test. As the results are given in Table 4, the IDH was eluted from the affinity column with 0.2 mM ATP, 5 mM $NADP^+$, or 5 mM DL-isocitrate with a recovery of 80-100%. NAD^+ or NADH at 1 mM were also effective but resulted in low recovery (20 to 30%). On the other hand, only NADH at 0.1-0.2 mM, among the ligands tested, could elute MDH. One of the examples of affinity column chromatography is shown in Fig. 2. IDH was eluted from the affinity column either by $NADP^+$ or isocitrate, while MDH could be eluted by NADH but not by oxalacetate. This may reflect a probable difference in their reaction mechanisms, since a rapid equilibrium random type mechanism was proposed for IDH (14, 15), and a compulsory ordered mechanism was proposed for MDH (16).

TABLE 4
Effects of Various Compounds on Elution of Malate Dehydrogenase and Isocitrate Dehydrogenase of T. flavus *from Blue Dextran-Sepharose Affinity Columns*

Dehydrogenase	*Compound*	*Concn. (mM)*	*Elution*
Malate dehydrogenase	$NADP^+$	10	-
	NADH	0.2	+
	Oxalacetate	10	-
	NAD^+	1	±
	KCl	150	+
Isocitrate dehydrogenase	$NADP^+$	5	+
	NAD^+	1	+
	DL-isocitrate	5	+
	ATP	0.2	+
	EDTA	10	-
	KCl	110	+

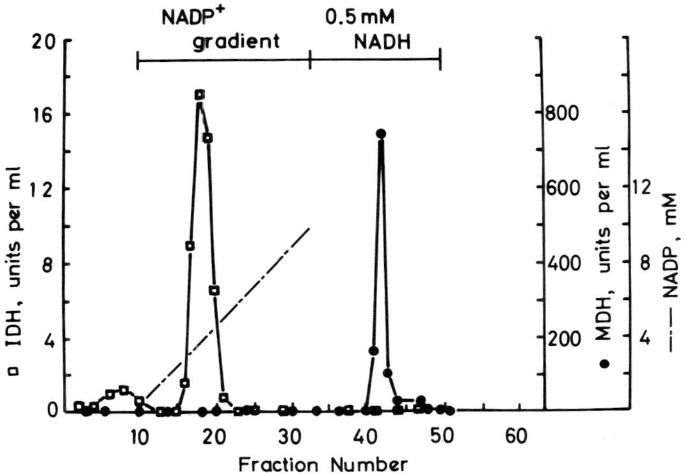

FIGURE 2. Elution pattern of MDH and IDH on blue dextran-Sepharose column. 0.5 ml of a partially purified preparation having MDH and IDH activities (see Table 1) was applied to the blue dextran-Sepharose column (1 x 5 cm) in the presence of 10 mM Tris-HCl buffer (pH 7.5). IDH activity was eluted with a linear gradient of $NADP^+$ from 0 to 10 mM, while MDH activity was eluted with 0.5 mM NADH.

EFFECTS OF CIBACRON BLUE F3GA ON DEHYDROGENASES

According to Thompson *et al.* (17), Cibracron blue F3GA is a potent competitive inhibitor (K_i of 10^{-6} to 10^{-7}) for NADH, the substrate of lactate dehydrogenase (Table 5). The effect of Cibacron blue F3GA on the MDH from *T. flavus* and the IDHs from *T. flavus* and *Escherichia coli* were examined at 30°C. The IDH of *E. coli* B was purified 10-fold with a DEAE-cellulose column using a buffer system containing 20% glycerol (18). Kinetic measurements of the initial rates for the three dehydrogenases in the presence and absence of Cibacron blue F3GA are shown in Fig. 3. The chromophore compound inhibited the IDH of *E. coli* and thermophile MDH in a competitive manner with K_i values of 4.0×10^{-6} M and 0.9×10^{-6} M, respectively, while the dye compound inhibited the thermophile IDH in a mixed manner with a K_i value of 1.1×10^{-4} M (Table 5). The results described above may indicate the presence of diverse dinucleotide fold structures. The thermophile IDH often failed to adsorb to the affinity column

TABLE 5
Inhibition of Various Dehydrogenases by Blue Chromophore

Dehydrogenase	K_i *(M)*	*Type of inhibition*
Malate dehydrogenase (*T. flavus*)	0.9×10^{-6}	Competitive
Isocitrate dehydrogenase (*E. coli*)	4.0×10^{-6}	Competitive
Isocitrate dehydrogenase (*T. flavus*)	1.1×10^{-4}	Mixed
*M_4 lactate dehydrogenase (rabbit muscle)	0.1×10^{-6}	Competitive
*H_4 lactate dehydrogenase (beef heart)	2.1×10^{-6}	Competitive

**Cited from reference (17).*

in the presence of excess MDH, whereas MDH activity could be adsorbed tightly. This finding may be ascribed to the lower affinity of IDH for the blue dye.

THERMAL AND SOME OTHER PROPERTIES OF PURIFIED DEHYDROGENASES

Final preparations of the dehydrogenases were proven to be homogeneous by polyacrylamide disc gel electrophoresis and SDS-polyacrylamide gel electrophoresis (Fig. 4). The MDH and IDH were shown by SDS-polyacrylamide gel electrophoresis to be dimers with molecular weights of 72,000 and 98,000, respectively. This MDH is considered to belong to the low molecular weight subgroup of MDHs (19) consisting of animal and several bacterial MDHs, including that of *T. aquaticus* (20). MDHs isolated from bacilli, on the other hand, have relatively high molecular weights (120,000 daltons) (21).

The isoelectric points were confirmed by electrofocusing to be pH 5.2 and 5.6 for MDH and IDH, respectively. The isoelectric point of the MDH is much higher than that of *T. aquaticus* MDH (pH 4.4) (20).

The amino acid composition of the *T. flavus* MDH (Tables

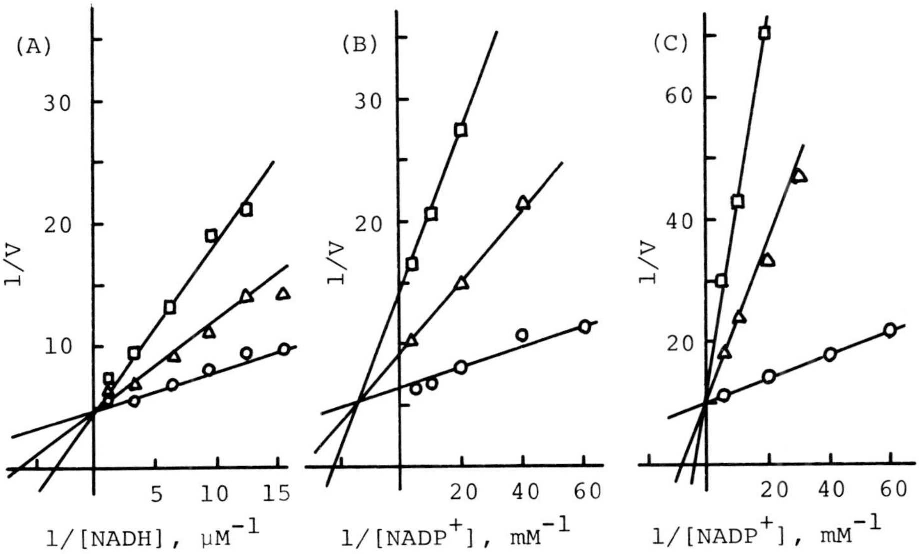

FIGURE 3. Double-reciprocal plots of the catalytic activity of dehydrogenases in the absence and the presence of Cibacron blue F3GA. The initial rates of catalysis, V, have the units ΔA_{340}/min. All measurements were made at 30°C. (A) T. flavus *MDH. The concentrations of blue chromophore were: o, none; Δ, 1.7 μM; □, 3.4 μM. (B)* T. flavus *IDH. The concentrations of blue chromophore were: o, none; Δ, 100 μM; □, 200 μM. (C)* E. coli *IDH. The concentrations of blue chromophore were: o, none; Δ, 20 μM; □, 100 μM.*

6 and 7) was not unusual, compared with the MDHs from other thermophilic and mesophilic organisms, in any of three parameters: frequency of non-polar side chain residues (22), polar ratio (23), and average hydrophobicity ($H\phi_{ave}$) (24) (Table 8). Numerous attempts have been made to relate protein structure to thermal stability by comparing amino acid compositions of mesophile and thermophile proteins, but this approach has not been fruitful. It is to be noted, however, that the thermophile enzymes, D-glyceraldehyde-3-phosphate dehydrogenases and phosphofructokinases, tend to have fewer cysteine residues than their mesophilic counterparts (25). It was suggested that the non-essential cysteine residues might have been replaced in enzymes from thermophiles so as to prevent oxidative inactivation of the enzymes under aerobic conditions at elevated temperatures. As to MDHs, there are 20 cysteines in the chicken heart mitochondrial enzyme (21), 11.4 in chicken heart supernatant (21), 7.9 in *Neurospora crassa* (21), 8.3 in *E. coli* (21), 6.8 in *T. aquaticus*

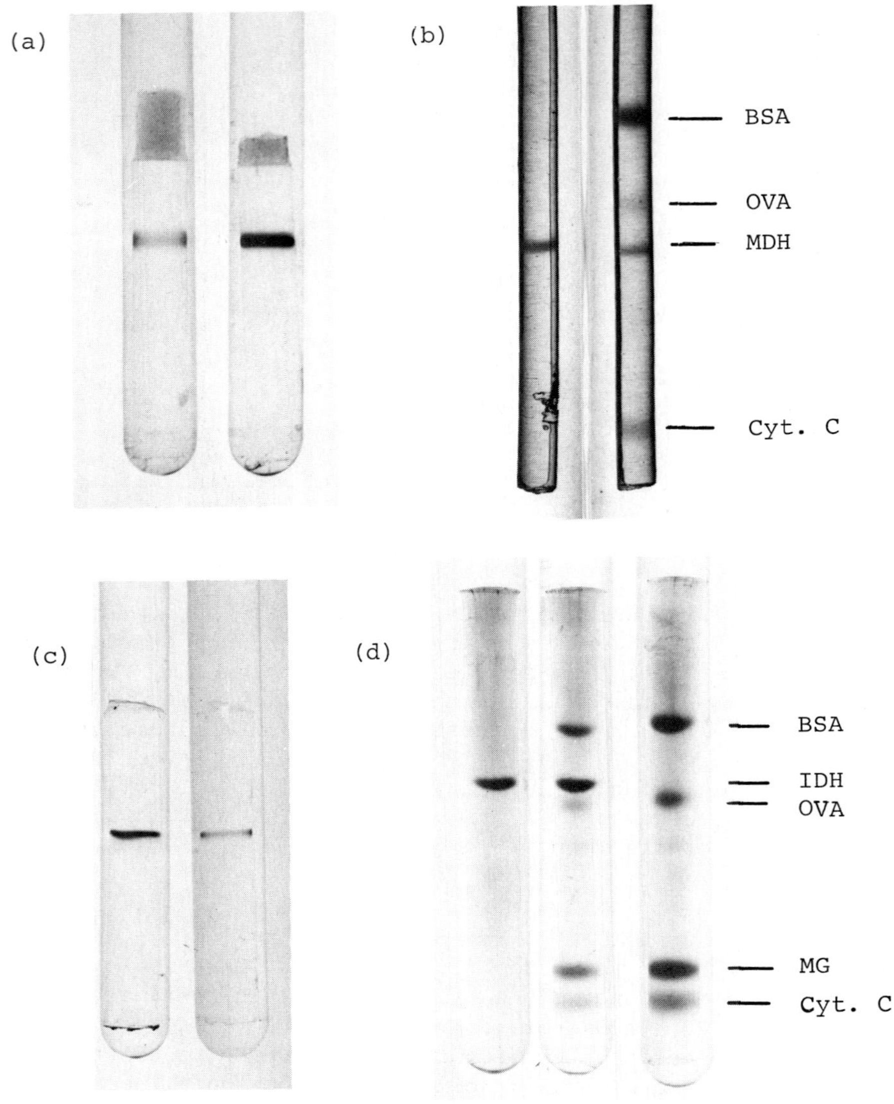

FIGURE 4. Homogeneity of MDH and IDH. Patterns of polyacrylamide disc gel electrophoresis at pH 9.4 of MDH (a, left: MDH activity specific staining; right: staining with Amido black) and IDH (c, left: staining with Amido black; right: IDH activity specific staining). Patterns of SDS-polyacrylamide gel electrophoresis (7.5% acrylamide containing 1% SDS) of MDH (b) and IDH (d) with standard proteins. BSA: bovine serum albumin, MG: myoglobin, OVA: ovalbumin and cyt. C: cytochrome C.

TABLE 6
Amino Acid Composition of T. flavus *MDH*

Amino acid	*Residues/mol*	*Amino acid*	*Residues/mol*
Asx	50.1	Val	45.2
Glx	60.9	Ile	27.5
Thr	24.0	Leu	55.9
Ser	14.7	His	6.8
Tyr	12.1	Pro	28.7
Phe	18.0	Met	20.3
Lys	32.3	Cys	4.0
Arg	35.1	Trp	13.0
Ala	88.8	Gly	49.2

(20) and 4.0 in *T. flavus*. Therefore, there is also a positive correlation between thermal stability of MDHs and a decreased content of cysteine residues. It is noteworthy, however, that MDH from the mesophile, *Bacillus subtilis*, contains no cysteine and is not inhibited by iodoacetate or p-hydroxymercuribenzoate (21). We must be cautious in arriving at a conclusion to this problem because (1) cysteine residues as a potential source of instability are closely related to their location in the 3-D structure and (2) thermophilic and mesophilic enzymes of different phylogenetic origins are compared with each other.

As with most enzymes from thermophiles so far studied, the MDH and IDH of *T. flavus* AT-62 showed remarkable heat stability and required no stabilizing factors. Both enzymes were not inactivated after 60 min at 70°C and the activities declined slowly at 80°C (Fig. 5). It should be noted that the homoserine dehydrogenase of the same organism is specifically stabilized against heat inactivation by monovalent metal ions, such as Na^+ or K^+ (2).

Using the melting-profile method developed by Fujita and Imahori (26), the melting temperature of MDH was observed to be 95°C, which is nearly 10°C higher than that of glyceralde-

TABLE 7
Amino Acid Composition of Various MDHs

Amino acid(s)	Organism			
	T. flavus	T. aquaticus*	E. coli**	*Pig mitochondria***
Asx Glx	18.9	20.4	18.5	14.3
Thr Ser	6.6	8.6	10.8	12.5
Tyr Phe	5.1	3.7	5.5	4.8
Lys Arg	11.5	11.7	9.2	10.6
Ala Val	22.8	36.9	37.0	35.4
Ile Leu	14.2			
His Pro	6.1	6.7	5.3	9.6
Met Cys	4.1	1.6	2.4	3.7
Gly	8.4	10.0	10.7	9.1
Trp	2.2	0.4	0.6	0.0

*Cited from reference (20).
**Cited from reference (21).

TABLE 8
*Calculated Values of Polarity for Malate Dehydrogenases from Various Sources**

MDH from	*Nonpolar side chains*	*Polarity ratio*	*Average hydrophobicity (Kcal/mol)*
T. flavus	0.35	0.74	1.17
T. aquaticus	0.328	0.828	1.029
B. subtilis	0.346	1.045	1.021
E. coli	0.370	0.719	1.128
N. crassa	0.320	1.01	1.050
Pig mito.	0.370	0.73	1.210
Pig sol.	0.360	0.87	1.183
Chick mito.	0.367	0.824	1.133
Chick sol.		0.834	1.185

Data on MDHs other than that of* **T. flavus are from reference (20).

hyde-3-phosphate dehydrogenase from *T. thermophilus* (26). At 100°C, decrease of the CD signal at 220 nm stopped at 50% of the original value (Fig. 6). Fig. 7 shows the comparison between the CD spectra at room temperature and at 100°C. After heat treatment by the melting profile procedure, the double minimums at 208 and 223 nm of the native enzyme disappeared and the broad negative band centered at 218-220 nm was observed. This structure was so stable as to be maintained for at least 20 min, suggesting the presence of an intermediate of heat denaturation.

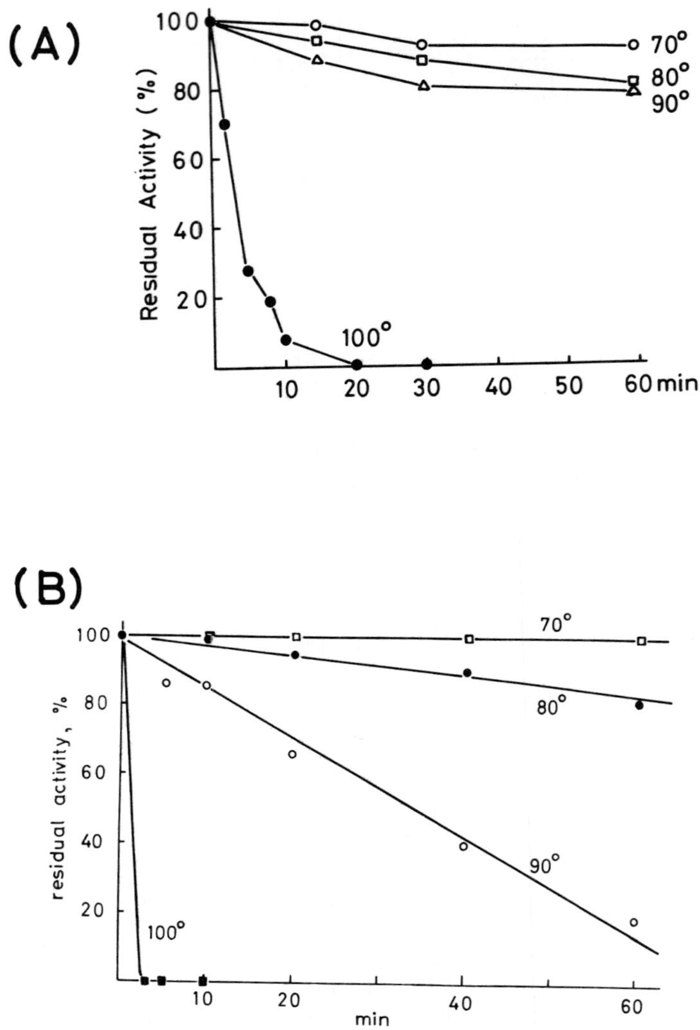

FIGURE 5. Thermostability of MDH (A) and IDH (B) of T. flavus. *The enzyme solutions in 20 mM potassium phosphate buffer (pH 7.0) were incubated at various temperatures and the residual activities were measured at the indicated intervals. Data for IDH (B) taken from reference (3) and reproduced with permission from The Japanese Biochemical Society.*

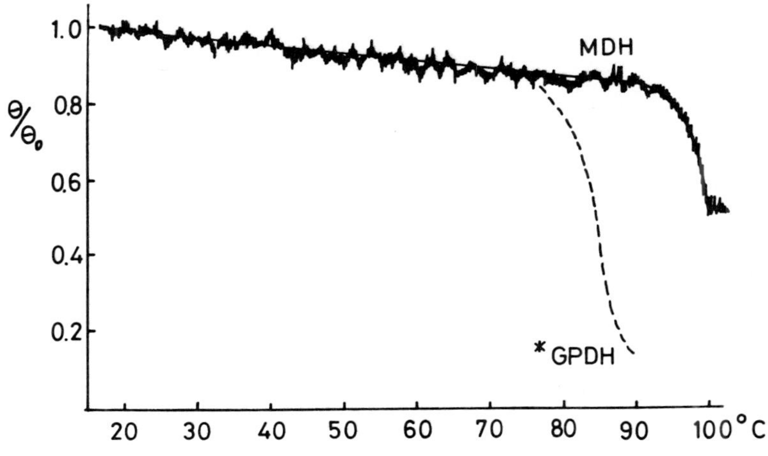

FIGURE 6. Melting-profile of T. flavus *MDH measured by CD spectroscopy.* θ/θ_o *indicates the fractional signal intensity (the intensity at any temperature divided by the intensity at room temperature). The heating rate was 1°C per min. Protein concentration was 0.13 mg/ml in 20 mM* K_2HPO_4*-*KH_2PO_4 *buffer (pH 7.0). *GPDH (glyceraldehyde-3-phosphate dehydrogenase) from* T. thermophilus. *Taken from reference (26) and reproduced with permission from John Wiley & Sons.*

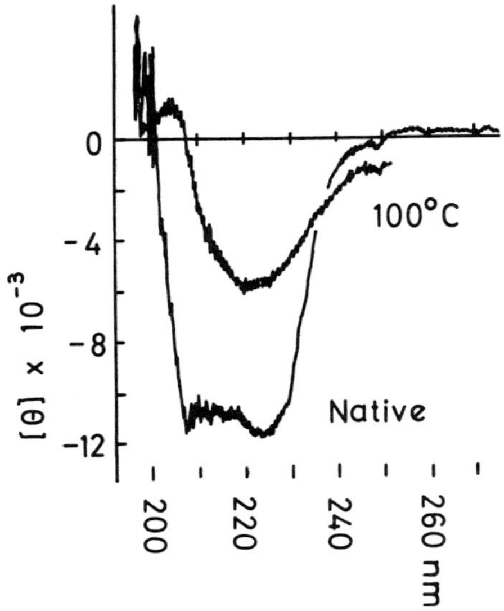

FIGURE 7. CD spectra of native MDH and the heat-denatured MDH of T. flavus. *Protein concentration was 0.13 mg/ml in 20 mM* K_2HPO_4*-*KH_2PO_4 *buffer (pH 7.0).*

SUMMARY

NAD^+-specific malate dehydrogenase and $NADP^+$-specific isocitrate dehydrogenase were purified to homogeneity from an extreme thermophile, *Thermus flavus*, by blue dextran-Sepharose affinity chromatography. The IDH was effectively eluted from the affinity column with 0.2 mM ATP, 5 mM $NADP^+$, or 5 mM DL-isocitrate, while only NADH (0.1-0.2 mM) among the ligands tested could elute MDH from the column.

The chromophore compound of blue dextran, Cibacron blue F3GA, inhibited MDH and IDH in a competitive manner (K_i value of about 1×10^{-6} M) and a mixed manner (K_i value of about 1×10^{-4} M), respectively, suggesting the presence of diverse dinucleotide fold structures.

While the amino acid composition of MDH was not unusual compared with MDHs from other thermophilic and mesophilic organisms, a decreased content of cysteine residues was observed. This may indicate replacement of non-essential cysteine residues so as to prevent oxidative inactivation.

The MDH and IDH exhibited remarkable heat stability, retaining full activity after 60 min at 70°C. The melting temperature of the MDH was observed to be 95°C by CD spectroscopy using the melting-profile method. After heat treatment by the melting profile method, the double minimums at 208 and 223 nm of the native enzyme disappeared and the broad negative band centered at 218-220 nm was observed, indicating the presence of an intermediate of heat denaturation.

ACKNOWLEDGEMENTS

The authors wish to express their thanks to the Central Research Laboratories of Ajinomoto Co., Ltd. (Kawasaki) for large-scale cultivation of the organism.

REFERENCES

1. Saiki, T., and Arima, K. (1970). Agric. Biol. Chem. *34*, 1762.
2. Saiki, T., Shinshi, H. and Arima, K. (1973). J. Biochem. *74*, 1239.
3. Saiki, T. and Arima, K. (1975). J. Biochem. 77, 233.
4. Saiki, T., Mahmud, I., Matsubara, N., Taya, K. and Arima, K. (1976). In *Enzymes and Proteins from Thermophilic Microorganisms* (H. Zuber, ed.), p. 169,

Birkäuser Verlag, Basel.

5. Saiki, T., Yamazumi, K. and Arima, K. (1977). Agric. Biol. Chem. *44*, 1651.
6. Saiki, T., Kimura, R. and Arima, K. (1972). Agric. Biol. Chem. *36*, 2357.
7. Oshima, T. and Imahori, K. (1971). J. Gen. Appl. Microbiol. *17*, 513.
8. Oshima, T. and Imahori, K. (1974). Intern. J. Syst. Bacteriol. *24*, 102.
9. Brock, T. D. and Freeze, H. (1969). J. Bacteriol. *98*, 289.
10. Brock, T. D. (1974). In *Bergey's Manual of Determinative Bacteriology*, 8th ed. (R. E. Buchanan and N. E. Gibbons, co-eds.), p. 285, The Williams and Wilkins Co., Baltimore.
11. Ramaley, R. F. and Hixon, J. (1970). J. Bacteriol. *103*, 527.
12. Thompson, S. T., Cass, K. H. and Stellwagen, E. (1975). Proc. Natl. Acad. Sci. USA *72*, 669.
13. Ryan, L. D. and Vestling, C. S. (1974). Arch. Biochem. Biophys. *160*, 279.
14. Marr, J. J. and Weber, M. M. (1973). Arch. Biochem. Biophys. *158*, 782.
15. Wicken, J. S., Chung, A. E. and Franzen, J. S. (1972). Biochemistry *11*, 4766.
16. Telegdi, M., Wolfe, E. V. and Wolfe, R. G. (1973). J. Biol. Chem. *248*, 6484.
17. Thompson, S. T. and Stellwagen, E. (1976). Proc. Natl. Acad. Sci. USA *73*, 361.
18. Reeves, H. C., Daumy, G. O., Lin, C. C. and Houston, M. (1972). Biochim. Biophys. Acta *258*, 27.
19. Murphey, W. H., Kitto, G. B., Everse, J. and Kaplan, N. O. (1967). Biochemistry *6*, 603.
20. Biffen, B. H. F. and Williams, R. A. D. (1976). In *Enzymes and Proteins from Thermophilic Microorganisms* (H. Zuber, ed.), p. 157, Birkhäuser Verlag, Basel.
21. Murphey, W. H., Barnaby, C., Lin, F. J. and Kaplan, N. O. (1967), J. Biol. Chem. *242*, 1548.
22. Waugh, D. F. (1954). Adv. Protein Chem. *9*, 326.
23. Fisher, H. F. (1964). Proc. Natl. Acad. Sci. USA *51*, 1285.
24. Bigelow, C. C. (1967). J. Theoret. Biol. *16*, 187.
25. Hocking, J. D. and Harris, J. I. (1976). In *Enzymes and Proteins from Thermophilic Microorganisms* (H. Zuber,

26. ed.), p. 121, Birkhäuser Verlag, Basel.

26. Fujita, S. C. and Imahori, K. (1974). In *Peptides, Polypeptides and Proteins* (E. R. Blout, F. A. Bovey, M. Goodman and N. Lotan, eds.), p. 217, John Wiley & Sons, New York.

THERMODYNAMIC STUDIES ON REVERSIBLE DENATURATION OF THERMOSTABLE PROTEINS FROM AN EXTREME THERMOPHILE

Hiroshi Nojima
Atsushi Ikai
Haruhiko Noda

Department of Biophysics and Biochemistry
University of Tokyo
Hongo, Tokyo

Koyu Hon-nami
Tairo Oshima

Mitsubishi-Kasei Institute of Life Sciences
Machida, Tokyo

Information on the stability of proteins of thermophiles has been mostly restricted to remaining activity after heating or the rate of irreversible denaturation at certain temperatures. Only in some cases, kinetic parameters, such as $\Delta H^{\ddagger}$ and $\Delta S^{\ddagger}$, of denaturation have been reported. In order to estimate thermodynamic parameters, it is necessary to test proteins that denature reversibly.

The authors found that phosphoglycerate kinase (PGK) and cytochrome C (cyt. *c*) would serve as favorable models for thermodynamic studies of denaturation because the unfolding reaction of these proteins from thermophilic and mesophilic organisms was reversible under suitable conditions. These proteins consist of a single polypeptide chain, so that interpretation of the data obtained can be restricted to the intrapeptide interactions.

This paper describes thermodynamic properties of reversible denaturation of PGK or cyt. *c* from an extreme thermophile, *Thermus thermophilus*, in the presence of guanidine hydrochloride (GuHCl).

ISBN 0-12-268250-5

T. THERMOPHILUS PHOSPHOGLYCERATE KINASE

PGK was purified from an extract of *T. thermophilus* using DEAE and Sephadex G-150 column chromatographies. Roughly 30 mg of a recrystallized preparation were obtained from 500 g of wet cell paste (Nojima *et al.*, in preparation).

The thermophile PGK is similar to counterparts from the moderate thermophile, *Bacillus stearothermophilus* (Suzuki and Imahori, 1974) and mesophilic yeast or animals (Scopes, 1973) in respect to subunit structure, molecular weight, Km's for ATP and glycerol 3-phosphate, and pH optimum. However, like other proteins from thermophiles, *T. thermophilus* PGK is highly thermostable. After heating at 90°C for 10 min, *T. thermophilus* PGK lost 50% activity, whereas 50% activity was lost when rabbit muscle PGK was heated at 55°C for 15 min or *B. stearothermophilus* PGK was heated at 70°C for 10 min.

The amino acid composition of *T. thermophilus* PGK was compared with those of *B. stearothermophilus* and yeast PGK's. Though Arg, Pro, Glu and Ala are increased and Lys, Asp and Ser are decreased in *T. thermophilus* PGK, no remarkable differences were found in average hydrophobicity defined by Bigelow (1967), secondary structure contents and number of helix forming and breaking residues (Nojima *et al.*, unpublished data).

T. THERMOPHILUS CYTOCHROME *C*-552

Purification procedures for cytochrome *c*-552 from *T. thermophilus* were published elsewhere (Hon-nami and Oshima, 1977). The molecular weight was determined to be 15,000 and the cytochrome contained a *c*-type heme and an iron atom per molecule. Coordinated binding between heme iron and a Met residue can be followed by measuring absorption at 690 nm. The binding was stable up to 80°C in a range between pH 1 to 12 (Kihara *et al.*, 1977) since the absorption band remained unchanged.

THERMAL DENATURATION AND ITS REVERSIBILITY

Reversibility of the denaturation by GuHCl of *T. thermophilus* and yeast PGK was confirmed by the complete recovery of their enzyme activities as well as the fluorescence spectra characteristics of their native forms by simply dialyzing out GuHCl after complete denaturation by GuHCl. Reversibility of the denaturation of *T. thermophilus* cyt *c* was also confirmed in a similar way. An example is illustrated in Fig. 1.

Thermal denaturation was followed by recording the molar ellipticity at 225 nm for *T. thermophilus* PGK and cyt *c* and at 220 nm for yeast PGK while changing the temperature of the sample solutions in the presence of various concentrations of GuHCl (Nojima *et al.*, 1977; Nojima *et al.*, 1978). Some of the results of thermal denaturation at various concentrations of GuHCl are shown in Fig. 2. The reversibility of thermal denaturation was also confirmed by the fact that the denaturation curves obtained by raising the temperature were superimposable with those obtained by lowering the temperature, as shown in Curve D of Fig. 2.

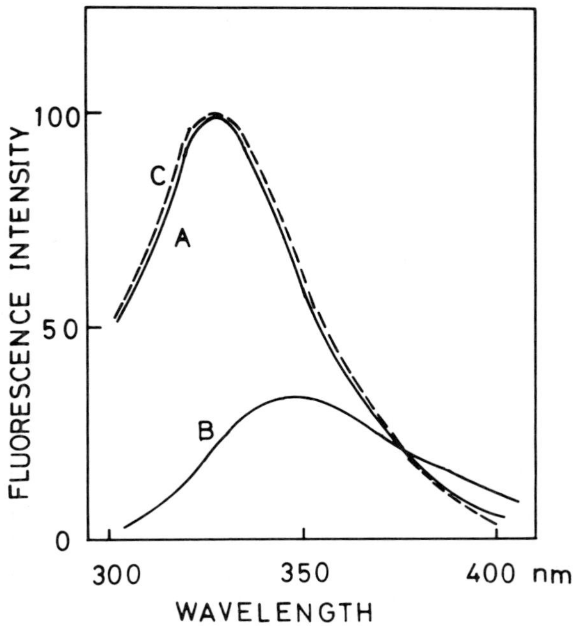

FIGURE 1. Reversibility of denaturation by GuHCl observed by fluorescence spectra at 20°C (excited at 295 nm). Corrected fluorescence spectra of T. thermophilus *PGK are shown. A, native; B, denatured by 4.2 M GuHCl; C, renatured by dialyzing out GuHCl after complete denaturation by 4.2 M GuHCl. Taken from Nojima* et al. *(1977) and reproduced with permission from Academic Press (London).*

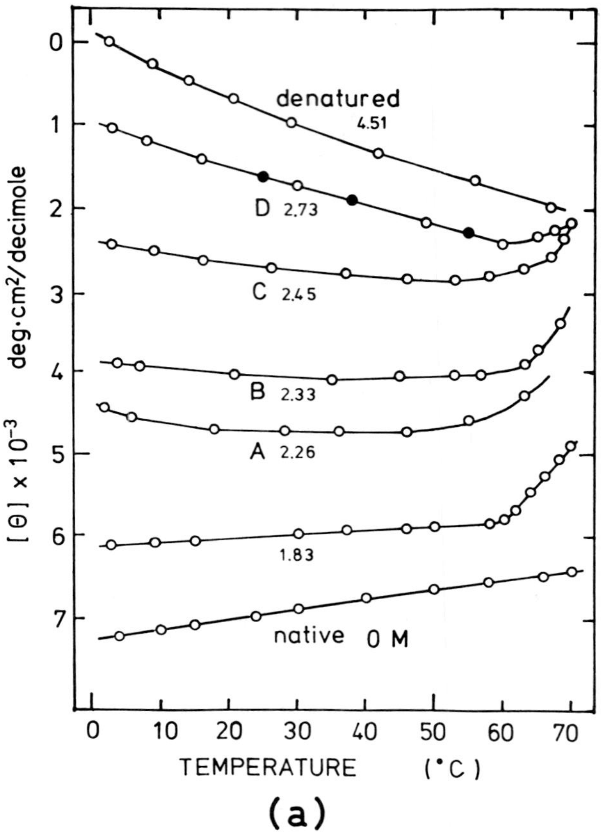

FIGURE 2. Denaturation curves of (a) T. thermophilus *PGK, (b) yeast PGK and (c)* T. thermophilus *cytochrome* c *followed by CD spectra. Molar ellipticity at 225 nm for* T. thermophilus *PGK and* T. thermophilus *cytochrome* c, *and at 220 nm for yeast PGK was plotted against temperature. The reversibility of thermal denaturation was confirmed by Curve D as described in the text. (o o). Molar ellipticity obtained by raising the temperature (o); molar ellipticity obtained by lowering the temperature (●). Numbers for each curve represent the molarities of GuHCl in each enzyme solution. Adopted from Nojima* et al. *(1977) and Nojima* et al. *(1978) and reproduced with permission from Academic Press (London).*

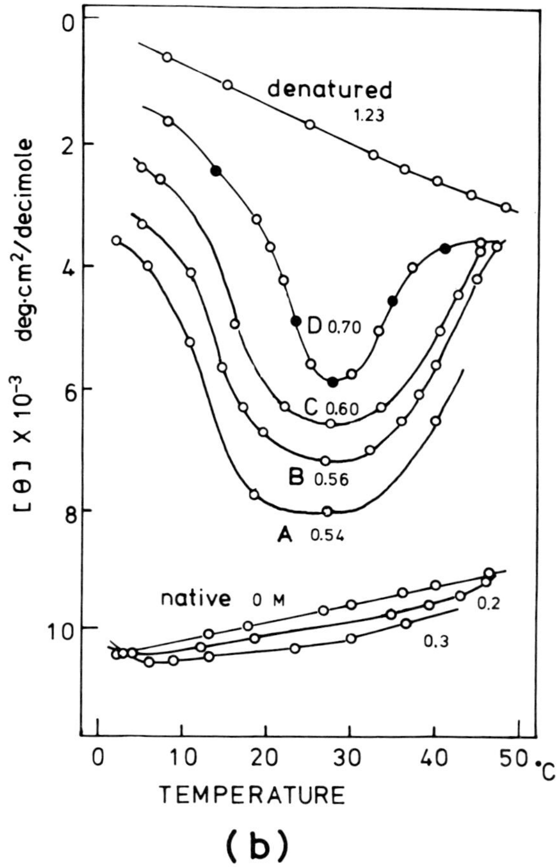

FIGURE 2b.

ISOTHERMAL DENATURATION WITH GuHCl

Isothermal denaturation induced by GuHCl at 25°C was carried out and the results obtained by recording circular dichroism or fluorescence are shown in Figs. 3-5. The circular dichroism spectrum at around 220 nm reflects the changes in the secondary structure of the protein, while the fluorescence spectrum reflects the conformations of tyrosine and tryptophan residue(s) in a protein molecule and gives more localized information.

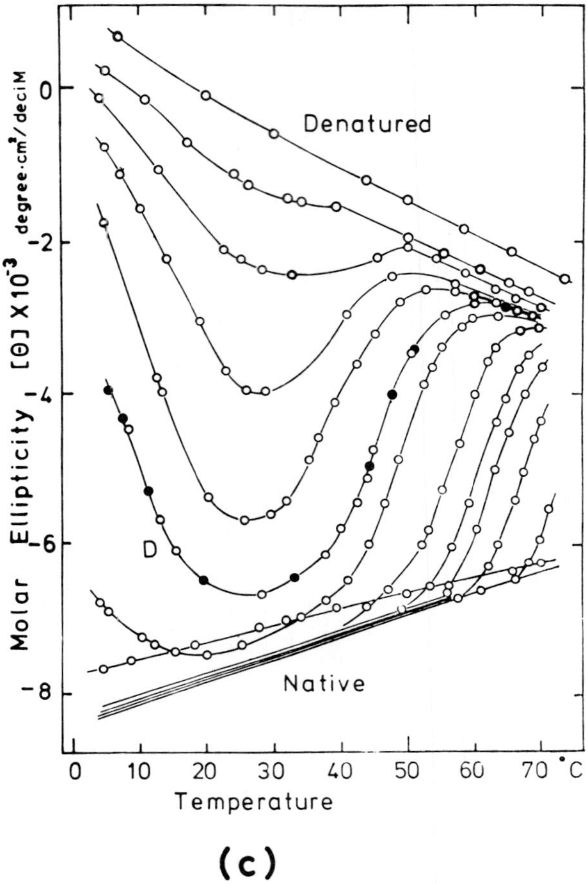

FIGURE 2c.

DISCUSSION

The transition curve at any stage of the denaturation is characterized by the fraction (f) of the enzyme molecules in the denatured state as follows:

$$f = \frac{y - y_N}{y_D - y_N} \tag{1}$$

Where y is the fluorescence intensity or the molar ellipticity, and y_N and y_D represent the value of y characteristic of the native and the denatured states, respectively. In

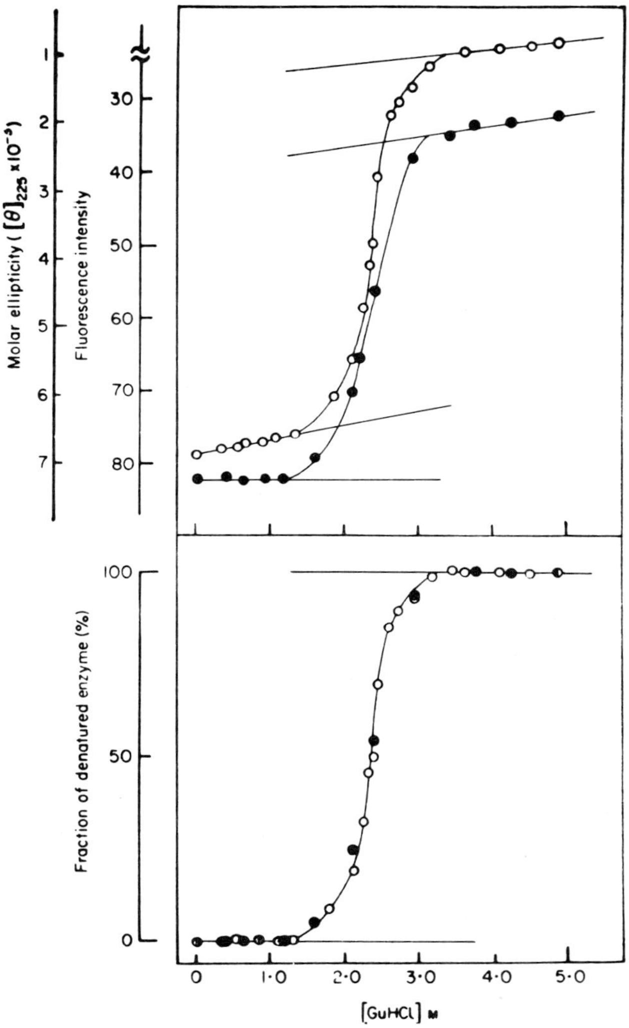

FIGURE 3. Isothermal denaturation profile of T. thermophilus *PGK at 25°C. Upper: Molar ellipticity at 225 nm (o) and fluorescence intensity at 328 nm (●) (excited at 295 nm). Lower: The percentages of the fraction of denatured enzyme versus GuHCl molarity. Taken from Nojima* et al. *(1977) and reproduced with permission from Academic Press (London).*

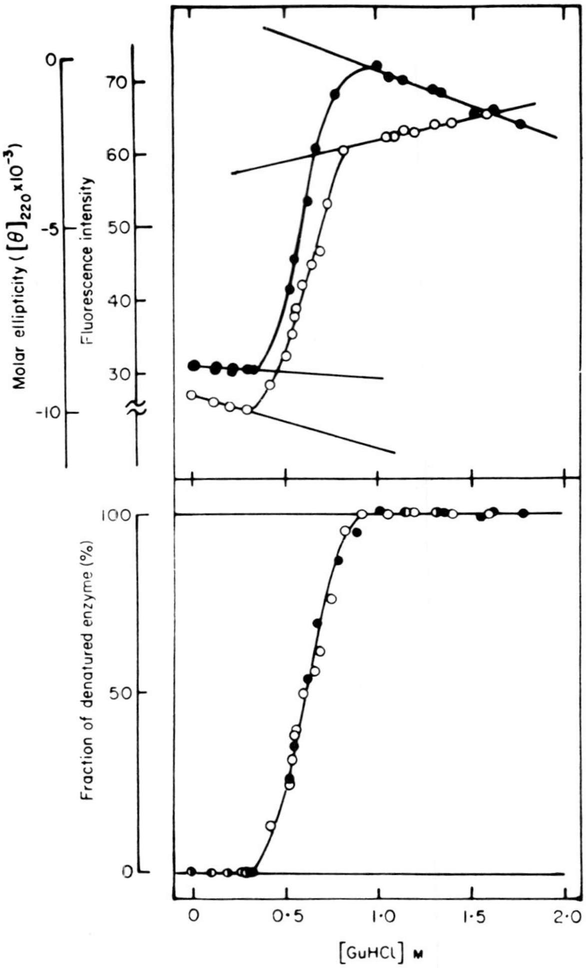

FIGURE 4. Isothermal denaturation profiles of yeast PGK at 25°C. Upper: Molar ellipticity at 220 nm (o) and fluorescence intensity at 340 nm (●) (excited at 295 nm). Lower: The fraction of denatured PGK versus *GuHCl molarity. Taken from Nojima* et al. *(1977) and reproduced with permission from Academic Press (London).*

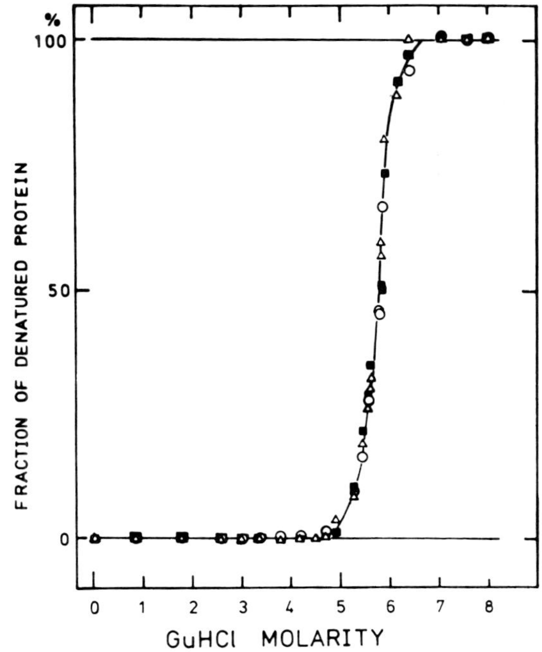

FIGURE 5. Isothermal denaturation of T. thermophilus *cytochrome* c *at 25°C. Data obtained by circular dichroism at 225 nm (Δ), fluorescence intensity at 302 nm (o) (excited at 280 nm), and fluorescence intensity at 345 nm (■) (excited at 280 nm) were plotted. Taken from Nojima* et al. *(1978) and reproduced with permission from Academic Press (London).*

Figs. 3 to 5, the fraction of the denatured protein calculated from circular dichroism or fluorescence results is plotted against GuHCl molarity. Since the isothermal denaturation profiles obtained by circular dichroism spectra and by fluorescence intensity were superimposable, thermodynamic analyses were done based on the two state model for denaturation, in which it was assumed that the solution contains a mixture of the native and the denatured protein and no other conformational state makes any significant contribution to the measured optical properties in the transition region.

The equilibrium constant K for the reaction $N \rightleftharpoons D$ can then be calculated as follows:

$$K = \frac{f}{1 - f} = \frac{y_N - y}{y - y_D} \qquad (2)$$

The Gibbs free energy change from the native to the denatured state is given by:

$$\Delta G = -RT\ln K \quad (3)$$

Aune and Tanford (1969) proposed several methods to estimate ΔG in the absence of a denaturant. One method is based on the assumption that denaturation results from the binding of the denaturant, and gives the following formula:

$$\Delta G = \Delta G^{\circ} - \Delta nRT\ln\,(1 + ka_{\pm}) \quad (4)$$

where k represents the binding constant of the denaturant to the protein and Δn is the difference between the number of denaturant molecules bound to the denatured and the native states respectively. The terms $a_{\pm}$ is the mean ionic activity of GuHCl at the given concentration (Aune and Tanford, 1969).

A value of k = 1.20 was employed in the following calculations, which has been used in the estimation of ΔG° values for other proteins whose reversible thermal denaturation experiments have been done heretofore (Aune and Tanford, 1969; Salahuddin and Tanford, 1970; Brandts, 1965; Knapp and Pace, 1974). Figs. 6 and 7 show ΔG plotted against GuHCl activity and the best-fit curve calculated by the least square analysis. The calculated ΔG° values are given in Table 1 together with those of other proteins. From these results it can be concluded that at 25°C, *T. thermophilus* PGK (11.87±0.21 kcal/mole) is more stable than yeast PGK (5.33±0.13 kcal/mole) by 6.54 kcal/mole and that *T. thermophilus* cyt. *c* (28.54±0.15 kcal/mole) is more stable than, for example, horse cyt. *c* by 15.84 kcal/mole.

The temperature dependence of ΔG at a constant concentration of GuHCl is shown in Figs. 8 and 9. For the evaluation of thermodynamic parameters, such as ΔH and ΔS, the free energy change (ΔG) has been expressed as a power series in the absolute temperature as follows (Brandts, 1969):

$$\Delta G = A + BT + CT^2 + DT^3 \quad (5)$$

By fitting equation (5) to each set of results, the best-fit curve can be determined by the least square analysis. The temperature of maximum stability (T*) where $\partial\Delta G/\partial T = 0$ is 25 to 28°C for yeast PGK, while that for *T. thermophilus* PGK cannot be uniquely determined. For *T. thermophilus* cyt. *c*, T* is 27.0±1.4°C, which is higher than those for horse and cow cytochromes *c* (12°C) or *Candida krusei* cyt. *c* (9°C) (Kawaguchi and Noda, 1976).

A remarkable difference between *T. thermophilus* PGK and yeast PGK is that the enthalpy change obtained by van't Hoff

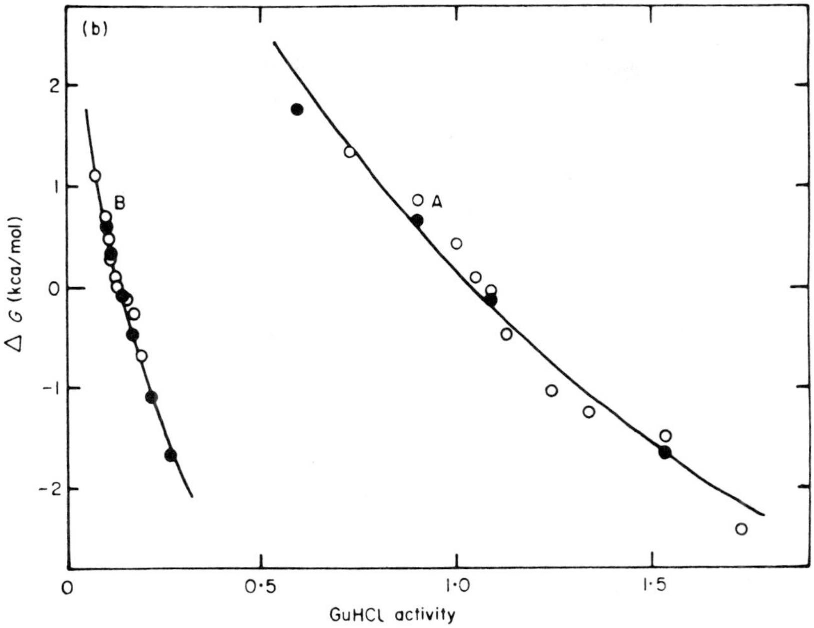

FIGURE 6. ΔG plots versus *GuHCl activity at 25°C. Activity of GuHCl was obtained from the following expression:* $\log a = -0.5191 + 1.4839 \log c - 0.2562(\log c)^2 + 0.5884 (\log c)^3$*, where c is the GuHCl molarity. ΔG values are calculated from the results obtained by CD spectra (o) and fluorescence spectra (●) for* T. thermophilus *PGK (Curve A) and for yeast PGK (Curve B), respectively. Taken from Nojima* et al. *(1977) and reproduced with permission from Academic Press (London).*

plots is very small in the case of the thermophile protein. Moreover, its dependence on temperature, and hence the heat capacity change on denaturation, is also very small over this temperature range (0 to 60°C). This property, however, is not seen in the case of *T. thermophilus* cyt. *c*.

Figs. 10-12 show the temperature dependence of ΔH and ΔS, respectively. The temperature dependence of ΔH and ΔS for *T. thermophilus* cyt. *c* (Fig. 12) are very similar to those of horse, cow and *Candida krusei* cytochromes *c* (Kawaguchi and Noda, 1977), whereas those for *T. thermophilus* PGK are very small compared with those for yeast PGK.

Considering these results, we wish to point out here that in terms of thermodynamics, there are at least three

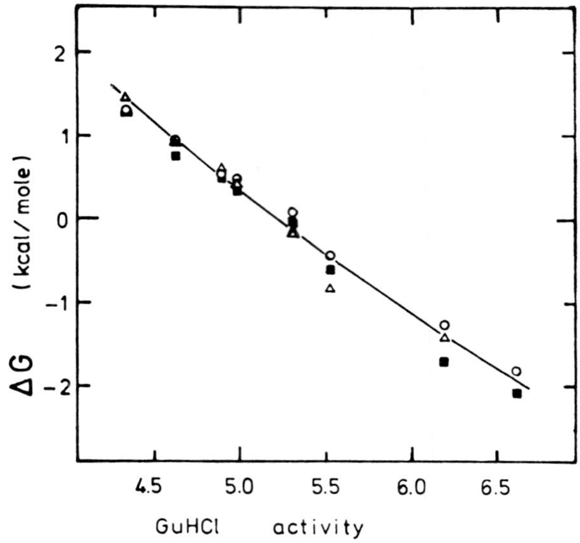

FIGURE 7. ΔG versus GuHCl activity for denaturation of T. thermophilus *cytochrome* c. *Data obtained by circular dichroism (Δ) and fluorescence (o, ■) were plotted. Taken from Nojima* et al. *(1978) and reproduced with permission from Academic Press (London).*

basic ways to make a protein thermostable (Nojima *et al.*, 1977). Curve 0 in Fig. 13 shows the change in free energy on denaturation of the three dimensional structure of ordinary proteins. The protein has maximum stability at T*, becoming unstable both at higher and lower temperatures. One way to make a protein more thermostable is to deepen curve 0 to curve 1, thus stabilizing the protein over a wider temperature range. Most of the discussion on the mechanism of protein thermostability has been centered around this type of mechanism, e.g., by trying to find more hydrophobic residues, hydrogen bonds, disulfide bridges, and the like. We would, however, like to suggest a second mechanism, which is to flatten curve 0 to curve 2 without increasing the maximum stability (ΔG at T*) of the protein. The stability at a low temperature is also expected to increase by this mechanism. The third possible mechanism is simply to shift curve 0 to curve 3. This mechanism predicts a loss of stability at low temperature for thermostable proteins which is contrary to common observation for thermostable proteins studied heretofore.

As is clearly shown in Fig. 6 and Table 1, *T. thermophilus* PGK is a very likely candidate for a protein attaining thermostability by the second mechanism. Since the maximum

TABLE 1
ΔG Values in the Absence of GuHCl at 25°C Calculated by the Least-square Method Assuming Equation (4)

Protein	*Source*	*ΔG°* *(kcal/mole)*	Δn[1]
PGK	*T. thermophilus*	11.87±0.21	25.1
	yeast	5.33±0.13	24.2
Cytochrome *c*-552	*T. thermophilus*	28.54±0.15	36.5
	Cow[2]	15.4	30.4
	Horse[2]	12.7	26.4
	Candida[2]	14.0	32.9
Myoglobin[3]	Horse	11.8	30
	Sperm whale	14.3	30
Ribonuclease[4]		13.0	24.4
Lysozyme[5]		14.2	21.5

[1]*From equation (4).*
[2]*Knapp and Pace, 1974.*
[3]*Puett, 1973.*
[4]*Salahuddin and Tanford, 1970.*
[5]*Aune and Tanford, 1969.*

stability (ΔG at T*) is also larger than that of the yeast enzyme, the first mechanism is also involved to some extent. In contrast to the results for *T. thermophilus* PGK, it is suggested that *T. thermophilus* cyt. *c* is stabilized by a combination of the first and third mechanisms since ΔG is larger and T* is higher than those of mesophilic counterparts.

The free energy change on denaturation is mainly the sum of the stabilizing contribution of hydrogen bonds and hydrophobic interactions, and the destabilizing contribution of conformational entropy change. In the case of chymotrypsinogen (Brandts, 1964), the free energy contribution of hydrogen bonds is mainly an enthalpic one of around 120 kcal/mole. The contribution from hydrophobic interactions will be about 105 kcal/mole at 0°C, and this value will increase at higher temperatures (for example, 145 kcal/mole at 75°C). These two contributions are approximately balanced at all tempera-

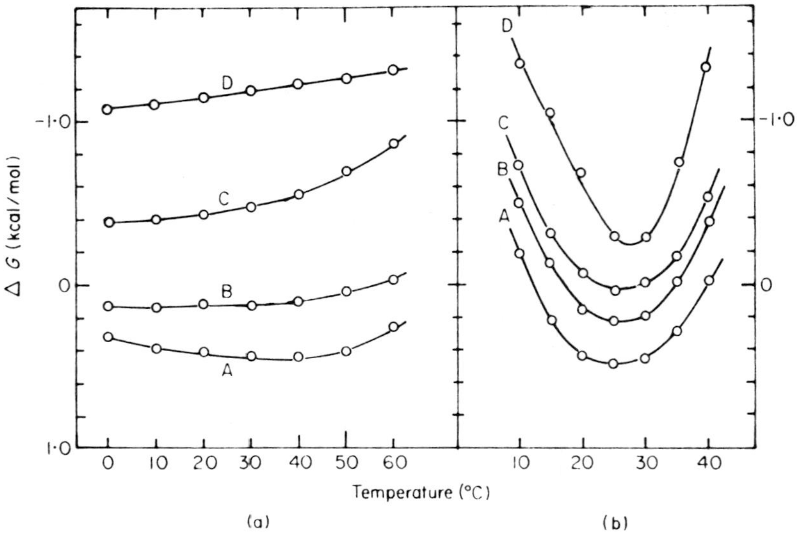

FIGURE 8. Relationship between ΔG values and temperature. (a) T. thermophilus *PGK, (b) yeast PGK. A-D correspond to those shown in Fig. 2. Taken from Nojima* et al. *(1977) and reproduced with permission from Academic Press (London).*

tures by the destabilizing contribution of the conformational entropy change. The net free energy of denaturation of ordinary proteins is around 10 kcal/mole (Table 1), resulting from a very delicate balance between the stabilizing and destabilizing factors for a protein.

Taking these facts into account, we can expect that the first mechanism in Fig. 13 can easily be attained by increasing a few hydrogen bonds or hydrophobic interactions through the replacement of some amino acid residues. The entropy change of denaturation comes predominantly from two sources (Brandts, 1969): (i) the changes in conformational entropy--a positive contribution at all temperatures, and (ii) the change due to the exposure of hydrophobic residues to the solvent -- a negative contribution at low temperatures, which becomes more positive at high temperatures. At the temperature of maximum stability (T*), where $\Delta S = -\partial \Delta G / \partial T = 0$, these two contributions exactly cancel each other. If we neglect a small change in conformational entropy caused by the increase of hydrophobic interactions, a protein with a higher content of hydrophobic interactions can be expected to exhibit maximum stability at a higher temperature. This means

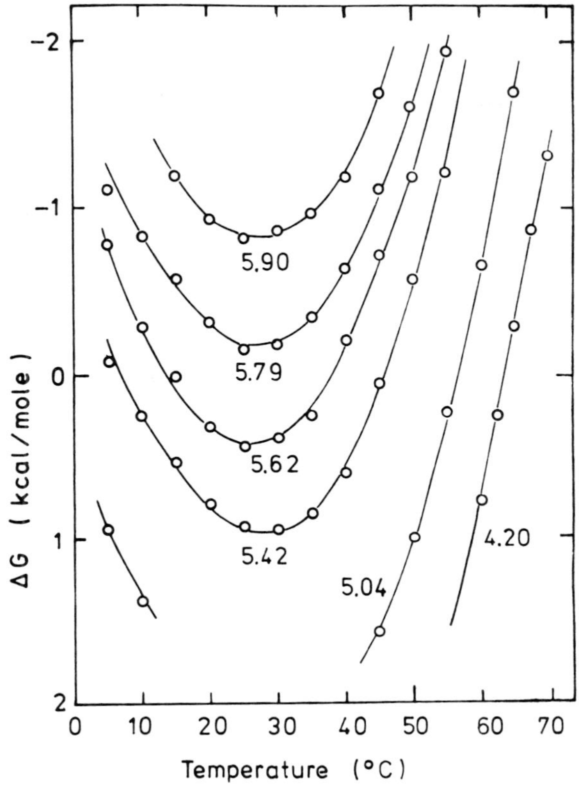

FIGURE 9. ΔG versus temperature for denaturation of T. thermophilus *cytochrome* c. *Numbers represent the GuHCl concentrations. From Nojima* et al. *(1978).*

that one way for attaining the third mechanism is by an increase in the content of hydrophobic residues.

It is not easy to understand the second mechanism from the known properties of the native form of proteins. We must recall, however, that ΔG values in Fig. 7 are the free energy *differences* between the native and the denatured states. Let us consider, for example, a protein that is left with considerable hydrophobic interactions among its nonpolar residues even in the denatured state. In this case we may expect that the nonlinear contribution of hydrophobic interactions to ΔG is lost in large part and the ΔG *versus* T curve becomes less quadratic. The presumed hydrophobic interactions in the denatured state may be different from those found in the native form. The amino acid residues participating in the hydrophobic interactions in the denatured form

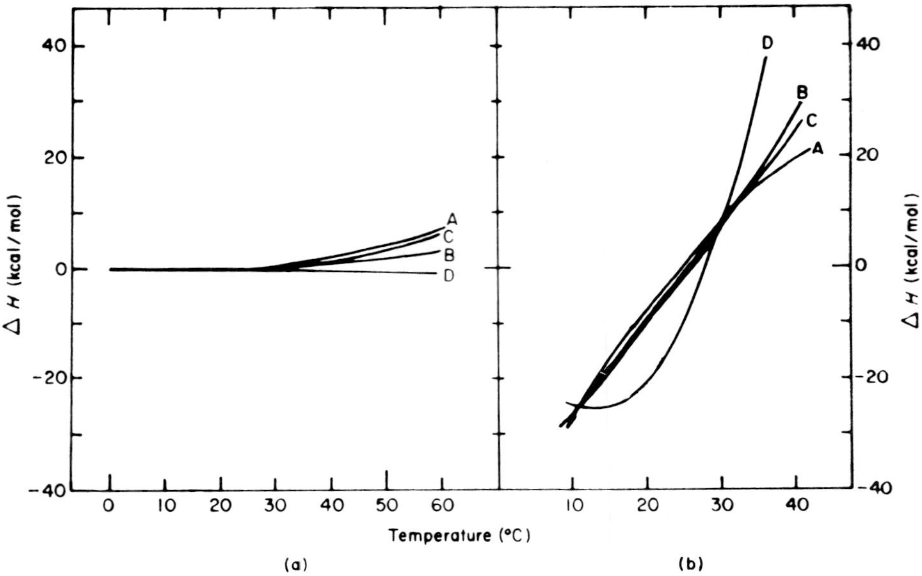

FIGURE 10. Temperature dependence of ΔH. (a) T. thermophilus PGK, (b) yeast PGK. A-D correspond to those shown in Fig. 2 (a) and (b). Taken from Nojima et al. (1977) and reproduced with permission from Academic Press (London).

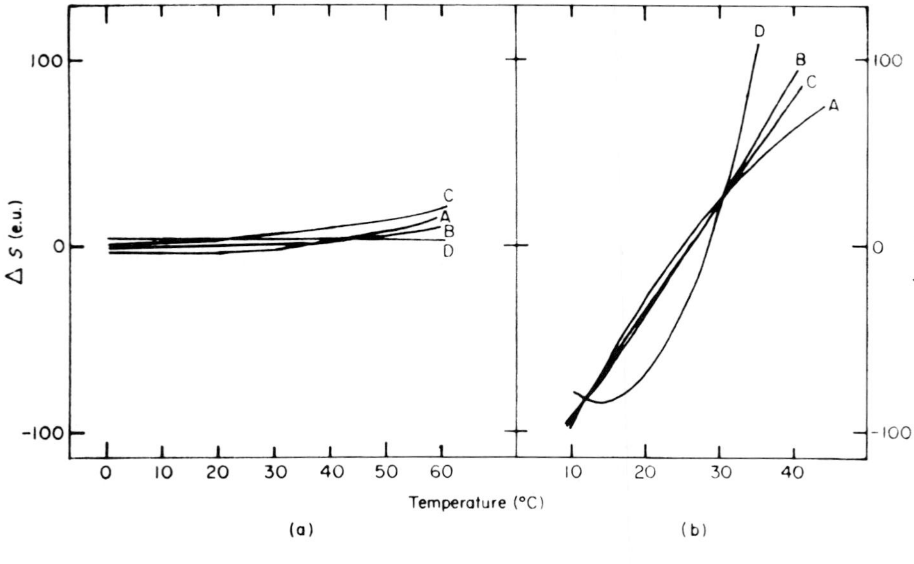

FIGURE 11. Temperature dependence of ΔS for denaturation of (a) T. thermophilus *PGK and (b) yeast PGK. A-D correspond to those in Fig. 2 (a) and (b). Taken from Nojima* et al. *(1977) and reproduced with permission from Academic Press (London).*

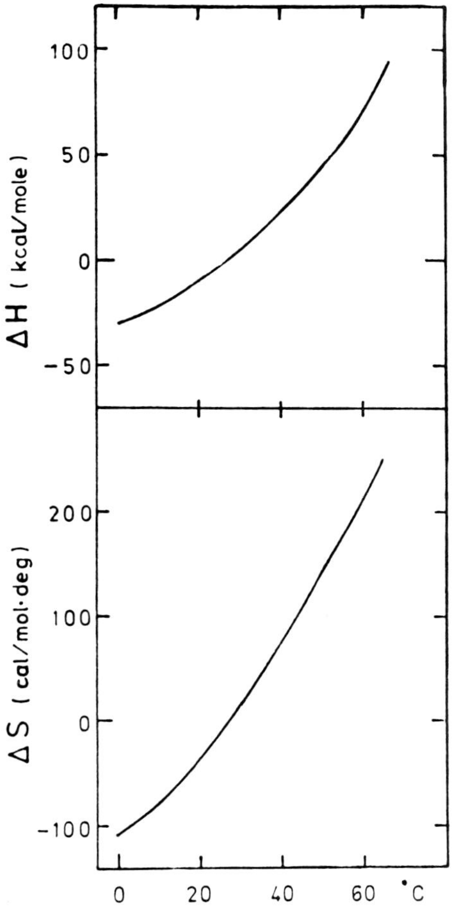

FIGURE 12. Temperature dependence of ΔH *and* ΔS *for denaturation of* T. thermophilus *cytochrome* c *in the presence of 5.42 M GuHCl. Taken from Nojima* et al. *(1978) and reproduced with permission from Academic Press (London).*

---

would be the neighboring residues in amino acid sequence, thus bearing little relation to the long range interactions responsible for the native three dimensional structure. The overall conformation and optical properties for such a denatured state might be still close to those of a random coil. Our present concern is to detect the structural elements in thermophile PGK which confer a very slight temperature dependence on ΔG of the denaturation.

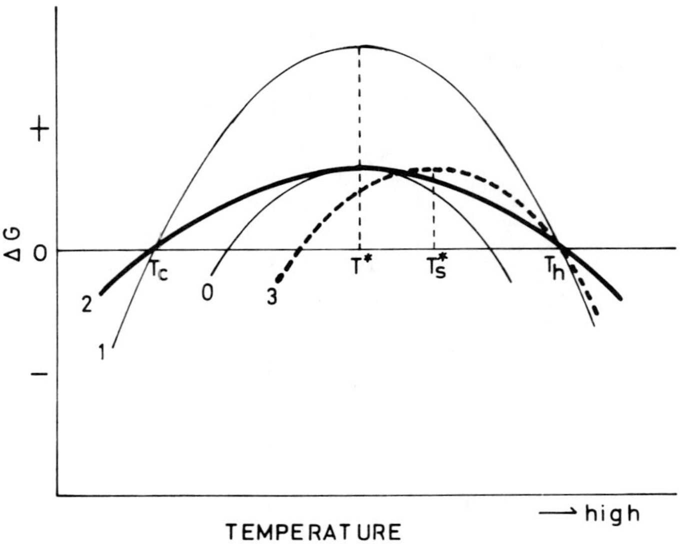

FIGURE 13. Hypothetical free energy (ΔG) curves of globular proteins. ΔG is defined as the difference of Gibbs free energy of the native enzyme from that of the denatured one. Curve 0; ΔG versus *temperature (T) for an ordinary globular protein. T* is the temperature of maximum stability. Curves 1-3; ΔG* versus *T for hypothetical thermophilic proteins. See text for details.*

REFERENCES

Aune, K. C. and Tanford, C. (1969). Biochemistry *8*, 4586.
Bigelow, C. C. (1967). J. Theoret. Biol. *16*, 187.
Brandts, J. F. (1964). J. Am. Chem. Soc. *86*, 4302.
Brandts, J. F. (1965). J. Am. Chem. Soc. *87*, 2759.
Brandts, J. F. (1969). *Structure and Stability of Biological Macromolecules*, Marcel Dekker, New York.
Hon-nami, K. and Oshima, T. (1977). J. Biochem. *82*, 769.
Kihara, H., Nakatani, H., Hiromi, K., Hon-nami, K. and Oshima, T. (1977). Biochim. Biophys. Acta *460*, 480.
Kawaguchi, H. and Noda, H. (1977). J. Biochem. *81*, 1307.
Knapp, J. A. and Pace, C. N. (1974). Biochemistry *13*, 1289.
Nojima, H., Ikai, A., Oshima, T. and Noda, H. (1977). J. Mol. Biol. *116*, 429.
Nojima, H., Hon-nami, K., Oshima, T. and Noda, H. (1978). J. Mol. Biol., in press.

Puett, D. (1973). J. Biol. Chem. *248*, 4623.
Salahuddin, S. and Tanford, C. (1970). Biochemistry *9*, 1342.
Scopes, E. K. (1973). In *The Enzymes* (P.D. Boyer, ed.) Vol. 8, p. 335, Academic Press, New York.
Suzuki, K. and Imahori, K. (1974). J. Biochem. *76*, 771

PROPERTIES AND REGULATION OF THERMOPHILIC GLUTAMINE SYNTHETASES

Frederick C. Wedler

Department of Chemistry
Biochemistry Program
Rensselaer Polytechnic Institute
Troy, New York

BACKGROUND

Review of Thermophilic Enzymes with Allosteric Properties

Of the many enzymes isolated from thermophilic microorganisms, studied in considerable detail, quite a few exhibit regulation (activation or inhibition) by non-substrate (allosteric) small molecules. Table 1 presents a reasonably comprehensive summary of such systems (Refs. 1-44). Some of these enaymes were studied at the temperature of cell growth, others below it, for reasons of practicality. The most important conclusion one may draw from this table is that thermophilic enzymes *closely* resemble their mesophilic counterparts with regard to regulatory properties. This implies comparable numbers of high specificity binding sites for substrates and for allosteric modifiers, i.e., "information" content of the protein chein. It also requires similar, although not necessarily identical, degrees of freedom or flexibility for polypeptide chains that interact between substrate and modifiers sites. The large number of enzymes that exhibit homotropic and/or heterotropic cooperative binding of ligands in both mesophilic and thermophilic examples supports this view. One may also point out that many of these enzymes exhibit non-linear Arrhenius plots, due either to effects induced in K_m or V_{max} or both. One is thus led to wonder (a) what the limits of functional information per protein subunit might be in a protein that is intrinsically stable at extreme temperatures, and (b) how this thermostabilization is

ISBN 0-12-268250-5

TABLE 1
Thermophilic Enzymes with Allosteric Properties

Enzyme	Source (Ref.)	Properties*
A. Catabolic		
1) Phosphofructokinase	*Flavobact. thermophilium* (1, 2)	(I) PEP - sigmoidal (A) ADP - antag. PEP inh.
	B. stearo., *T. aquaticus* (3)	(I) PEP; (A) ADP
	Thermus X-1 (4)	(I) PEP (c); (A) ADP (c $\sim$ F-6-P) sigmoidal F-6-P binding ($n_H = 2$)
2) F-1,6-diP'ase (FBPase)	*F. thermophilium* HB8 (5)	(I) AMP (incr. S-inh. by FBP); (ADP, ATP, IMP, ...) (A) PEP (decr. S-inh. by FBP)
3) FBP Aldolase	*B. stearo.* (6-8)	(A) K^+ (Na^+); Zn^{++}, Co^{++}, Fe^{++}, Mn^{++}, Cd^{++} (I) EDTA; S-inactivation by FBP, DHAP, G-3-P
	T. aquaticus (9)	(A) NH_4^+; Fe^{++}, Co^{++}, Cys
4) D-Glyc.-3-P DHase	*T. thermophilus* HB-8 (10)	(A) NH_4^+ (Na^+, C_2H_5OH)

	Enzyme	Organism	Regulation
5)	Lactate DHase	*B. caldolyticus* (11)	(I) ADP, PEP, sigmoidal S-binding
6)	Citrate Synthase	*B. stearo.* (12)	(I) ATP ($C \sim$ AcCoA, $NC \sim$ OAA); (ADP) $\pm$ KCl not AMP
7)	[$NADP^+$] Isocit. DHase	*T. aquaticus* YT-1 (13) *T. flavus* AT-62 (14)	(I) glyoxalate, OAA (oxomalate) stronger > 60° (concerted?)
8)	Fumarase	*B. stearo.* *Thermus* X-1 (15)	(I) fumarate, S-inh.
9)	6-P-Gluconate DHase	*B. stearo.* (16)	(I) NADPH ($C \sim NADP^+$, strong prod. inh.)
10)	Isocitrate lyase	*B. stearo.* BS-1 (17,18)	(A) K^+, Na^+, Li^+, NH_4^+; Mg^{++}, Ba^{++}, Cd^{++}

B. Anabolic

	Enzyme	Organism	Regulation
1)	Acetate kinase	*C. thermaceticum* (19)	sigmoidal M_gATP binding ($n_H = 1.8$)
2)	Formyl-THF synthetase	*C. thermaceticum* (20)	(A) K^+, NH_4^+ (decr. K_m (formate))
3)	Acetohydroxy-acid synthetase	*T. aquaticus* *Bacilus sp.* (21)	(I) Val, Ile (not concerted) req. Mg^{++}, TPP or FAD
4)	PEP carboxylase	*T. aquaticus* (22)	(A) AcCoA (abs. req.); Mg^{++}

TABLE 1 - Continued
Thermophilic Enzymes with Allosteric Properties

Enzyme	*Source (Ref.)*	*Properties*
B. Anabolic - Cont'd.		
5) Pyruvate carboxylase	*B. stearo.* (23, 24)	(I) Asp ($n_H = 4$) (A) AcCoA ($n_H = 2$) (I) chlorothricin (C ∿ AcCoA, NC ∿ ATP, Pyr) $n_H = 3$, binds coop., induces ATP S-inh., blocks (A) pyr.
6) Pyrimidine ribonucleotide kinase	*B. stearo.* (25)	(I) CTP (inh. strength incr. with T)
7) Ribonucleotide reductase	*Thermus* X-1 (26)	(A) dGTP
8) Deoxythymidine kinase	*B. stearo.* (27)	(I) dTTP (C ∿ dT)
9) Aspartokinase	*B. stearo.* (28,29) E-I	(I) (Lys (hyperbolic)) concerted (Thr (sigmoidal)) (NC ∿ ATP, M ∿ Asp) (n_H and K_i incr. with T)
	B. stearo. (30) E-II *Thermus sp.* (31)	(I) *meso* - diamino pimelate (I) Lys, Thr (synergistic)

10)	Homoserine DHase	*B. stearo.* (32)	(I) Thr (C ∿ Thr) (A) K^+
		T. flavus AT-62 (33)	(I) L-Cys (NC), not other end prods. (A) K^+, Na^+ Homoser. neg. coop. binding (n_H = 0.45)
11)	Threonine deaminase	*B. stearo.* (34, 35)	(I) Ile (sigmoidal), Val; weaker inh. as T incr. induces $n_H > 1$ for L-Thr (A) Na^+; K^+, Li^+
12)	Glutamine synthetase	*B. stearo.* (36-39)	(I) many amino acids, nucleotides, C-P, GA-6-P, etc. sigmoidal MnATP binding (n_H = 2.2), complex S-I, I-I site interactions
		B. caldolyticus YTP (39, 40)	(I) many amino acids, nucleotides, GA-6-P, etc. (A) NMP's hyperbolic MnATP binding complex S-I, I-I site interactions *isozymes*: complementary inh. sensitiv.
13)	Met-tRNA synthetase	*B. stearo.* (41)	neg. coop. S-binding ($n_H < 1$)
14)	DNA polymerase	*T. aquaticus* (42)	(A) Mg^{++}, abs. requirement

TABLE 1 - Continued
Thermophilic Enzymes with Allosteric Properties

Enzyme	*Source (Ref.)*	*Properties*
B. Anabolic - Cont'd.		
15) RNA polymerase	*T. aquaticus* T-2 (43, 44)	(I) K^+, NH_4^+; Heparin, rifampicin

*Abbreviations, Symbols: (I) = inhibition by, (A) = activation by, inh. = inhibits, inhibition; incr. = increased, increases; decr. = decreased, decreases; abs. = absolute; req. = requires; C = competitive, NC = non-competitive, M = mixed; S = substrate; I = inhibitor; prod. = product; n_H = Hill constant accomplished. Several recent in-depth reviews of the latter question should be cited, the first by Singleton and Amelunxen (45), the second by Ljungdahl and Sherod (46). All present data indicate that one is faced with the problem of distinguishing small subtle differences against a background of larger, similar effects, regardless of what parameter is considered.

Experimental Rationale: Glutamine Synthetase

In dealing with the above questions, it seemed reasonable to study an enzyme system which in the mesophilic case appears to contain nearly maximal levels of functional information as ligand binding sites. The elegant work of Stadtman and co-workers (47) indicates that in micro-organisms the activity of glutamine synthetase is rigorously regulated by 6-10 metabolic and product feedback modifiers derived from glutamine. The mode of regulation includes adenylylation-deadenylylation in gram-negative organisms such as *E. coli*, but involves synergistic inhibition in gram-positive mesophiles, e.g., *Bacillus sp.* Since the regulation of the *E. coli* enzyme is cumulative, separate sites for two metal ions, three substrates, and 6-10 feedback inhibitors per subunit (MW 50,000) are indicated. Thus, it occurred to us to ask how much of this functional information could be and actually is maintained in closely related thermophilic organisms living at 55-70°C, exactly which binding interactions were most important for regulation of nitrogen metabolism, and how the maintenance of high information content and thermo-stabilization are accomplished. Our previous studies with the enzyme from a moderate thermophile, *B. steraothermophilus* (36, 37) showed that minimal simplifications in responses to feedback regulators were required at 55°C. Obviously, more extreme conditions were needed to amplify the most relevant effects. An extreme thermophile, *B. caldolyticus YTP*, capable of growth at 85-90°C, was then selected for study. Preliminary reports of our experimental results have appeared earlier (39), in which it was assumed that a single enzyme had been isolated. We have since discovered that two enzymes exist in *B. caldolyticus*, and that these are probably regulatory isozymes (40). These isozymes now have been partially characterized. What follows is a discussion of the implications of these data for both this specific key regulatory enzyme and thermophily in general.

GLUTAMINE SYNTHETASES OF *B. CALDOLYTICUS*

Microbiology, Cell Growth Conditions

Optimal cell growth was obtained in the limiting defined medium of Heinen (48, 49) composed of: 3.0 mM K_2HPO_4, 0.9 mM KH_2PO_4, 4.3 mM NaCl, 0.25 mM $MnCl_2.6H_2O$ and $MgCl_2.6H_2O$, 0.36 mM $FeSO_4.7H_2O$, 0.034 mM $CaCl_2$, 15 mM glucose, 1.5-2.5 g/L Bacto-tryptose (Difco) or trypticase-peptone (BBL) and 0.15 ml/L trace elements solution that contained (per L): 2.37g $ZnCl_2$, 5.0g H_3PO_4, 0.25g $CuSO_4.5H_2O$, 0.25g $Na_2MoO_4.2H_2O$, 0.48g $CoCl_2.6H_2O$ and 5.0 ml H_2SO_4. All growth was carried out at 68-72°C with high aeration in volumes up to 300L. Cultures were maintained on 3% nutrient agar dissolved in this minimal medium. Addition of either NH_4^+ or L-glutamine above 2 mM reduced the specific activity of total glutamine synthetase (measured by GHS assay, pH 6.5, with $MnCl_2$) to 25-30% of the original value. The molecular basis for this effect is not yet known. Enzyme activity levels were not sensitive to addition of 20 mM L-Glu, 20 mM L-Asp, or 2.5g/L Tryptose. Well-aerated cultures of cells in the defined media with no added NH_4Cl typically doubled their A_{650} about every 30 min at 70°C up to $A_{650} \sim 1.5$ with a cell yield of 5g wet weight/L culture.

Recent investigations reveal that *Bacillus caldolyticus*, strain YTP, as isolated and supplied by NASA Ames Labs (48, 49), contains two "variants" that can be distinguished with regard to morphology of colonies, growth on various defined media, and other characteristics (50, 51). If grown on nutrient agar, one of these variants produces large translucent yellowish colonies, while the other grows more slowly and forms small, opaque, white colonies. Pure cultures of each were separately grown on our defined media, harvested, the cells broken, and enzyme purified through step 3 of the purification scheme in Table 2. *Each* variant was found to contain *both* electrophoretically separable enzymes E_I and E_{II} (see below), determined by slicing the polyacrylamide gels into 3 mm pieces and assaying for glutamine synthetase activity. This argues that the two strains are very closely related and are most likely variants of the same species.

Assays for Enzyme Activity

Assays were carried out at 70°C at pH 6.50, 25 mM Im-OA_c, with 50 mM L-glutamate, 10 mM $MnCl_2$-ATP, and 10 mM of eith NH_4Cl or NH_2OH. The "biosynthetic" activity (with NH_4Cl) was measured by Pi release colorimetrically (A_{660}) by

TABLE 2
Purification of B. caldolyticus *Glutamine Synthetases*

Step		*Specific Activity (U/mg)*	*Yield (%)*
1) Crude extract, streptomycin SO_4		0.132	100
2) 60%-35% Ammonium sulfate back-extraction		0.442	112
3) Affinity chromatog.-MSOX		2.61	135
4) Bio-Gel A 1.5m		4.10	110
5) DEAE-cellulose		5.50	100
6) Bio-Gel A 0.5m		8.50	94
7) Affinity chromatog.-MSOX	E_I	24.2	77
	E_{II}	11.3	

addition of $FeSO_4$-acid molybdate complex formation and the γ-glutamyl-hydroxamate synthetase (GHS, with NH_2OH) activity was measured colorimetrically (A_{540}) by addition of acidic $FeCl_3$ (36, 37, 52).

Purification of Glutamine Synthetases

Purification was carried out according to the procedures outlined in Table 2. Since L-Met-sulfoximine (MSOX) inhibits glutamine synthetase from *B. caldolyticus* and a variety of other microorganisms in the micro-molar range (53) we have used an affinity column as a very effective purification procedure (steps 3 and 7). MSOX was coupled to CN-Br activated Bio-Gel A-15 m through ε-aminocaproic acid, and final separation of E_I and E_{II} (step 7) was accomplished with a 0-0.3 M KCl gradient. Bio Rad Affigel Blue with a KCl gradient (0-1.0 *M*) also is a very effective final purification and separation procedure. The rather striking result is that *B. caldolyticus* contains two separable, distinctly different glutamine synthetases.

Criteria of Homogeneity

Electrophoresis procedures included (a) migration of native enzyme in 5% polyacrylamide gels with the pH 7.2 buffer system of Chrambach (54); (b) isoelectric focusing in 4% gels containing 8 *M* urea and 1.7% Nonidet NP-40 with pH 3-6 LKB Ampholytes; (c) SDS (0.1%)-polyacrylamide (55) gels. E_I migrates faster than E_{II} on pH 7.2 PAGE, due to isoelectric point differences of 5.2 (E_I) and 5.5 (E_{II}). SDS-PAGE techniques reveal a subunit molecular weight very close to 50,000 for each enzyme. All three techniques indicated pure, homogeneous enzyme from the procedures of Table 2. Of special note is the fact that the two enzymes are not distinguishable on PAGE with the "standard" pH 8.9 Tris-Gly buffer system or on gels of >5% polyacrylamide. Our preliminary experiments with electron microscopy (39) and sedimentation velocity (53) in the ultracentrifuge do not distinguish or separate the two forms in any way. The dodecameric structures seen in electron micrographs clearly indicate a native enzyme MW = 600,000 for both E_I and E_{II}. The two enzymes purified from low-NH_4^+ grown cells differ from each other in virtually every parameter *except* quaternary structure, size, shape and molecular weight, as will be described. Table 3A summarizes and compares certain physical properties of the glutamine synthetases from *B. subtilis* (37°C), *B. stearothermophilus* (55°C), and *B. caldolyticus* (70°C).

Kinetic Properties

Kinetic properties in terms of substrate Km values, specific activity, and substrate specificity (assessed by reaction rates with structural analogs of substrates), are summarized in Table 3B. Obviously, the *B. caldolyticus* enzymes differ from each other rather markedly by these criteria, but have maintained a relatively high degree of specificity under the extreme condition of 70°C. E_I and E_{II} differ strongly with regard to relative degrees of activation of catalysis by Mn^{2+} and Mg^{2+}. The absolute magnitudes of the specific activity values do not increase markedly as one proceeds from the mesophilic enzymes to the thermophilic. This was found to occur in similar comparisons with other thermophilic enzymes (46) and is interpreted as reflecting a conformationally less flexible protein for thermophilic enzymes, i.e., a sacrifice of catalytic efficiency for structural stability. Certainly, the absence of cooperative binding of MnATP, seen clearly in the 37°C and 55°C enzymes (36, 60), with the 70°C *B. caldo* enzymes is consistent with this hypothesis.

Regulatory Properties

The existence of two glutamine synthetases in the case of the extreme thermophile, *B. caldolyticus*, begins to gain significance when one screens a wide variety of small molecules likely to be biosynthesized in part, directly or indirectly, from L-glutamate and L-glutamine. The inhibitions one observes, even at relatively high levels of substrates, are presented in Fig. 1. E_I and E_{II} differ markedly in their responses to these inhibitors, and the divalent cation present also has a pronounced effect on % inhibition in a number of cases. In separate experiments (not shown), % inhibition was also observed with each substrate concentration decreased in turn to twice the Km value. This also produced inhibition patterns that differed in complex but interesting ways: inhibition by certain modifiers was sometimes strengthened, sometimes weakened by high *vs.* low substrate levels. Despite the obvious complexity of these inter-related effects, the data of Fig. 1 clearly indicate that (a) E_I is inhibited significantly by Gly, L-Ala, L- and D-Ser, and L-Trp, but not by L-Gln or AMP; (b) E_{II} is not inhibited significantly by amino acids except L-Trp but strongly by L-Gln and AMP; (c) CTP, NAD and glucosomine-6-P (GA-6-P) inhibit both enzymes rather strongly at 10 mM; (d) nucleotide monophosphates other than AMP may slightly stimulate or inhibit activity, whereas mucleotide diphosphates inhibit activity moderately; and (e) inhibition is generally stronger with Mn^{2+} than with Mg^{2+} present. Overall, glutamine synthetase activity is seen to respond to the full spectrum of metabolic end products derived from L-Gln in the cell. However, differences in the regulatory responses of E_I and E_{II} cited above suggest an isozymic "division of labor" to accomplish this goal. This strategy may result from the necessary evolution of more stable, less flexible proteins capable of survival at 70°C or above.

At this point, our understanding of the regulatory behavior of the *B. caldolyticus* enzymes is rather incomplete. We have only preliminary, suggestive data on substrate-inhibitor site interactions. In an effort to probe modifier-modifier site interactions, experiments with all possible *pairs* of inhibitors with E_I and E_{II}, with either Mn^{2+} and Mg^{2+} present, have been carried out as done earlier by

TABLE 3

Summary of Properties of Bacillus sp. *Glutamine Synthetases*

Property *Cell growth T:*	*B. subtilis* *(37°C)*	*B. stearo.* *(55°C)*	*B. caldo.* *(70°C)*	
			E_I	E_{II}
A. Physical				
SU-MW	50,000	50,000	50,000	50,000
No. SU's	12	12	12	12
pI	-	4.6	5.2	5.6
$S^{\circ}_{20,W}$	19.3	19.75	19.4	19.4
B. Kinetic				
Km(NH_3), mM	0.4[d,e]	2.0	0.5, 6.0[b]	0.3, 3.5[b]
specificity[c]	high	high	moderate	moderate
Km(L-Glu), mM	0.80[d,e]	1.5	2.2, 18[b]	4.9, 18[b]
specificity[c]	high	high	high	high
Km(ATP-Mn), mM	0.36[a]	0.30[a]	3.0	1.5
(ATP-Mg)	-	-	2.9	0.5
specificity[c]	high	high	moderate	v. high
pH-optimum Mn^{2+}:	7.0	6.0	6.5	6.0
Mg^{2+}:	7.0	7.5	7.5	7.5

Specific Activity (μmoles P_i/min/mg)	Mn^{2+}	7[e]	32	14	6
	Mg^{2+}	18[e]	22	40	4
	Co^{2+}	-	8	3	0.1
Arrhenius plot:		NL[f]	NL[f]	NL[f]	NL[f]
References:		(60 - 64)	(48)	(50, 51)	(50, 51)

[a] *Cooperative, Hill n = 2.3*

[b] *Biphasic substrate binding*

[c] *As judged by relative reaction rates with substrate analogs (not shown)*

[d] *S-inhibition with Mn^{2+} present*

[e] *Determined at 25°C*

[f] *NL = non-linear*

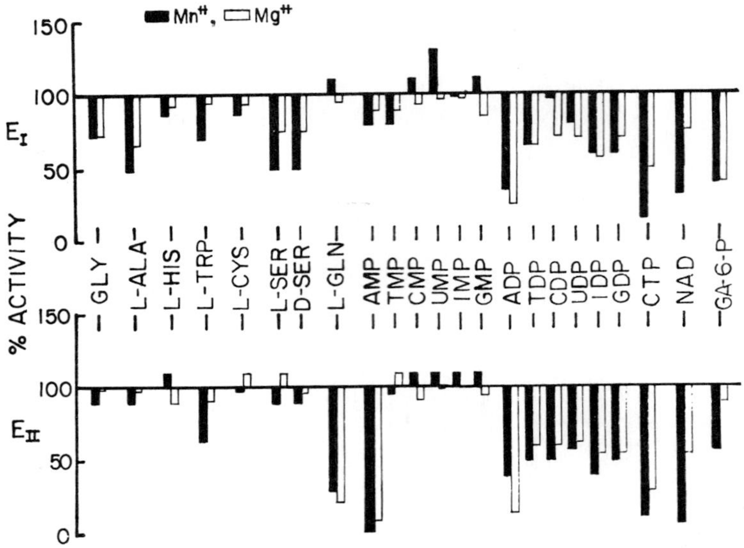

FIGURE 1. Inhibition of biosynthetic activity of B. caldolyticus *glutamine synthetases E_I (top) and E_{II} (bottom) by feedback modifiers derived from L-Gln. Each modifier was used at 10 mM, except that L-Trp and TMP were used at 5 mM. Nucleotide di- and tri-phosphates were studied with divalent metal ion at 1 mM in excess of the* total *nucleotide level; (filled bars) Mn^{2+}, pH 6.20; (open bars) Mg^{2+}, pH 7.50 in the biosynthetic assay with substrate levels at 5 x Km values. Taken from Wedler* et al. *(40) and reproduced with permission from Academic Press.*

Woolfolk and Stadtman with the *E. coli* enzyme (61). The effect of each possible pair of inhibitors, together, was judged to be either antagonistic, cumulative, additive, or synergistic. Synergistic or cumulative effects demand separate modifier binding sites, whereas antagonistic or additive effects may result from separate sites or competition for the same site. Fig. 2 summarizes the relative number of each type of interaction seen for E_I and E_{II} with Mn^{2+} or Mg^{2+} present, along with those modifiers most frequently involved in each type of site interaction. The number of separate sites for modifiers, as well as the complexity of their interactions, is apparently quite high for each enzyme.

Whether adenylylation-deadenylylation occurs with E_I (or E_{II}) is not yet known definitively, despite preliminary observations (40) that suggested it may occur. Present research is directed at this key question, along with completion of detailed studies of modifier action.

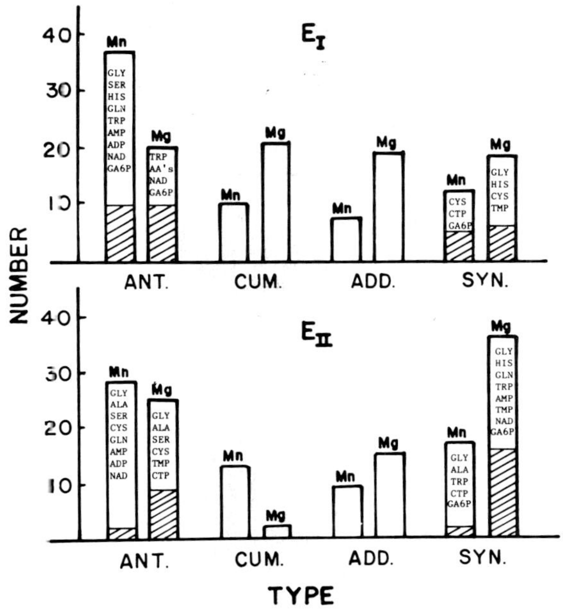

FIGURE 2. Summary of modifier - modifier site interactions with B. caldolyticus *glutamine synthetases* E_I *and* E_{II} *isozymes, dependent on activating divalent metal ion. The height of the bar represents the total number of interactions of a given type, while the shaded heights are for the number that are very strong. The % inhibition with each inhibitor was determined separately (cf. Fig. 1) at a concentration that gave 20-30% inhibition. Then the % inhibition with all possible* pairs *of inhibitors at those concentrations was determined. The % cumulative inhibition was calculated according to Woolfolk and Stadtman (61). Antagonistic effects were those below the theoretical cumulative value, and synergistic effects were those above the additive inhibition effects calculated.*

Thermostabilization

With our discovery of *B. caldo.* E_I and E_{II}, isozymes, we now estimate that our earlier data (39), presented at the Zürich conference in 1975, on thermostabilization of the *B. caldo.* "enzyme" involved a *ca.* 1:1 mixture of E_I and E_{II}. In-depth studies with the separated, purified enzymes have not carried very far yet; however, we do know that E_I is inherently less stable than E_{II}. *Intrinsic* thermostabilization of one or both enzymes seem to correlate with increases in polar amino acids and very likely by disulfide cross-links. *Extrinsic* stabilization effects occur by rather potent substrate protection with L-Glu and MnATP and some modifiers, and by mild stabilization with a few divalent cations. Aggregation (driven by heat, due to hydrophobic interactions) of dodecameric native enzyme to higher polymers may also be important *in vivo*. Taken together, these effects must provide considerable net stabilization.

Overview

The observation of two forms of the same enzyme in the same cell is not totally new. There are two forms of ribulose-1,5-bisphosphate carboxylase in *R. spheroides* (62), of triosephosphate isomerase in *Euglena gracilis* (63), and both class I and class II aldolases in *Euglena* (63) and *L. casei* (64). Very recently, two forms of glutamine synthetase were observed in *Rhizobia* (65). In several of the mesophilic organisms, the isozymes either occur with compartmentalization or are expressed under different growth conditions. With the extreme thermophile, *B. caldolyticus*, it apparently is necessary to both reduce the number and interactions of ligand sites per subunit to provide a more thermostable protein. Thus, the evolution of isozymes ensures a complete response by glutamine synthetase to the full spectrum of feedback modifiers, and regulation of L-glutamine production is no less rigorous.

A number of very interesting questions remain unanswered at this point. Does either E_I or E_{II} undergo covalent modification (adenylylation or other changes) *in vivo*? If so, how does this alter regulation of this form? Is it possible that *B. stearothermophilus*, *B. subtilis*, or *B. licheniformis* also contain regulatory isozymes that were overlooked previously? What are the most potent feedback modifiers among those studied so far, and how do their sites interact with each other and with substrate sites? Can we really unravel the very subtle mechanisms of thermostabilization for this complex but intriguing enzyme?

REFERENCES

1. Yoshida, M., Oshima, T., and Imahori, K. (1971). Biochem Biochem. Biophys. Res. Comm. *43*, 36.
2. Yoshida, M. (1972). Biochemistry *11*, 1087.

3. Hengartner, H., Kolb, E., and Harris, J. I. (1976). In *Enzymes and Proteins from Thermophilic Microorganisms* (H. Zuber, ed.), p. 199, Birkhauser Verlag, Basel.
4. Cass, K. H. and Stellwagen, E. (1975). Arch. Biochem. Biophys., *171*, 682.
5. M. Yoshida, M., Oshima, T., and Imahori, K. (1973). J. Biochem. *74*, 1183.
6. Sugimoto, S. and Nosoh, Y. (1971). Biochem. Biophys. Acta., *235*, 210.
7. Hill, H. A. O., Lobb, R. R., Sharp, S. L. and Stokes, A. M. (1976). Biochem. J. *153*, 551.
8. Howard, R. L. and Becker, R. R. (1972). Biochem. Biophys. Acta, *268*, 249.
9. Freeze, H. and Brock, T. D. (1970). J. Bacteriol. *101*, 541.
10. Fujita, S. C., Oshima, T. and Imahori, K. (1976). Eur. J. Biochem. *64*, 57.
11. Weerkamp, A. and MacElroy, R. D. (1972). Arch. Mikrobiol. *85*, 113.
12. Higa, A. I. and Cazzulo, J. J. (1976). Experientia *32*, 1373.
13. Ramaley, R. F. and Hudock, M. O. (1973). Biochem. Biophys. Acta, *315*, 22.
14a. Saiki, T., Mahmud, I., Matsubara, N., Taya, K. and Arima, K. (1976). In *Enzymes and Proteins from Thermophilic Microorganisms* (H. Zuber, ed.), p. 169, Birkhauser Verlag, Basel.
14b. Saiki, T. and Arima, K. (1975). J. Biochem. *77*, 233.
15. Cook, W. R. and Ramaley, R. F. (1976). In *Enzymes and Proteins from Thermophilic Microorganisms* (H. Zuber, ed.), p. 207, Birkhauser Verlag, Basle.
16. Veronese, F. M., Baccu, E. and Fontana, A. (1976). Biochemistry *16*, 4026.
17. Griffiths, M. W. and Sundaram, T. K. (1973). J. Bacteriol. *116*, 1160.
18. Sundaram, T. K., Libor, S. and Chell, R. M. (1976). In *Enzymes and Proteins from Thermophilic Microorganisms* (H. Zuber, ed.), p. 263, Birkhauser Verlag, Basel.
19. Ljungdahl, L. G., Sherod, D. W., Moore, M. R. and Andreesen, J. R. (1976). In *Enzymes and Proteins from Thermophilic Microorganisms* (H. Zuber, ed.), p. 237, Birkhauser Verlag, Basel.
20. O'Brien, W. E., Brewer, J. M. and Ljungdahl, L. G. (1976). In *Enzymes and Proteins from Thermophilic Microorganisms* (H. Zuber, ed.), p. 249, Birkhauser Verlag, Basel.
21. Chin, H. W. and Trela, J. M. (1973). J. Bacteriol. *114*, 674.

22. Bridger, G. P. and Sundaram, T. K. (1976). J. Bacteriol. *125*, 1211.
23. Sundaram, T. K., Cazzulo, J. J. and Kornberg, H. L. (1971). Biochem. J. *122*, 663.
24. Schindler, P. W. and Zahner, H. (1973). Eur. J. Biochem. *39*, 591.
25. Orengo, A. and Saunders, G. F. (1972). Biochemistry *11*, 1761.
26. Sando, G. N. and Hogenkamp, H. P. C. (1973). Biochemistry *12*, 3316.
27. Kobayashi, S. H., Hubbell, H. R. and Orengo, A. (1974). Biochemistry *13*, 4537.
28a. Kuramitsu, H. K. (1968). Arch. Biochem. Biophys. *167*, 643.
28b. Kuramitsu, H. K. (1970). J. Biol. Chem. *245*, 2991.
29. Cavari, B. Z., Arkim-Shlank, H. and Grossowicz, N. (1972). Biochem. Biophys. Acta *261*, 161.
30. Kuramitsu, H. K. and Yoshimura, S. (1972). Biochem. Biophys. Acta *264*, 152.
31. Saiki, T. and Arima, K. (1970). Agr. Biol. Chem. (Tokyo) *34*, 1762.
32. Cavari, B. Z. and Grossowicz, N. (1973). Biochem. Biophys. Acta *203*, 183.
33. Saiki, T., Shinshi, H. and Arima, K. (1973). J. Biochem. *74*, 1239.
34. Thomas, D. A. and Kuramitsu, H. K. (1971). Arch. Biochem. Biophys. *145*, 96.
35. Muramatsu, N. and Nosoh, Y. (1976). J. Biochem. *80*, 485.
36. Wedler, F. C. and Hoffmann, F. M. (1974). Biochemistry *13*, 3207.
37. Wedler, F. C., Carfi, J. and Ashour, A. E. (1976). Biochemistry *15*, 1749.
38. Hachimori, A., Matsunaga, A., Shimizu, M., Samejima, T. and Nosoh, Y. (1974). Biochem. Biophys. Acta *350*, 461.
39. Wedler, F. C., Hoffmann, F. M., Kenney, R. and Carfi, J. (1976. In *Enzymes and Proteins from Thermophilic Microorganisms* (H. Zuber, ed.), p. 187, Birkhauser Verlag, Basel.
40. Wedler, F. C., Kenney, R. M., Ashour, A. E. and Carfi, J. (1978). Biochem. Biophys, Res. Comm. *81*, 122.
41. Mulvey, R. S. and Fersht, A. R. (1976). Biochemistry *15*, 243.
42. Chien, A., Edgar, D. B. and Trela, J. M. (1976). J. Bacteriol. *127*, 1550.
43. Air, G. M. and Harris, J. I. (1974). FEBS Letters *38*, 277.
44. Fábry, M., Suemegi, J. and Venetainer, F. (1976). Biochem. Biophys. Acta *435*, 228.

45. Singleton, R., Jr. and Amelunxen, R. E. (1973). Bacteriol. Rev. *37*, 320.
46. Ljungdahl, L. G. and Sherod, D. (1976. In *Extreme Environments: Mechanisms of Microbial Adaptation* (M. R. Heinrich, ed.), p. 147, Academic Press, New York.
47. Stadtman, E. R. and Ginsburg, A. (1973). In *The Enzymes* (P. D. Boyer, ed.), Vol. 10, p. 755, Academic Press, New York.
48. Heinen, W. (1971). Arch. Mikrobiol., *76*, 2.
49. Heinen, U. J. and Heinen, W. (1972). Arch. Mikrobiol. *82*, 1.
50. Ramaley, R., personal communication.
51. Johnson, E., personal communication.
52. Shapiro, B. M. and Stadtman, E. R. (1971). *Methods in Enzymology* *17*, 910.
53. Unpublished results, this laboratory.
54. Rodbard, D. and Chrambach, A. (1971). Analyt. Biochem. *40*, 95.
55. Weber, K. and Osborn, M. (1969). J. Biol. Chem. *244*, 440.
56. Deuel, T. F., Ginsburg, A., Yeh, J., Shelton, E. and Stadtman, E. R. (1970). J. Biol. Chem. *245*, 5195.
57. Deuel, T. F. and Stadtman, E. R. (1970). J. Biol. Chem. *245*, 5206.
58. Deuel, T. F. and Turner, D. C. (1972). J. Biol. Chem. *247*, 3039.
59. Deuel, T. F. and Prusiner, S. (1974). J. Biol. Chem. *249*, 257.
60. Wedler, F. C. (1974). Biochem. Biophys. Res. Comm. *60*, 737.
61. Woolfolk, C. A. and Stadtman, E. R. (1964). Biochem. Biophys. Res. Comm. *17*, 313.
62. Gibson, J. L. and Tabita, F. R. (1977). J. Biol. Chem. *252*, 943.
63. Mo, Y., Harris, B. G. and Gracy, R. W. (1973). Arch. Biochem. Biophys. *157*, 580.
64. London, J. (1974). J. Biol. Chem. *249*, 7977.
65. Darrow, R. A. and Knotts, R. R. (1977). Biochem. Biophys. Res. Comm. *78*, 554.

ON THE THERMOSTABILITY OF AMINO GROUP MODIFIED GLUTAMINE SYNTHETASE FROM *BACILLUS STEAROTHERMOPHILUS*

T. Sekiguchi
A. Matsunaga
A. Shimamura
Y. Nosoh
A. Hachimori
T. Samejima

Laboratory of Chemistry of Natural Products
Tokyo Institute of Technology
Meguro, Tokyo

Many studies have recently been made to elucidate the mechanism of thermostability or thermophilicity of thermophilic microorganisms, especially the mechanism of thermostability of macromolecules such as proteins, enzymes and nucleic acids from thermophiles. Oshima (1976) proposed three different types of mechanisms for the thermostability of thermophiles. Thermostability of the first type is intrinsic, and the enzymes of this type are thermostable without any stabilizer. Most of intracellular enzymes belong to this type. The second type involves protective factors, and extracellular enzymes such as thermolysin, a proteolytic enzyme from *Bacillus thermoproteolyticus* (Endo, 1962; Feder *et al.*, 1971; Drucker and Borchers, 1971), and α-amylase from *B. stearothermophilus* (Ogasawara *et al.*, 1970) are known to belong to this type. These enzymes are stabilized by metal ions such as Ca^{2+}. The thermostability of macromolecules of the third type is thermally induced, as exemplified by tRNA from *Thermus thermophilus* (Watanabe *et al.*, 1974; Oshima, 1976). When the bacterium was grown at elevated temperatures the 5-methyl-2-thiouridine content of tRNA increased, and concomitantly the content of ribothymidine decreased. A correlation between thermostability and 5-methyl-2-thiouridine content of tRNA and the thermal induction of tRNA thermostability was suggested.

ISBN 0-12-268250-5

In order to reveal the mechanism of thermostability of thermophile enzymes of the first type, almost all the studies have been made by comparing the conformations of thermophile enzymes with those of the counterparts from mesophiles. As is well known, the conformations of mesophile enzymes catalyzing the same reaction are more or less different from each other, and similarly the thermophile proteins showing the corresponding enzyme activity have different conformations. In order to obtain a clear conclusion on peculiar conformation relating to the thermostability of thermophile enzyme we have to accumulate much information on the conformations of mesophile and thermophile enzymes, and compare the conformations of the two groups of enzymes (Ljungdahl and Sherod, 1976; Singleton, 1976).

Investigations on the enzymes or other macromolecules of the second type may therefore be commendable because the difference in conformation between the enzyme samples in the presence and absence of stabilizer may be considered to be mainly related to the thermostability of the enzyme. Similarly, if an enzyme from a thermophilic bacterium grown at high temperatures is thermostable and the corresponding enzyme from the organism grown at low temperatures is thermolabile, the difference in conformation between the two enzymes may be related to the thermostability of the thermophile enzyme. Campbell (1955) reported that α-amylase from *B. stearothermophilus* grown at 55°C was thermostable, but the enzyme from the bacterium grown at 37°C was thermolabile. Isono (1970) examined the amino acid compositions of the amylases from the 55°C- and 37°C-grown cells of another strain of *B. stearothermophilus* and found that the compositions of the two enzymes were different.

A similar but more interesting approach to reveal the mechanism of thermostability of thermophile enzymes has recently been devised by Haberstisch and Zuber (1974). They have succeeded in growing thermophilic bacteria such as *B. caldotenax* at 37°C. The bacteria are unable to grow at 37°C unless subjected to an adaptation procedure. Enzymes from the cells grown at 65°C were thermostable, while the corresponding enzymes from the cells grown at 37°C were thermolabile. Zuber and his co-workers have been studying the mechanism of thermostability of thermophile enzymes by comparing the conformations of the 65°C enzymes with those of the counterparts from the 37°C cells. Some of the results are presented by Zuber in these proceedings.

Chemical modification of amino acid residues of enzymes has been used to reveal the relationship between the structure and function of enzymes (Cohen, 1970). The modifications of amino acid residues of an enzyme may cause some conformational change in the protein. Loss of thermostability of thermophile enzymes may therefore be expected to result from chemical modification of the enzyme. If a thermophile enzyme can be modified without any loss of enzyme activity and the modified enzyme becomes thermolabile, the comparison of the conformations of the native and modified enzymes may afford a new approach to reveal the thermostability mechanism.

The modified thermolysin with 3 out of 28 tyrosyl and 2 out of 3 tryptophanyl residues reacted with tetranitromethane was less thermostable than the native enzyme, but the modified enzyme exhibited only 50% of the activity of the native enzyme (Ohta *et al.*, 1974). Glutamine synthetase from *B. stearothermophilus* contains 12 subunits (Hachimori *et al.*, 1974), and 1 out of 3 SH groups/subunit has been shown to be essential for its activity by the modification method with 5,5'-dithiobis(2-nitrobenzoic acid) (DTNB) (Hachimori *et al.*, 1975). The modified enzyme with 0.7 SH group/subunit reacted with DTNB was shown to exhibit only 25% activity compared to the original level and also lost thermostability. These modified enzymes, although thermolabile, were considerably inactivated. In the present study, therefore, chemical modification of thermophile glutamine synthetase was attempted to obtain thermolabile modified enzymes without any loss of enzyme activity. Among the amino acid residues which have usually been subjected to modification, tryptophan, histidine and lysine were chosen to be modified.

As reported previously (Hachimori *et al.*, 1974), the enzyme contains 5 tryptophan, 13 histidine and 27 lysine residues/subunit. When the enzyme was treated with 50 moles of 2-hydroxy-5-nitrobenzyl bromide per subunit according to the procedure of Horton and Koshland (1965), about 2 out of 5 tryptophan residues/subunit were modified without any loss of activity. The thermostability of the modified enzyme was the same as that of the native enzyme. Modification of additional tryptophan residues resulted in decreased activity and also thermostability. When the enzyme was treated with 150 moles of diethyl pyrocarbonate/subunit according to the procedure of Thomé-Beau *et al.* (1971), about 4 out of 13 histidine residues/subunit were modified. The modified enzyme exhibited only 50% activity compared to the original level and was less thermostable than the native enzyme. When 10 histidine residues/subunit were modified, the activity was completely lost. On treating the enzyme with 100 moles of succinic anhydride/subunit according to the method of Bethune *et al.* (1967), 4 out of 27 lysine (amino groups)/subunit were succinylated. The modified enzyme retained completely the native enzyme activity, but became thermolabile. The effects of succinylation of amino groups of the enzyme on the activity and thermostability were then examined in more detail.

When the enzyme was treated with increasing amounts of succinic anhydride, increasing amounts of amino groups were modified. As shown in Fig. 1, on succinylation of up to 4.6 amino groups/subunit, no activity change was observed. When more amino groups were succinylated, the activity gradually decreased, and on succinylation of 12 amino groups the activity was almost completely lost.

The thermostability of the enzyme gradually decreased on succinylation of increasing amounts of amino groups (Table 1). The succinylated enzyme of which 4.6 amino groups/subunit were modified with succinic anhydride (4.6 NH_2-modified enzyme) was less thermostable than the native enzyme. As in the case of the native enzyme (Hachimori *et al.*, 1974), Mg^{2+} showed a remarkable protection against thermal inactivation of the modified enzymes. For example, the 4.6 NH_2-modified enzyme exhibited almost the same thermostability as that of the native enzyme in the presence of 50 mM $MgCl_2$. Similar protection against thermal inactivation of the native and modified enzymes was observed with 7.5 mM $MnCl_2$.

The pH and temperature profiles for the γ-glutamyl transfer reaction of the 4.6 NH_2-modified enzyme were similar to those of the native enzyme (Hachimori *et al.*, 1974), and the optimum pH and temperature were 7.8-8.0 and 68°C, respectively. The values for the Michaelis constant and maximum

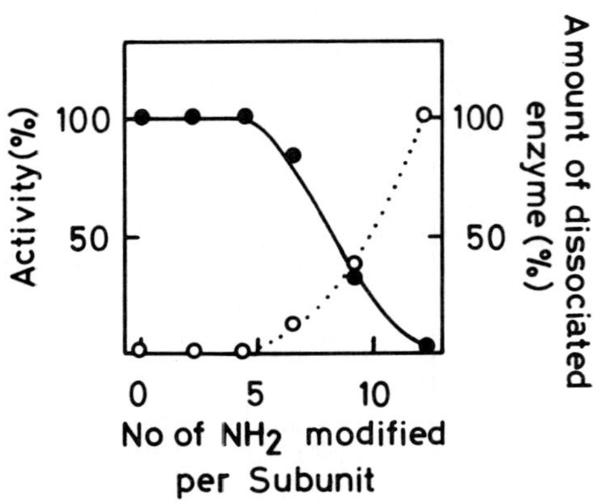

FIGURE 1. Effect of succinylation on the activity of glutamine synthetase. The activity (●) was measured by the γ-glutamyl transfer method (Hachimori et al.*, 1974). The ultracentrifugal patterns of succinylated enzyme samples exhibited one fast sedimenting component of about 18S or a fast sedimenting component of about 18S and a slowly sedimenting component of about 3S. The amount of dissociated molecules in each enzyme sample (o) was calculated from the areas of the fast and slow sedimenting peaks.*

TABLE 1
Effect of Succinylation on Thermostability of Glutamine Synthetase

No. of NH_2 groups modified/subunit	$MgCl_2$ (50 mM)	Activity[a] 30 min	Activity[a] 60 min
0	-	86	77
	+	99	97
2.3	-	60	46
	+	95	91
4.6	-	42	20
	+	96	89
6.3	-	7	6
	+	59	45
9.1	-	5	4
	+	16	9

[a]*The native or modified enzyme (1.0 mg protein/ml) was heated at 65°C in 20 mM Tris-HCl buffer (pH 7.2) containing 0.1 M NaCl. The remaining activity was measured by the γ-glutamyl transfer reaction method.*

velocity for glutamine in the transfer reaction of the 4.6 NH_2-modified enzyme at 65°C were 66.7 mM and 0.54 μmole/min/mg protein with 50 mM $MgCl_2$, respectively. These values for the native enzyme were the same as those for the modified enzyme. The enzyme has been shown to be cumulatively inhibited by some amino acids and nucleotides (Hachimori *et al.*, 1974). As shown in Table 2, the transfer reaction of the 4.6 NH_2-modified enzyme was cumulatively inhibited by alanine, glycine ans serine, as in the case of the native enzyme. The 4.6 NH_2-modified enzyme may be similar to the native enzyme in catalytic property, but different in thermostability. The difference in protein conformation between the native and 4.6 NH_2-modified enzymes can be considered to be related to the thermostable property of the enzyme.

As shown in Fig. 1, on succinylating up to 4.6 amino groups/subunit, no subunit-dissociation of the enzyme was observed. The value of the sedimentation constant ($s^{\circ}_{20,w}$) of the 4.6 NH_2-modified enzyme (17.9S), however, was slightly

TABLE 2
Effect of Inhibitors on the 4.6 NH_2-modified Enzyme

Inhibitor (20 mM)	*Activity (%)*[a]	
	Native	*Modified*
None	100	100
Alanine	50	54
Glycine	64	64
Serine	59	62
Ala + Gly + Ser	22 (19)[b]	23 (22)[b]

[a]*The activity was measured by the biosynthetic reaction method at 65°C (Hachimori* et al., *1974).*

[b]*Calculated for cumulative inhibition according to the method of Stadtman (1970).*

smaller than that of the native enzyme (18.5S), indicating a slight change in molecular shape of the enzyme on succinylation of amino groups. As in the case of the native enzyme (18.5 → 19.0S), the value of $s^{\circ}_{20,w}$ of the modified enzyme increased, although slightly, on addition of Mg^{2+} (17.9 → 18.6S). This indicates a slight change in molecular shape of the native and succinylated enzymes on addition of Mg^{2+}.

It has been shown that the α-helix content of the enzyme increased from 7 to 20% on addition of Mg^{2+} (Hachimori *et al.*, 1974). The α-helix contents of the succinylated enzymes (2.3 to 13.6 NH_2-modified enzymes) were calculated to be 20%, even in the absence of Mg^{2+}, and no change in α-helix content of the modified enzymes was observed on addition of Mg^{2+}.

Fig. 2 shows the CD spectra in the near ultraviolet region of the native and succinylated enzymes in the presence and absence of Mg^{2+}. As shown in the figure, the state of the aromatic amino acid residues in the enzyme molecule (Hachimori *et al.*, 1974) was not changed on succinylation of up to 6.3 amino groups/subunit, and the changes in the state of amino acid residues of the modified enzymes on addition of Mg^{2+} were the same as that of the native enzyme.

The native enzyme, when excited at 290 nm, exhibited a fluorescence spectrum with an emission maximum at 337 nm, indicating that the spectrum may be due to the tryptophan residues of the enzyme (Matsunaga and Nosoh, 1974). As shown in Fig. 3, the state of tryptophan residues manifested as the fluorescence spectrum of the enzyme was not changed

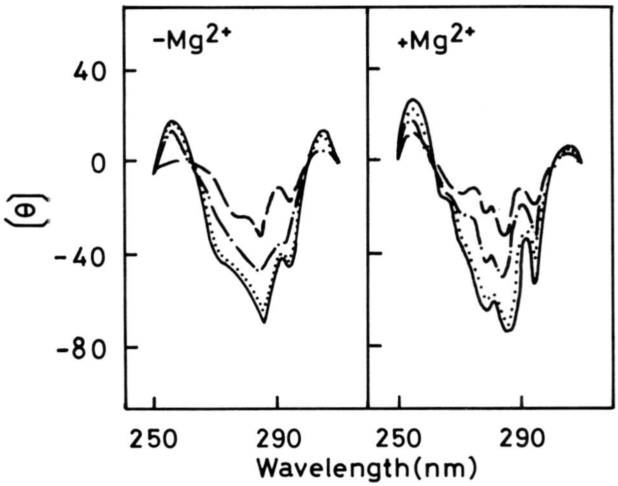

FIGURE 2. CD spectra in the near ultraviolet region of NH_2-modified enzymes in the presence and absence of 50 mM $MgCl_2$. ——— native; ···· , 2.3, 4.6 and 6.3 NH_2-modified; —·—·—, 9.1 NH_2-modified; and - - -, 12.5 NH_2-modified.

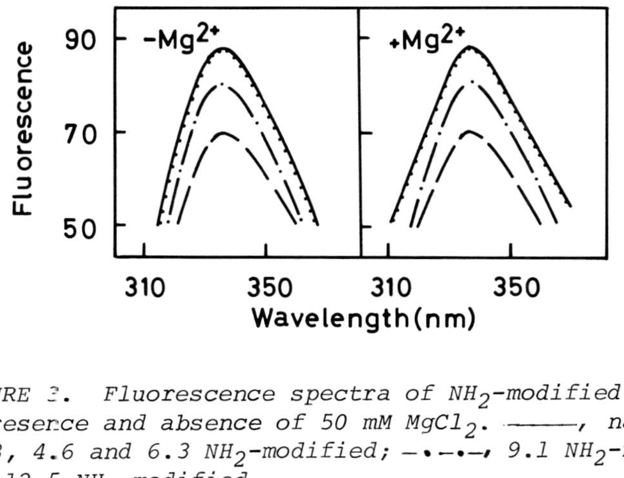

FIGURE 3. Fluorescence spectra of NH_2-modified enzymes in the presence and absence of 50 mM $MgCl_2$. ———, native; ···· , 2.3, 4.6 and 6.3 NH_2-modified; —·—·—, 9.1 NH_2-modified; and ---, 12.5 NH_2-modified.

on succinylation of up to 6.3 amino groups, and the states of the tryptophan residues of the succinylated enzymes were not affected on addition of Mg^{2+}.

The fluorescence spectrum of *p*-toluidinylnaphthalene sulfonic acid (TNS) in aqueous solution, when excited at 330 nm, exhibited an emission maximum at 470 nm. The emission maximum of TNS, when excited with the enzyme, was shifted to 450 nm with a considerable increase in intensity, suggesting that TNS was non-covalently bound to a hydrophobic site on the enzyme (Matsunaga and Nosoh, 1974). As shown in Fig. 4, the TNS-binding property of the enzyme gradually changed on succinylation of increasing amounts of amino groups. This property of the enzyme was considerably affected by Mg^{2+}. On addition of Mg^{2+} to the 2.3, 4.6 and 6.3 NH_2-modified enzymes, the TNS-binding properties of these modified enzymes seemed to change to the level of the native enzyme in the presence of Mg^{2+}.

The results shown in Table 1 and Figs. 2, 3 and 4 may indicate that among the conformations of the native and succinylated enzymes examined in the present study, only the TNS-binding properties (probably hydrophobic properties) of the enzymes are related to the thermostabilities of the proteins. Any clear conclusion on the relationship between hydrophobicity and thermostability of thermophile enzymes has not yet been deduced from the comparison of the amino acid compositions of thermophile enzymes with those of the counterparts from mesophiles (Singleton and Amelunxen, 1973; Bull and Breese, 1973; Ljungdahl and Sherod, 1976; Middaugh and MacElroy, 1976). As presented in this proceedings and

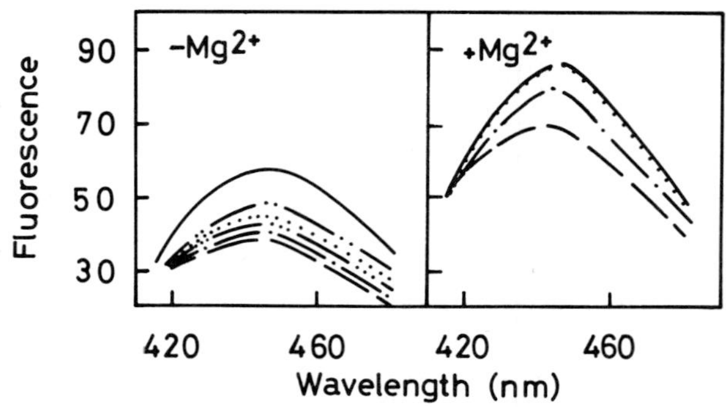

FIGURE 4. Fluorescence spectra of TNS bound to NH_2-modified enzymes in the presence and absence of 50 mM $MgCl_2$. ——, native; — •• —, 2.3 NH_2-modified; ····, 4.6 NH_2-modified; — ••• —, 6.3 NH_2-modified; — • —, 9.1 NH_2-modified; and ---, 12.5 NH_2-modified.

recently reported by Yutani *et al.* (1977), however, the thermostability of an enzyme may result from a subtle change in hydrophobicity of a small part of the protein molecule. Succinylation of few amino acid residues of thermophile glutamine synthetase may have caused some conformational change resulting in such a subtle change in hydrophobic property of the enzyme. The possibility that salt bridges between amino and carboxyl groups may partly be attributed to thermostability of the enzyme, as suggested for glyceraldehyde-3-phosphate dehydrogenase from *B. stearothermophilus* (Biesecker *et al.*, 1977), however, should also be considered. Succinylation of amino groups of the enzyme may have disrupted such salt bridges, resulting in decreased thermostability.

Thermophile enzymes have been shown to be not only more thermostable but also more resistant against urea denaturation than the corresponding mesophile enzymes (Ohta, 1967; Ogasawara *et al.*, 1970; Amelunxen *et al.*, 1970; Suzuki and Imahori, 1973). This may indicate that thermophile enzymes have more firmly folded protein structure than the corresponding mesophile enzymes. We may therefore expect that the stabilities to urea denaturation of the native and 4.6 NH_2-modified enzymes in the presence and absence of Mg^{2+} would decrease in their order of decreased thermostability. The native and modified enzymes were treated with increasing concentrations of urea at 35°C for 16 hrs in the presence and absence of 50 mM $MgCl_2$. The ultracentrifugal analysis and CD measurement of each of the urea treated enzyme samples were carried out in the same medium as used for the urea treatment after appropriate dilution of the sample, if necessary.

When the four enzyme samples were treated with urea, the amount of a component with an $s_{20,w}$ value of 18S in each enzyme sample decreased, as shown in Fig. 5. Concomitantly, a small component with an $s_{20,w}$ value of about 3S appeared and the amount of this component increased. The results shown in the figure indicate that the stabilities to urea denaturation of the quaternary structure of the four enzymes decreased as the thermostabilities of the samples decreased, except in the case of the native enzyme without Mg^{2+}.

The states of the tryptophan residues manifested as the CD spectrum in the near ultraviolet region of the four enzymes were found to be changed, as shown in Fig. 6. The changes are represented as those of the values of mean residues ellipticity, (θ), at 293 nm. It may be considered from the results that the tertiary structures related to the state of tryptophan residues of the enzymes were more easily unfolded by urea as the thermostabilities of the samples decreased.

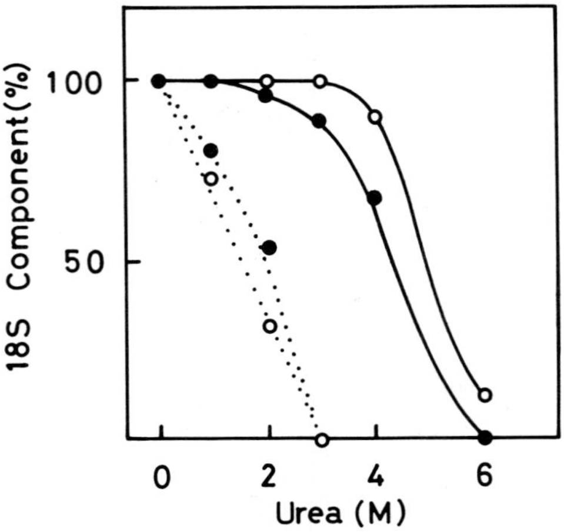

FIGURE 5. Effects of urea on the subunit structure of the native (o) and 4.6 NH_2-modified (●) enzymes in the presence (——) and absence (····) of 50 mM $MgCl_2$.

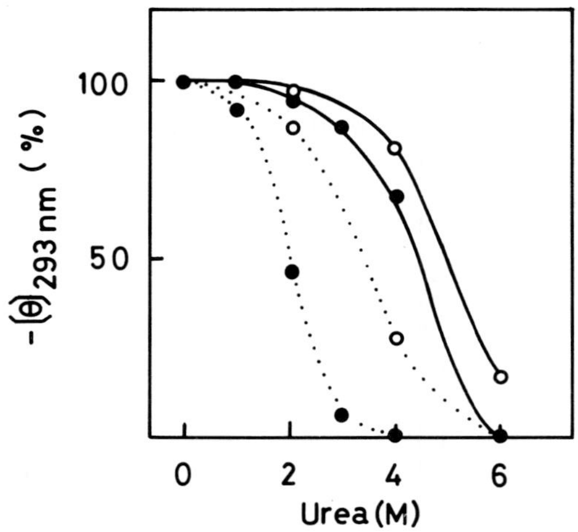

FIGURE 6. Effects of urea on the CD spectra in the near ultraviolet region of the native (o) and 4.6 NH_2-modified (●) enzymes in the presence (——) and absence (····) of 50 mM $MgCl_2$.

Fig. 7 shows the changes of the values of (θ) at 220 nm of the four enzymes on treating with increasing concentrations of urea. The results represent the changes in the content of the ordered structures in the secondary conformation of the enzymes. The change in stability to urea denaturation of the secondary structure of the four enzymes seemed not to parallel the changes in thermostability of the samples.

The change in thermostability of the four enzyme samples, the native and 4.6 NH_2-modified enzymes in the presence and absence of Mg^{2+}, may be parallel to the change in stability of the quaternary and tertiary structures of the sample proteins. Urea has been considered to disrupt hydrophobic bonding (Nozaki and Tanford, 1963) and/or hydrogen bonding (Robinson and Jencks, 1965). Thermophile enzymes may therefore be more thermostable and, at the same time, more resistant against urea denaturation than the counterparts from mesophiles, partly due to increased hydrophobic properties.

The results obtained in the present study indicate that chemical modification can be used as one of the effective approaches to reveal the mechanism of thermostability of thermophile enzymes. If the modification of a thermophile enzyme is carried out carefully with modifiers specific for a given type of functional group, we may obtain thermolabile modified enzymes with different amino acid residues reacted

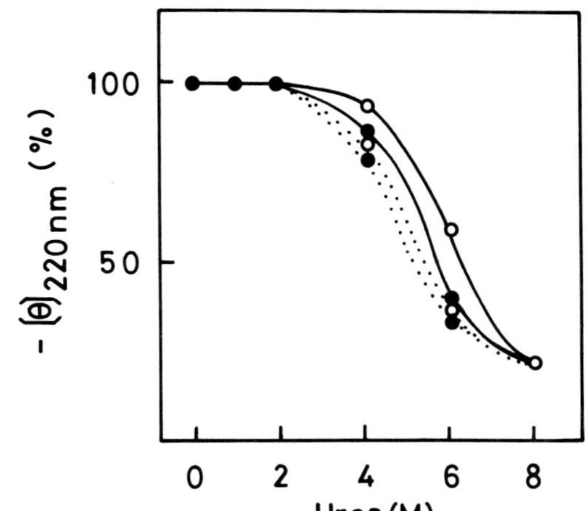

FIGURE 7. Effects of urea on the CD spectra in the far ultraviolet region of the native (o) and 4.6 NH_2-modified (•) enzymes in the presence (——) and absence (····) of 50 mM $MgCl_2$.

with the modifiers. If the modified enzymes exhibit the activity of the native enzyme, the accumulated information on the differences in conformation between the native and modified enzymes will help us to deduce any conclusion on the mechanism of thermostability of the enzyme. In contrast to thermostable thermophile enzymes, chemical modification of thermolabile mesophile enzymes can be expected to yield thermostable enzymes. Comparison of the conformations of the native and modified mesophile enzymes will also be useful to reveal the mechanism of thermostability. In addition, it seems worthwhile to examine the effects on thermophile and mesophile enzymes of bifunctional modifiers specific for amino groups, such as dimethyladipoimidate. The thermostable property will be increased or endowed with thermophile or mesophile enzymes, respectively, upon bridge formation between two amino acid residues in the protein molecule.

REFERENCES

Amelunxen, R. E., Noelken, M. and Singleton, R. Jr. (1970). Arch Biochem. Biophys. *141*, 447.

Bethune, J. L., Ulmer, D. D. and Vallee, B. L. (1967). Biochemistry *6*, 1955.

Biesecker, G., Harris, J. L., Thierry, J. G., Walker, J. E. and Wonacott, A. J. (1977). Nature *266*, 328.

Bull, H. B. and Breese, K. (1973). Arch. Biochem. Biophys. *158*, 681.

Campbell, L. L. (1955). Arch. Biochem. Biophys. *54*, 154.

Cohen, L. A. (1970). In *The Enzymes* (P. D. Boyer, ed.), Bol. 1, p. 147, Academic Press, New York.

Drucker, H. and Borchers, S. L. (1971). Arch. Biochem. Biophys. *147*, 242.

Endo, S. (1962). J. Ferment. Technol. *40*, 346.

Feder, J., Garrett, L. R. and Will, B. S. (1971). Biochemistry *10*, 4552.

Haberstisch, H.-U. and Zuber, H. (1974). Arch. Microbiol. *98*, 275.

Hachimori, A., Matsunaga, A., Shimizu, M., Samejima, T. and Nosoh, Y. (1974). Biochim. Biophys. Acta *350*, 461.

Hachimori, A., Tanabe, A., Nagaoka, T., Suzuki, H., Nosoh, Y. and Samejima, T. (1975). J. Biochem. *78*, 1235.

Hartman, F. C. and Wold, F. (1967). Biochemistry *6*, 2439.

Horton, H. R. and Koshland, D. E. (1965). J. Am. Chem. Soc. *87*, 1126.

Isono, K. (1970). Biochem. Biophys. Res. Comm. *41*, 852.

Ljungdahl, J. G. and Sherod, D. (1976). In *Extreme Environments: Mechanism of Microbial Adaptation*

(M. R. Heinrich, ed.), p. 147, Academic Press, New York.

Matsunaga, A. and Nosoh, Y. (1974). Biochim. Biophys. Acta *365*, 208.

Middaugh, C. R., and MacElroy, R. D. (1976). In *Extreme Environments: Mechanisms of Microbial Adaptation* (M. R. Heinrich, ed.), p. 201, Academic Press, New York.

Nozaki, Y. and Tanford, C. (1963). J. Biol. Chem. *238*, 4074.

Ogasawara, K., Imanishi, A. and Isemura, T. (1970). J. Biochem. *67*, 77.

Ohta, Y. (1967). J. Biol. Chem. *242*, 509.

Ohta, Y., Tanabe, S., Hachimori, A. and Samejima, T. (1974). J. Biochem. *76*, 871.

Oshima, T. (1976). Seikagaku *48*, 895.

Robinson, D. A. and Jencks, W. P. (1965). J. Am. Chem. Soc. *87*, 2462.

Singleton, R., Jr. and Amelunxen, R. E. (1973). Bacteriol. Rev. *37*, 320.

Singleton, R., Jr. (1976). In *Extreme Environments: Mechanisms of Microbial Adaptation* (M. R. Heinrich, ed.), p. 189, Academic Press, New York.

Stadtman, E. R. (1970). In *The Enzymes* (P. D. Boyer, ed.), Vol. 1, p. 397, Academic Press, New York.

Suzuki, K. and Imahori, K. (1973). J. Biochem. *74*, 955.

Thomé-Beau, F., Lê-Thi-Lan, Olomucki, A., and VanThoai, N. (1971). Eur. J. Biochem. *19*, 270.

Watanabe, K., Oshima, T., Saneyoshi, M. and Nishimura, S. (1974). FEBS Letters *43*, 59.

Yutani, K., Ogasawara, K., Sugino, Y. and Matsushiro, A. (1977). Nature *267*, 274.

PROPERTIES OF RNA POLYMERASES FROM *THERMUS THERMOPHILUS* HB8 AS CHARACTERIZED BY THEIR SUBUNITS

Shuichi Tsuji
Seiichi Kawashima
Kazutomo Imahori

Department of Biochemistry
Faculty of Medicine
University of Tokyo
Bunkyoku, Tokyo

INTRODUCTION

The existence of two forms of RNA polymerases (ribonucleoside triphosphate: RNA nucleotidyltransferase (DNA-dependent), EC 2.7.7.6) in *Thermus thermophilus* HB8 has been reported (Date *et al.*, 1975a, b). One form (polymerase A) can utilize various DNAs as templates, while the other (polymerase B) is not active in using ordinary double or single stranded DNAs but is active only in using poly[d(A-T)] as a template. The subunit structures of polymerases A and B are 2I.II.III.IV and 2I.III.IV(X.Z), respectively. The content of subunits X and Z varied depending on the preparations, but

ABBREVIATIONS. SDS: sodium dodecyl sulphate; LIS: 3,5-diiodosalycylic acid, Li salt; EDTA: ethylenediaminetetraacetic acid; DTT: dithiothreitol; NTP: ribonucleoside 5' -triphosphate; BPB: bromophenol blue; CT-DNA: calf thymus DNA; poly[d(A-T)]: alternating copolymer of deoxyadenylic and deoxythymidylic acids; Buffer A(X): 10 mM Tris-HCl, pH 8.0, 10 mM $MgCl_2$, 0.1 mM EDTA, X mM KCl, 10 mM β-mercaptoethanol; Buffer C(X:Y): 50 mM Tris-HCl, pH Y, 0.5 mM DTT, X mM KCl, 8 M urea; Buffer R: 10 mM Tris-HCl, pH 8.0, 10 mM $MgCl_2$, 0.1 mM EDTA, 0.5 mM DTT, 0.5 M KCl; n.d.: not determined.

ISBN 0-12-268250-5

the ratio of X to Z was always fixed at a value of about 1. Although the content of X plus Z seemed to increase at the expense of subunit III, the content of subunit X and Z has no effect on the activity or template specificity. Thus, the basic subunits of polymerase B are thought to be I, III and IV.

From the molecular weight of each subunit, it is deduced that subunits I, II, III and IV correspond to α, σ, β and β' of the *E. coli* enzyme. If subunit II corresponds to the σ subunit, polymerase B is the core enzyme of polymerase A (holo enzyme) as found in other bacteria (Chamberlin, 1974). In the latter cases, however, the core enzyme shows high activity when using poly[d(A-T)] and also natural DNAs as templates. It has been believed that the core enzyme possesses only a catalytic site and is therefore able to transcribe single or double stranded DNAs and poly[d(A-T)], while the σ subunit specifically stimulates initiation of RNA synthesis. The template specificity of polymerase B is therefore more limited than that of other bacterial core enzymes. Thus, polymerase B may be different from the core enzyme in function.

In eukaryotes, RNA synthesis is known to be carried out by different forms of RNA polymerase (Chambon, 1975), whereas in prokaryotes multiple forms of RNA polymerase have been reported only in *Bacillus subtilis*. The different forms of polymerase are supposed to play a role depending upon whether the cell is undergoing vegetative growth or sporulation (Fukuda *et al.*, 1975). Chao and Speyer (1973) reported the existence of a new form of RNA polymerase in stationary phase cells of *E. coli* which is almost inactive toward natural DNAs as templates and is active only with poly[d(A-T)]. This enzyme, however, has not been purified and it is therefore uncertain whether it contains the same core unit as RNA polymerase B. Two or three RNA polymerase activities with different template specificities have also been found in crude extracts of *E. coli* (Travers and Buckland, 1973). Such functional differences, however, appear to be mediated by protein factors which complex with a common RNA polymerase holo enzyme. In contrast, Fukuda *et al.* found another σ subunit, called σ', which can be complexed to the same core enzyme unit (Fukuda *et al.*, 1974). Judging from the molecular weights, it is possible that subunit II and subunit X correspond to σ' and σ, respectively.

In the present study, we examined the reconstitution of RNA polymerase A and B to learn more about the relationship between these two forms and the functions of subunits II, X and Z. The results obtained indicate that (i) RNA polymerase B is the core enzyme of RNA polymerase A (holo enzyme), (ii) the function of subunit II is similar to that of the σ subunit of other bacterial RNA polymerases and subunit II could restore narrow template specificities to those of holo types, and (iii) subunit X plus subunit Z can substitute for subunit III (β subunit of other bacterial RNA polymerases).

RESULTS

Mutual Conversion of RNA Polymerases A and B

RNA polymerase A was treated with 7 M urea overnight at 15°C in order to dissociate it completely into subunits. However, when the resulting sample was subjected to electrophoresis on gels containing 7 M urea, only three bands appeared. Analysis of each band by SDS gel electrophoresis revealed that the three bands corresponded to subunit II, a complex of 2I and III and a complex of 2I, III and IV, respectively. This result indicated that subunit II can be easily separated from the rest of the molecule by treatment with urea. Subunit II was prepared by extraction from the gel according to the method of Jovin *et al.* (1964). Similar treatment of RNA polymerase B gave four bands, corresponding to subunit I (or Z), a complex of 2I and X, a complex of 2I and III, and a complex of 2I, III and IV, respectively.

Although such partially dissociated samples of both enzymes were completely inactive, about 90% of the control enzyme activity was restored by dialysis against buffer A(200) containing 20% glycerol for 27 hrs at 4°C. This indicated that the native enzyme was reconstituted from the partially dissociated mixtures.

Next we tried mutual conversion of RNA polymerases A and B. The results are shown in Table 1. As described previously, polymerase A can transcribe both poly[d(A-T)] and ϕYS-40 DNA, but polymerase B can transcribe only poly[d(A-T)]. One of the partially dissociated components of polymerase A with the subunit composition of 2I.III.IV exhibited the same template specificity as polymerase B. This indicated that elimination of subunit II from polymerase A reduced the enzyme to polymerase B. Next, polymerase B and subunit II were mixed in an 8 M urea solution and dialyzed against the buffer cited above. The resulting sample displayed the same template specificity as polymerase A. This shows that polymerase A was formed from polymerase B and subunit II. The reconstituted enzyme showed the same sedimentation profile as polymerase A when examined by glycerol gradient centrifugation (Fig. 1). The addition of subunit II to polymerase A had, of course, no effect. These results suggest that polymerase B is the core enzyme of polymerase A and subunit II is

TABLE 1
Summarized Results of Mutual Conversion of RNA Polymerases A and B.

	Specific activity (x 10^{-3} unit/mg) Template DNA	
	ϕYS-40 DNA	poly[d(A-T)]
RNA polymerase A + subunit II[a]	12.0	27.0
RNA polymerase B + subunit II[a]	17.0	35.0
RNA polymerase A[b]	12.8	30.4
RNA polymerase B[b]	0.1	9.2
RNA polymerase A - subunit II[c] (2I.III.IV)	0.1	8.0

[a] *Purified subunit II (10 μg) and RNA polymerase A (40 μg) or RNA polymerase B (40 μg) were mixed in 0.25 ml of buffer A(200) containing 8 M urea, dialyzed against buffer A(200) containing 20%(V/V) glycerol for 20 hr at 4°C, and assayed using poly[d(A-T)] and ϕYS-40 DNA as templates.*

[b] *RNA polymerase A or B in the absence of subunit II was subjected to the same treatment and assay as described in a.*

[c] *After exposure to 7 M urea, RNA polymerase A was subjected to electrophoresis on polyacrylamide gels containing 7 M urea. The complex of 2I.III.IV was extracted and treated and and assayed as described in a.*

"equivalent" to the σ subunit of other bacterial RNA polymerases.

Complete Dissociation and Reconstitution of RNA Polymerase A and B

In the previous experiments we succeeded in reconstituting the native enzyme from a partially dissociated sample. However, in order to clarify the role of subunits X and Z, which are the minor components of polymerase B, we had to establish systems for completely dissociating the polymerases into subunits and reconstituting the enzymes from mixtures of individual subunits.

The complete dissociation of polymerase A or B could be achieved by treating the enzyme with (i) 0.3 M LIS or (ii) 4 M LiCl plus 8 M urea. Next we examined the optimal condition for reconstitution of the enzyme from the mixture of subunits. The reconstitution was influenced by several factors, including pH, enzyme concentration, ionic strength and temperature. Some of these results are shown in Fig. 2. Fig. 2a indicates that the optimal salt concentration was around 0.5 M. The optimal temperature was determined to be 65°C as shown in Fig. 2b. The optimal protein concentration was fixed at 0.15 mg/ml for the following reasons. If the concentration was higher than this value, the recovery of the activity decreased. When the concentration was lower than this value, recovery was unaffected but the reconstitution rate decreased. Other factors, including the concentrations of DTT, EDTA, and $MgCl_2$ were determined as indicated in buffer R.

In the case of RNA polymerase A, the recovery or kinetics of reconstitution was not influenced by the addition of ribonucleoside triphosphates, but was slightly stimulated by the addition of DNA.

Thus, the reconstitution was carried out by procedures as shown in Fig. 3. Under these conditions, the recovery of both enzymes was around 70-90% judging from the activities measured with ϕYS-40 DNA, CT-DNA and poly[d(A-T)] as templates (Table 2).

Purification of Each Subunit

Since the conditions for reconstitution were now established, we proceeded to purify each subunit of polymerase B. The isolation procedures are outlined in Fig. 4. Methods for the preparation of subunit II have been described previously.

Passage through a column of Sephadex G-200 divided the mixture of subunits into two fractions. The first fraction contained subunits III, IV and X, while the second fraction contained subunits I and Z (Fig. 5a). Subunits I and Z were separated from each other by subjecting the second fraction to phosphocellulose column chromatography as shown in Fig. 5b. The first fraction was also chromatographed on a phosphocellulose column. As shown in Fig. 5c, the run through fractions contained only subunit X, while the fractions eluted by buffer C (500:8.4) contained subunits III and IV. By applying this second peak fraction to a heparin-Sepharose column, subunits III and IV were separated from each other. The identification as well as the homogeneity of each of the five subunit preparations was determined by SDS gel electrophoresis as shown in Fig. 6. The recoveries of subunits I, II, III, IV, X and Z were 50-65%, 35-50%, 25-40%, 10-20%, 50-60% and 55-90%, respectively. Each subunit preparation could be stored in buffer A(500) containing 20% glycerol at -20°C without any appreciable decrease in activity for several weeks.

FIGURE 1. Glycerol gradient centrifugation of RNA polymerases and their derivatives. Each of RNA polymerase A (A), RNA polymerase B (B), the mixture of subunit II and RNA polymerase B (2:1 in molar ratio) (C), the mixture of subunit II and RNA polymerase A (2:1 in molar ratio) (D), and RNA polymerase A devoid of subunit II (E) was treated with 8 M urea solution, dialyzed against buffer A(200) containing 20% (V/V) glycerol for 24 hr at 4°C and finally dialyzed against buffer A(100) for 6 hr at 4°C. An aliquot of each dialyzed sample (0.1 ml) was layered on 4.7 ml of a glycerol gradient (10-30%) containing buffer A(100) and centrifuged for 13 hr at 32,000 rev./min in a Hitachi RPS 50 rotor. Catalase (20 μg) and β-galactosidase (15 μg) were added to each tube as reference markers. After centrifugation, the solution in each tube was divided into 25 fractions and each fraction was assayed for CT-DNA-dependent activity (●), poly[d(A-T)]-dependent activity (■), catalase activity (Δ), and β-galactosidase activity (o). Assay of catalase and β-galactosidase activities were carried out essentially according to the methods of Beers and Sizer (1952) and Maruo et al. *(1969), respectively.*

Another aliquot of each dialyzed sample (0.1 ml) was also spun in the same manner but without reference markers. A portion of the fraction which had maximal activity of RNA polymerase A or B was analyzed by SDS gel electrophoresis. The results are shown in the inset of each panel.

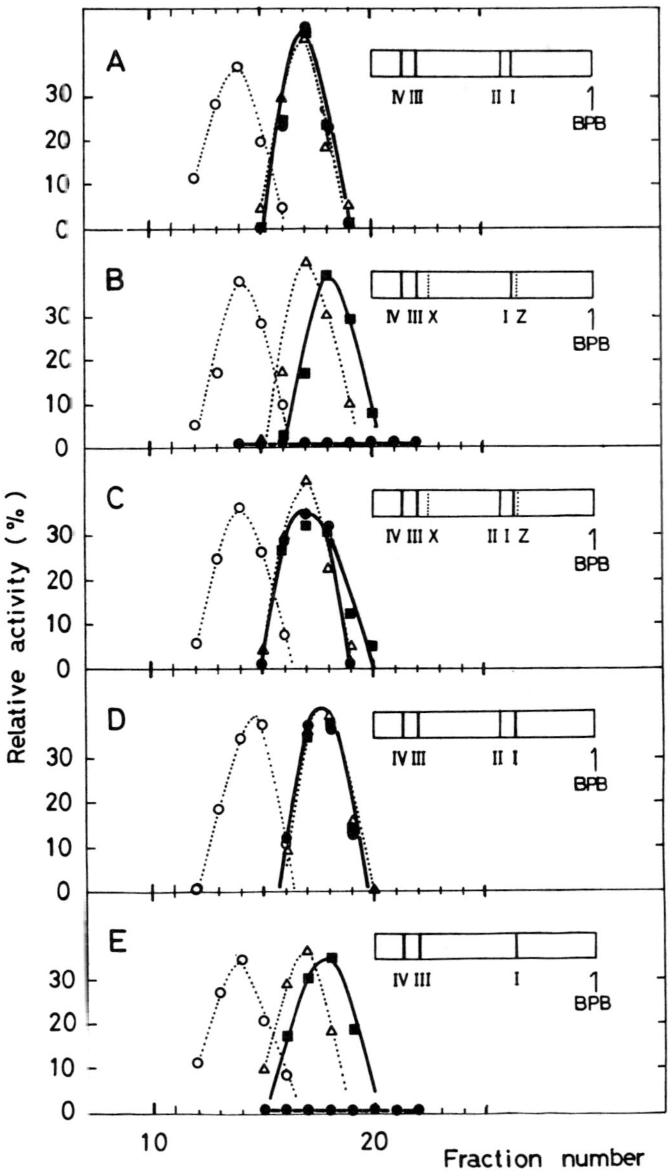
A
B
C
D
E
IV III
II I
BPB
IV III X
I Z
II I Z
I
Relative activity (%)
Fraction number
0
10
20
30
10
20

Figure 1.

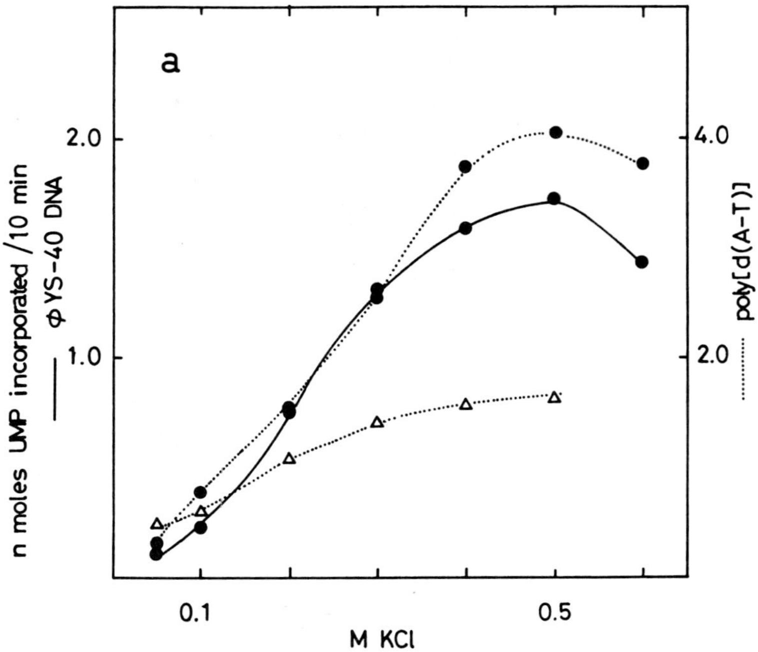

FIGURE 2. Reactivation of Dissociated RNA Polymerases A and B. RNA polymerases A and B were dissociated into subunits by dialyzing against a solution containing 8 M urea, 4 M LiCl, 0.2 M Tris-HCl, pH 8.0, and 0.5 mM DTT for 4 hr at 25°C. The dissociated enzyme solution was dialyzed against 0.05-0.6 M KCl, 10 mM Tris-HCl, pH 8.0, 0.5 mM EDTA, 0.5 mM DTT for 14 hr at 4°C. Dialyzed enzymes were incubated for various times at various temperatures.

a) Effect of KCl concentrations. RNA polymerase A (●) and RNA polymerase B (Δ) activities were assayed using poly[d(A-T)] (·····) and ϕYS-40 DNA (———) as templates, respectively.

b) Kinetics at various temperatures. This figure shows the results of RNA polymerase B activity using poly[d(A-T)] as template. 4°C (●), 37°C (■), 50°C (Δ), 65°C (o).

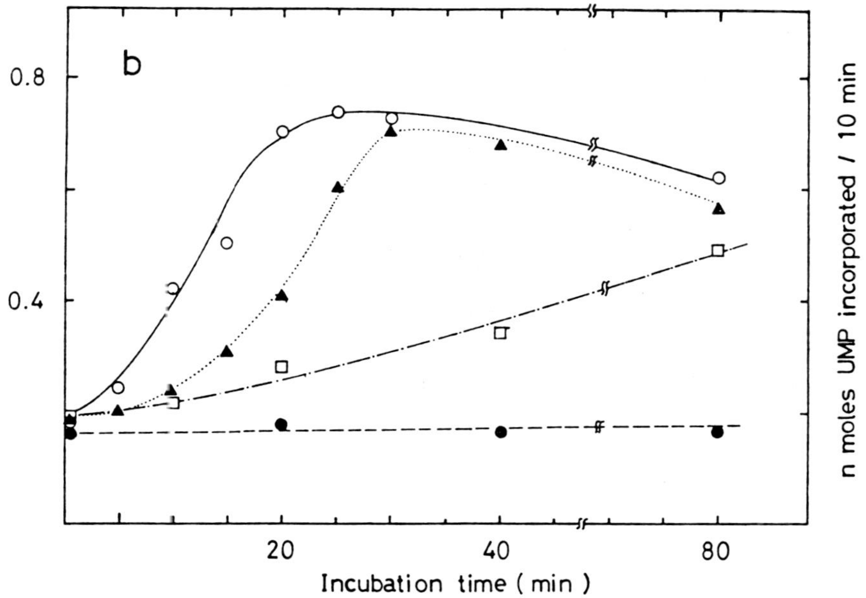

FIGURE 2b.

Subunits
Mixture:

Step I — Dialyzed against buffer A(500) containing 8 M urea, for 4 hr at 25°C.
↓
Step II — Dialyzed against buffer A(500) containing 6 M urea, for 2 hr at 4°C.
↓
Step III — Dialyzed against buffer R for 14 hr at 4°C.
↓
Step IV — Incubated for 25 min at 65°C.

FIGURE 3. Procedure of Reconstitution from Isolated Subunits.

TABLE 2

Recovery of Enzyme Activities under Optimum Conditions for Dissociated RNA Polymerases A and B

	RNA polymerase A			RNA polymerase B		
Template DNA:	*φYS-40*	*CT-DNA*	*poly[d(A-T)]*	*φYS-40 DNA*	*CT-DNA*	*poly[d(A-T)]*
Renatured enzyme from 0.3 M LIS treatment (units/mg)	10.4	--	24.0	0.1	--	9.6
Control[a] (units/mg)	15.0	--	32.0	0.1	--	12.0
Recovery (%)	69.0	--	75.0	--	--	80.0
Renatured enzyme from 4 M LiCl + 8 M urea treatment (units/mg)	9.8	5.2	25.0	0.1	0.1	11.7
Control[b] (units/mg)	13.1	8.7	28.0	0.1	0.1	14.0
Recovery (%)	75.0	60.0	89.0	--	--	84.0

The control enzyme was treated by the same renaturation procedures except that 0.3 M LIS (a) or 4 M LiCl (b) was omitted.

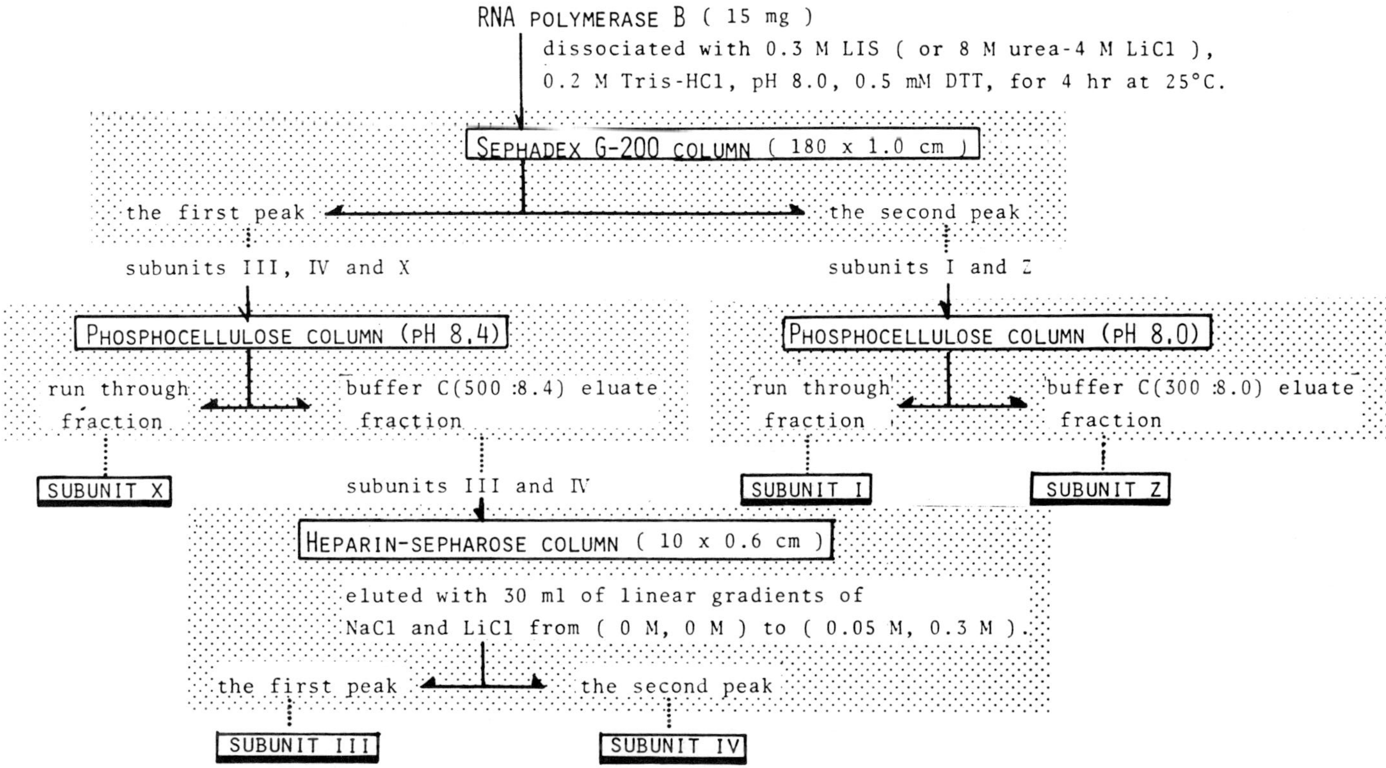

FIGURE 4. Isolation procedure for Subunits I, III, IV, X and Z. Further details were described in the legend to Fig. 5.

Reconstitution from Separated Subunits

Each of the purified subunits or various combinations of them were treated according to the reconstitution procedure outlined in Fig. 3. After incubation, each sample was tested for activity using φYS-40 DNA, CT-DNA and poly[d(A-t)] as templates. Once each of the subunits was treated separately by the reconstitution procedures and then mixed together with any combination of subunits, no activity dould be detected (data not shown). Next we conducted the experiments by mixing subunits in various combinations and treating the mixture by the procedures indicated in Fig. 3. The results are shown in Table 3. The best combination was 2I, II, III and IV, and the resulting product showed the same template specificity as polymerase A. These results correlate with the subunit composition of the two polymerases.

In order to determine the roles of subunits X and Z, we replaced either one or both of them with one of the other four subunits. These data are also shown in Table 3. Judging from the recovery of activity, subunit III can be replaced by subunit X, but substitution of X plus Z for subunit III gave much higher activity. This result indicates that a combination of X and Z could play the same role as subunit III.

Next we examined the assembly sequence in subunit

FIGURE 5. Separation of Subunits of RNA Polymerase B after Treatment with 0.3 M LIS.

a) Elution pattern of LIS-treated RNA polymerase B from a Sephadex G-200 column equilibrated with buffer C(0:8.4). Subunit composition of each peak was identified by SDS gel electrophoresis.

b) Chromatography of the second peak fraction of Fig. 5a on a phosphocellulose column (5 x 0.9 cm) equilibrated with buffer C(0:8.0). The sample solution was dialyzed against buffer C(0:8.0) before applying to the column. After washing with buffer C(0:8.0), elution was performed with buffer C(300:8.0).

c) Chromatography of first peak fraction of Fig. 5a on a phosphocellulose column (10 x 0.9 cm) equilibrated with buffer C(0:8.4).

d) Chromatography of the second peak fraction of Fig. 5c on a heparin-Sepharose column (10 x 0.6 cm) equilibrated with buffer C(0:7.5). The sample solution was dialyzed against buffer C(0:7.5) before applying to the column. Elution was carried out by double linear gradients of NaCl and LiCl (O M, O M) to (0.05 M, 0.3 M). Heparin-Sepharose was prepared according to the method of Sternbach et al. *(1975).*

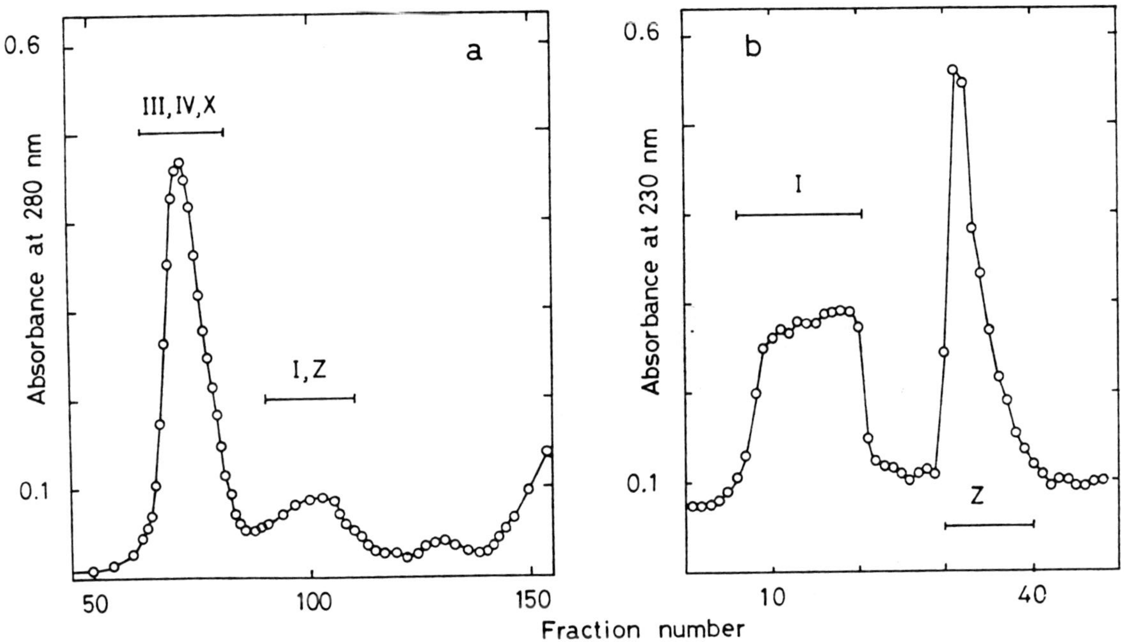
a
b
III,IV,X
I,Z
I
Z
0.6
0.1
Absorbance at 280 nm
Absorbance at 230 nm
50
100
150
10
40
Fraction number

FIGURE 5a, b.

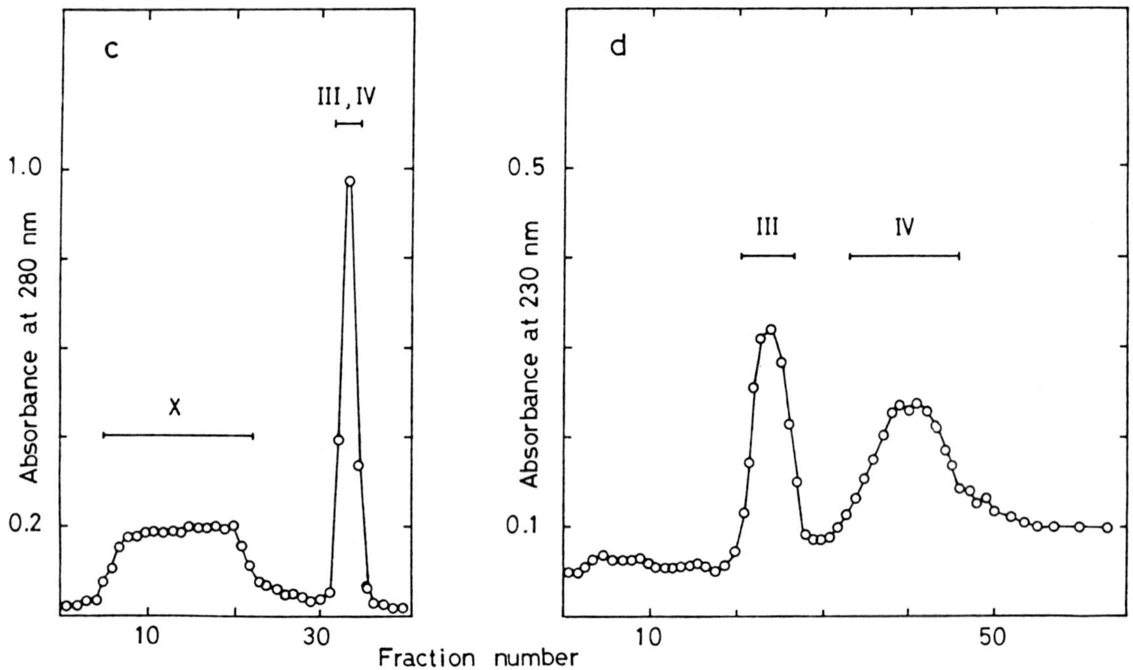
c
III, IV
X
1.0
0.2
Absorbance at 280 nm
10
30
d
III
IV
0.5
0.1
Absorbance at 230 nm
10
50
Fraction number

FIGURE 5c, d.

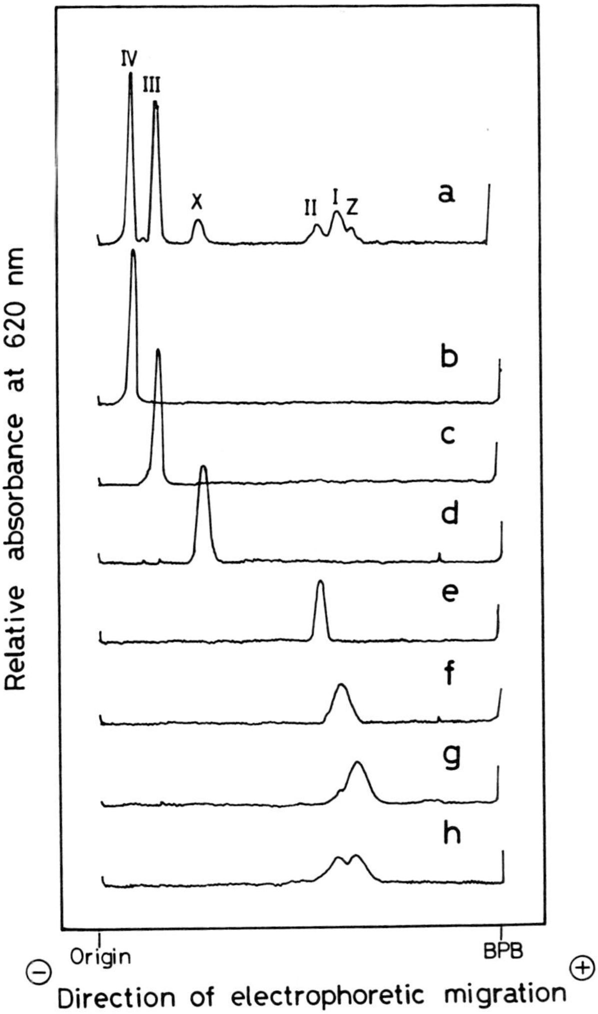

FIGURE 6. Electrophoresis of RNA Polymerase Subunits from T. thermophilus *HB8 on Polyacrylamide Gels Containing 0.1% SDS. Electrophoresis was carried out according to the methods of Weber and Osborn (1969) by applying 10-20 µg of each subunit. The gels were stained with 0.05% Coomasie brilliant blue R250 and recorded with a Joyce Loebel densitometer, Chromoscan 200. a: RNA polymerase A + RNA polymerase B; b: purified subunit IV; c: purified subunit III; d: purified subunit X; e: purified subunit II; f: purified subunit I; g: purified subunit Z; h: purified subunit I + purified subunit Z.*

TABLE 3
Reconstitution of RNA Polymerases A and B from Isolated Subunits

Subunit combination	Specific activity Template		
	Poly[d(A-T)]	φYS-40 DNA	CT-DNA
(2I, III, IV)	420	n.d.	0.1
(2I, III, IV, II)	730	n.d.	80.0
(2I, X, IV)	75	n.d.	0.1
(2I, X, IV, II)	190	n.d.	10.0
(2I, Z, IV)	10	n.d.	0.1
(2I, Z, IV, II)	10	n.d.	0.1
(2I, III, X)	10	n.d.	0.1
(2I, III, X, II)	15	n.d.	0.1
(2I, III, Z)	10	n.d.	0.1
(2I, III, Z, II)	10	n.d.	0.1
(Z, III, IV)	10	n.d.	0.1
(Z, III, IV, II)	10	n.d.	0.1
(X, III, IV)	10	n.d.	0.1
(X, III, IV, II)	10	n.d.	0.1
(2I, III, IV, Z)	280	20.0	0.1
(2I, III, IV, Z, II)	450	300.0	40.0
(2I, III, IV, X)	200	n.d.	5.0
(2I, III, IV, X, II)	240	n.d.	20.0
(2I, X, IV, Z)	240	10.0	10.0
(2I, X, IV, Z, II)	540	140.0	50.0
(2I, III, IV, X, Z)	50	10.0	n.d.
(2I, III, IV, X, Z, II)	270	240.0	n.d.
(III, X), (IV, X),...etc.	10	0.1	0.1
(2I, III)	30		
+ (X)	30		
+ (IV)	250		
(2I, X)	20		
+ (III)	50		
+ (IV)	80		
(2I, X, Z)	10		
+ (III)	20		
+ (IV)	180		
(2I, X)	20		
+ (III)+(Z)	40		
+ (IV)+(Z)	90		

reconstitution. In this experiment, a combination of subunits was mixed together, treated up to step II of Fig. 3 and then the following steps were carried out with or without supplementing additional subunits. The results are shown on the last twelve lines of Table 3. In the first parenthesis of Table 3, the combination of subunits for the first treatment is indicated while in the second parenthesis the subunit or combination of subunits supplemented before the second treatment is shown. When the second parenthesis is missing, no additional subunits were added before the second treatment. The results clearly indicate that only (2I, III)+(IV) and (2I, X, Z)+(IV) show substantial activities. The sequence of reconstitution thus deduced is the same as that for *E. coli* RNA polymerase (Saitoh and Ishihama, 1976) provided that subunits I, III and IV correspond to α, β and β', respectively.

Analysis of the Complexes

In the previous experiments we examined the reconstitution process only by tracing recovery of activity. However, we had no evidence indicating that reconstitution actually took place. In order to analyze the complexes, we layered the incubation products (after step III of Fig. 3) on glycerol gradients. After centrifugation, a hole was pierced at the bottom of the tube and 25 fractions were collected. Each fraction was analyzed by SDS gel electrophoresis in order to determine the amount and composition of subunits present. The incubation products before and after step IV gave the same results, suggesting that reconstitution is completed at step III and step IV is required only for activation of the complex.

The results are summarized in Fig. 7. The four panels (A, B, C and D) indicate that the combination of subunits in the initial mixtures were (2I, III, IV), (2I, X, IV), (2I, X, Z, IV) and (2I, III, IV, Z), respectively. In the case of A or C, the core enzyme was reconstituted in large amounts as expected, while another peak corresponded to a complex lack-

TABLE 3 Note. Subunits were isolated by chromatography as described in Fig. 4. The subunits were combined as indicated in parentheses, denatured and reconstituted according to the process indicated in Fig. 3. Reconstituted samples were assayed as described by Date et al. *(1975a) using ϕYS-40 DNA, CT-DNA and poly[d(A-t)] as templates. In these cases, specific activity is expressed as ^{3}H-UMP incorporated per 10 min per p mole of subunit IV, III or X.*

ing subunit IV. In the case of B, only one peak corresponding to 2I.X could be detected. In combination with the results of C, this indicates that complex 2I.X cannot bind subunit IV unless subunit Z is present. The pattern of D is essentially the same as that of A indicating that subunit Z cannot be incorporated into a core enzyme of (2I.III.IV) or a complex of (2I.III). Subunit Z was recovered only from a heavy aggregate sedimenting at the bottom of the gradient. These results indicate that subunits X and Z behave as a unit and that the combination of X and Z can sbustitute for subunit III.

DISCUSSION

Relation between RNA Polymerases A and B

The subunit compositions of polymerases A and B suggest that these correspond to holo and core enzymes of *E. coli*, respectively. As shown in Table 1, the addition of subunit II to polymerase B coverted the latter to polymerase A, while polymerase A devoid of subunit II showed the same template specificity as polymerase B. However, more work is required to prove this correspondence. The most convincing evidence would be to show that subunit II can play the same role as that of the σ subunit. However, all the antibiotics and inhibitors which have been used to characterize the σ subunit of *E. coli* were not effective for polymerases of *T. thermophilus*. The so-called σ-cycle could not be observed in the thermophilic organism. Nevertheless, the following experimental results provide support for the above idea.

Crude extracts of *T. thermophilus* HB8 were subjected to molecular sieve chromatography on Ultro-gel ACA 34 (2 x 120 cm) and the run through fractions were pooled. This fraction contained neither free RNA polymerase nor subunit II, but rather a ternary complex of DNA, RNA polymerase and product (RNA) (unpublished data). The activity of this complex for incorporation of UTP was inhibited by nucleases (DNase I, RNase IA and RNase T_1) but did not require exogeneous template (Table 4). Run through fractions were incubated at 65°C with NTP (50 μM each) to mature the nascent RNA. After the reaction, enzyme and product (RNA) were liberated from the complex. The liberated enzyme could transcribe poly[d(A-T)] but not CT-DNA. The liberated enzyme was purified by chromatography on DEAE-cellulose, hydroxyapatite, heparin-Sepharose and Ultro-gel ACA 34. The purified enzyme thus obtained was the same as polymerase B in respect to both subunit composition and template specificity (Table 4). Subunit

II could not be detected at any step in the purification process.

These results indicate that at the intermediate stage of transcription, the thermophile RNA polymerase assumes the B form, analogous to the core enzyme of *E. coli*. It would also appear that subunit II corresponds to the σ subunit, which is necessary for the initiation step but is released from the complex at the elongation stage. However, the template specificity of polymerase B is more narrow than that of the *E. coli* core enzyme. Further characterization of subunit II is in progress.

Purification of Subunits

Both RNA polymerase A and B are stable not only against heat but also against several denaturing reagents. The polymerases could hardly be dissociated into subunits even in the presence of 8 M urea. Thus we applied more drastic conditions, such as (i) 0.3 M LIS or (ii) 0.4 M LiCl plus 8 M urea for dissociation. Although these methods were suitable for complete dissociation, subunits were rather unstable in such solutions and recovery of the native enzyme was poor. However, if we use freshly prepared LIS or urea and if we conducted the purification of each subunit promptly (within a few days), the resulting subunits allowed for satisfactory reconstitution.

Although the assignment of subunits III and IV was difficult, the following facts suggest that they correspond to β and β', respectively: (i) Similarity of elution pattern from a phosphocellulose column in the presence of urea. When a mixture of subunits III and IV was applied to a phosphocellulose column (pH 8.0) in the presence of 8 M urea, subunit III was eluted first. It has been reported that subunit β is eluted before β' under the same conditions (Yarbrough and Hurwitz, 1974), (ii) a substrate analog (oxidation product of 8-^{3}H-ATP) was bound selectively to subunit III or X, and (iii) the assembly sequence of III and IV was exactly the same as β and β' (Saitoh and Ishihama, 1976).

Reconstitution

As shown in Fig. 2b, the reconstitution rate depended remarkably on temperature. However, since the reconstituted products before and after step IV of Fig. 3 gave the same results, the rate-limiting or temperature-dependent step should be step IV, where the inactive pre-enzyme complex is

TABLE 4

Characterization of the Run Through Fraction and Liberated Enzyme Fraction

Assay condition	Run through fraction[a] (CPM)	Liberated enzyme fraction[b,c]	Purified liberated enzyme[d]	RNA polymerase A[d]	RNA polymerase B[d]
		($\times 10^{-4}$ units/mg protein)			
-DNA	4582	0.01	0.01	0.01	0.01
$-Mg^{2+}$	45	0.01	0.01	0.01	0.01
-ATP	90	0.01	0.01	0.01	0.01
-GTP	70	0.01	0.01	0.01	0.01
-CTP	80	0.01	0.01	0.01	0.01
+CT-DNA	4290	0.01	0.01	2.26	0.013
+poly[d(A-T)]	5192	0.80	1.65	6.17	2.21
Incorporated ratio (template specificity) poly[d(A-T)] / CT-DNA	--	80	165	2.73	170

[a]*A crude extract (5.0 ml) of* T. thermophilus *HB8 prepared according to Date* et al. *(1975a) was applied to an Ultrogel ACA 34 column (120 x 2 cm). Run through fractions were collected and assayed. Assay of RNA polymerase activity (exogeneous DNA-independent) was carried out according to the procedure of Date* et al. *(1975a) with the following modifications: a reaction mixture (final volume: 0.2 ml) contained 100 μl of run through fraction, 10 μmoles glycine-NaOH buffer, pH 8.2 at 65°C, 1.0 μmole of β-mercaptoethanol, 0.02 μmoles EDTA, 2 μmoles $MgCl_2$, 20 μmoles KCl, 0.08 μmoles each of the unlàbelled NTP, 0.08 μmoles ^{3}H-UTP (50 x 10^4 cpm). The reaction was carried out for 10 min at 65°C, and the incorporation of UMP into acid-insoluble*

material was counted.

[b]Run through fraction (10 ml) was incubated for 30 min at 65°C with NTP (50 μM each), and applied to a DEAE-cellulose column (4 x 1 cm) equilibrated with buffer A(100) in order to remove nucleic acids. After washing with 10 ml of buffer A(100), stepwise elution was performed with buffer A(250) and A(500). No fraction showed activity unless exogeneous DNA was supplemented. Enzyme activity was only detected in the buffer A(250) eluate fraction (liberated enzyme fraction).

[c]The liberated enzyme was purified by chromatography on DEAE-cellulose, hydroxyapatite, heparin-Sepharose and Ultrogel ACA 34.

[d]Assay was carried out by the method of Date et al. (1975a).

converted to the active enzyme probably through a conformational change.

In our reconstitution experiments the addition of either subunit II or template DNA to the reconstitution mixture did not influence the rate or extent of reconstitution. This result is not consistent with reports of other laboratories (Heil and Zillig, 1970; Ishihama *et al.*, 1973; Fukuda and Ishihama, 1974; Palm *et al.*, 1975) which indicated stimulatory effects of DNA and σ subunit on reconstitution of the *E. coli* enzyme. Although this discrepancy is difficult to explain, it might be due to (i) differences in the organism, or (ii) incubation at high temperatures, which may substitute for addition of the σ subunit. In any event, the recovery under optimal conditions was 30-50%.

Subunits X and Z

As indicated previously, the subunit composition of polymerase A was stoichiometric and the molar ratios of subunits I, II, III and IV were 2:1:1:1. While polymerase B is missing subunit II, it contained the extra subunits X and Z, which were fractional and varied depending on the preparation. The functions of X and Z remain obscure. One possibility is that subunit X behaves as a kind of regulatory subunit and restricts the template specificity of polymerase B. Stetter and Zillig (1974) found a new subunit, y, in RNA polymerase from *Lactobacillus curvatus*. The molecular weight of subunit y (89,000) is close to that of subunit X. They reported that subunit y enhanced the activity of the σ subunit in a synergistic fashion. However, we could not demonstrate such an effect with subunit X (unpublished data). A more likely possibility is that subunits X and Z are the digestion products of subunit III and the combination of these two subunits can substitute for subunit III. As shown above, the enzyme reconstituted from subunits I, IV, X and Z revealed the same template specificity as that reconstituted from I, III and IV. Both of these reconstituted enzymes displayed the same specificity as native polymerase B. As shown in Fig. 7, existence of both X and Z is required for the binding of subunit IV to the complex and the complex thus formed showed the same sedimentation property as polymerase B. These results clearly indicated that a combination of X and Z can replace subunit III as both a functional and structural constituent of the complex. It should be added that the sum of the molecular weights of X and Z is very close to that of subunit III. Furthermore, when we treated polymerase A with trypsin, new fragments appeared which had the same mobilities as X and Z, respectively, on SDS gel electrophoresis (Fig. 8).

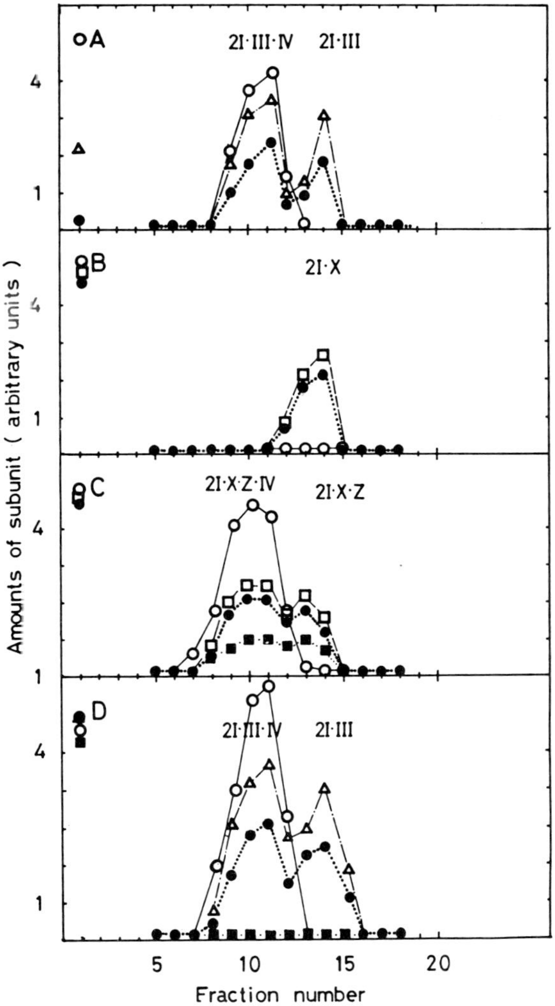

FIGURE 7. Analysis of Reconstituted Complexes. Subunits were mixed in proportions as indicated below and treated up to step III of Fig. 3. The resulting solution (0.5-1.0 ml) was condensed to 0.1 ml and layered on 4.7 ml of a 10 to 30% glycerol gradient in buffer A(100). The centrifugation was carried out in a Hitachi RPS 50 rotor for 13 hr at 32,000 rev./min. Each fraction (except Nos. 2, 3 and 4) was applied to SDS polyacrylamide gels. After electrophoresis, the gels were analyzed by a gel scanner. The content of each subunit was determined by weighing the corresponding peak on the tracing. The composition of subunits in each of the initial mixtures are as follows: (A) 2I, III and IV, (B) 2I, X and IV, (C) 2I, X, A and IV, and (D) 2I, III, IV and Z. I (o), III (Δ), IV (o), X (□), Z (■).

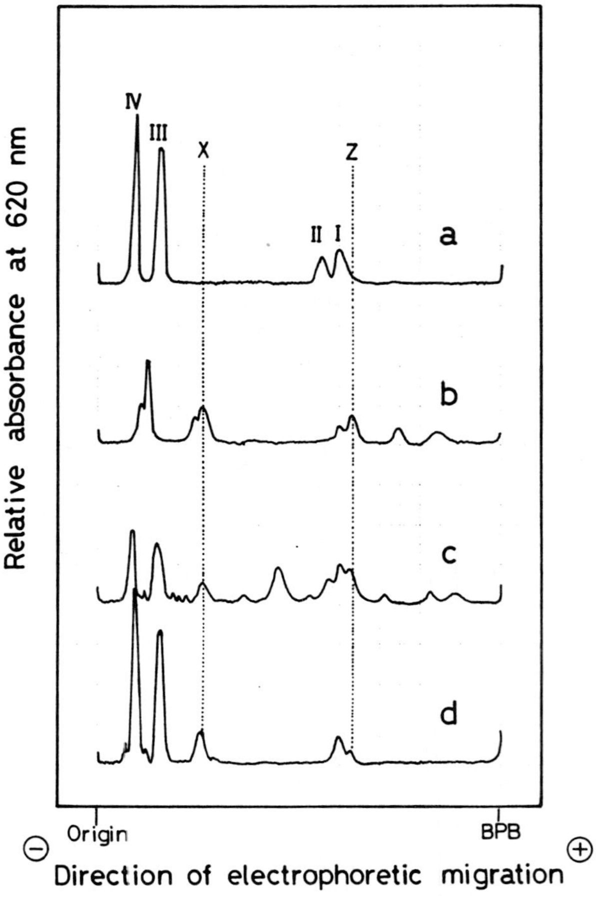

FIGURE 8. Proteolysis of RNA Polymerase A by Pronase and Trypsin. RNA polymerase A was digested by the proteases (10^{-2} parts W/W) at 37°C for 60 min. The reaction was stopped by heating the mixture for 2 min at 100°C after addition of 2% SDS and 1% β-mercaptoethanol. 30 μg of each product was applied to SDS-polyacrylamide gels.

a: RNA polymerase A.

b: Pronase-treated RNA polymerase A.

c: Trypsin-treated RNA polymerase A.

d: RNA polymerase B.

Similar fragments have been reported in other organisms. King *et al*. (1974) showed that the β subunit is proteolytically cleaved into two fragments, β_1 and β_2. The molecular weights of β_1 and β_2 were 124,000 and 52,000, respectively. Fragments of similar size were also found in *B. subtilis* (MW = 110,000) (Millet *et al*., 1972) and *B. thuringensis* (MW = 130,000) (Klier *et al*., 1973). Thus, we may conclude that X and Z are the proteolytic digestion products of subunit III. The latter two fragments are reported to have a special function in controlling transcription. However, at present it is not clear whether fragments X and Z have special functions or whether they are merely proteolytic products of subunit III. Further work is required to elucidate these problems.

ACKNOWLEDGEMENTS

The authors express their deep appreciation to Dr. T. Oshima of The Mitsubishi-Kasei Institute of Life Sciences for his generous gift of ϕYS-40 DNA. We thank Drs. T. Ohta and K. Suzuki for valuable advice and helpful discussions.

REFERENCES

Beers, R. F., Jr. and Sizer, I. W. (1952). J. Biol. Chem. *195*, 133.
Chamberlin, M. J. (1974). Ann. Rev. Biochem. *43*, 721.
Chambon, P. (1975). Ann. Rev. Biochem. *44*, 613.
Chao, L. and Speyer, J. F. (1973). Biochem. Biophys. Res. Comm. *51*, 399.
Date, T., Suzuki, K. and Imahori, K. (1975a). J. Biochem. *78*, 845.
Date, T., Suzuki, K. and Imahori, K. (1975b). J. Biochem. *78*, 955.
Fukuda, R., Iwakura, Y. and Ishihama, A. (1974). J. Mol. Biol. *83*, 353.
Fukuda, R. and Ishihama, A. (1974). J. Mol. Biol. *87*, 523.
Fukuda, R., Keilman, G., McVey, F. and Doi, R. H. (1975). In *Spores IV* (P. Gerhardt, R. N. Costilow and H. L. Sadoff, eds.), p. 226, American Society for Microbiology, Washington, D.C.
Heil, A. and Zillig, W. (1970). FEBS Letters *11*, 165.
Ishihama, A., Fukuda, R. and Ito, K. (1973). J. Mol. Biol. *79*, 127.
Jovin, T., Chambach, A. and Naughton, M. A. (1964). Anal. Biochem. *9*, 351.
King, A. M. Q., Lowe, P. A. and Nicholson, B. H. (1974). Biochem. Biophys. Res. Comm. *59*, 38.
Klier, A. F., Lecadet, M. and Dedonder, R. (1973). Eur. J. Biochem. *36*, 317.
Maruo, B., Seto, H. and Nagata, Y. (1969). J. Bacteriol. *100*, 209.
Millet, J., Kerjan, P., Aubert, J. P. and Szulmajster, J. (1972). FEBS Letters *23*, 47.
Palm, P., Heil, A., Boyd, D., Grampp, B. and Zillig, W. (1975). Eur. J. Biochem. *53*, 283.
Saitoh, T. and Ishihama, A. (1976). J. Mol. Biol. *104*, 621.
Stetter, K. O. and Zillig, W. (1974). Eur. J. Biochem. *48*, 527.
Travers, A. and Buckland, R. (1973). Nature New Biol. *243*, 257.
Weber, K. and Osborn, M. (1969). J. Biol. Chem. *244*, 4406.
Yarbrough, L. R. and Hurwitz, J. (1974). J. Biol. Chem. *249*, 5400.

based on growth yield studies (Andreesen *et al.*, 1973) and the presence of electron carriers such as cytochrome (Gottwald *et al.*, 1975). Electrons generated during glycolysis are accepted by NAD and those generated by oxidation of pyruvate, by ferredoxin. The electron donors in the reduction of CO_2 to acetate are NADPH in the formate dehydrogenase and methylenetetrahydrofolate dehydrogenase reactions and, most likely, $FADH_2$ in the methylenetetrahydrofolate reductase reaction. Thus, it is obvious that an electron transfer must occur from NADH and reduced ferredoxin to NADP and FAD. The possibility exists that it may be coupled with production of ATP. The feasibility of phosphorylation coupled with the reduction of CO_2 has been pointed out and discussed by Thauer *et al.* (1977).

Homoacetate fermentation and the synthesis of acetate from CO_2 and pyruvate is not an exclusive property of *C. thermoaceticum.* An almost identical fermentation and acetate synthesis is carried out by *Clostridium formicoaceticum* (Andreesen *et al.*, 1970; O'Brien and Ljungdahl, 1972) and, perhaps, also by the newly isolated *Acetobacterium woodii* (Balch *et al.*, 1977; Tanner *et al.*, 1978). These microorganisms are strict anaerobes and mesophiles. Both ferment fructose, but not glucose, to acetate as the only product and, as in *C. thermoaceticum*, one-third of the acetate appears to be formed from CO_2 and pyruvate.

C. thermoaceticum and *C. formicoaceticum* neither produce nor use molecular hydrogen. They require an organic substrate which is the source of electrons for the reduction of CO_2. *Acetobacterium woodii* differs from the two clostridia in that it has hydrogenase. The presence of hydrogenase in *A. woodii* enables this bacterium to use molecular hydrogen as a source of electrons in the reduction of CO_2 to acetate according to the following equation (Balch *et al.*, 1977):

$$4H_2 + 2HCO_3^- + H \rightarrow CH_3COO^- + 4H_2O \qquad (4)$$

What is perhaps the most interesting property of *A. woodii* is its capacity to grow autotrophically on a mineral salt medium requiring the addition of only pantothenate. Carbon dioxide and H_2 serve as the sources of carbon and energy. Thus, energy must be generated in the reduction of CO_2 to acetate (reaction 4) (Balch *et al.*, 1977; Schoberth, 1977).

The synthesis of acetate from CO_2 in the homoacetate-fermenting microorganisms is a major pathway. Consequently, it can be expected that they contain high levels of enzymes of this pathway. This has been confirmed for *C. thermoaceticum* by Andreesen *et al.* (1973), for *C. formicoaceticum* by O'Brien and Ljungdahl (1972), and for *A. woodii* by Tanner *et al.* (1978). The levels of formate dehydrogenase and tetrahydrofolate enzymes are 100- to 1000-fold higher than what is normally found for corresponding enzymes in other bacteria, plants, or mammalian tissues. Thus, the homoacetate-fermenting microorganisms are excellent sources for enzymes of the tetrahydrofolate pathway. In our studies of thermophilicity, we have been interested in comparing thermostabilities and other properties of enzymes and proteins from *C. thermoaceticum*, grown at 60°C, and *C. formicoaceticum*, grown at 37°C.

GENERAL PROPERTIES OF PROTEINS FROM *C. THERMOACETICUM*

As was mentioned in the introduction, we noticed in 1963 (Ljungdahl and Wood) that enzymes in extracts of *C. thermoaceticum* have reasonable activities only at temperatures above 40°C. It was subsequently found that such extracts efficiently catalyze the synthesis of acetate from CO_2 and pyruvate at 55°C (Ljungdahl *et al.*, 1965). The results clearly indicated that all enzymes required for acetate synthesis are stable in extracts at what can be considered unusually high temperatures.

Several enzymes and other proteins have now been obtained from *C. thermoaceticum* and *C. formicoaceticum*. Among these are formyltetrahydrofolate synthetase (O'Brien *et al.*, 1976), methylenetetrahydrofolate dehydrogenase (Ljungdahl *et al.*, 1976), formate dehydrogenase (Andreesen *et al.*, 1973), acetate kinase (Schaupp and Ljungdahl, 1974), ferredoxin (Yang *et al.*, 1977), rubredoxin and rubredoxin reductase (Yang and Ljungdahl, 1977), as well as, a 5-methoxybenzimidazolylcobamide-containing protein (Ljungdahl *et al.*, 1973). The purified proteins from *C. thermoaceticum* exhibit thermostabilities much higher than corresponding proteins from *C. formicoaceticum*. Thus, the two homoacetate-fermenting clostridia synthesize proteins with thermostabilities, which reflect the respective growth temperatures of the two bacteria.

Although the corresponding enzymes from *C. thermoaceticum* and *C. formicoaceticum* have different thermostabilities, other properties are very similar. Results of an extensive comparison between formyltetrahydrofolate synthetase from *C. thermoaceticum* and this enzyme from other clostridia have been published (O'Brien *et al.*, 1976). In all properties (physical, chemical, as well as kinetic), except for thermostability, the clostridial synthetases are very similar. We came to the conclusion that the high thermostability of the *C. thermoaceticum* synthetase is inherent in the protein structure and that it depends on the amino-acid composition or sequence. Similarly, methylenetetrahydrofolate dehydrofolate dehydrogenase (Ljungdahl *et al.*, 1976) and ferredoxin (Yang *et al.*, 1977) from *C. thermoaceticum* have been compared with equivalent proteins from other clostridia. Again, the higher thermostabilities of the *C. thermoaceticum* proteins as compared with the proteins from mesophilic clostridial sources clearly are inherent in the protein structures. Thus, it is apparent that *C. thermoaceticum* is able to grow at 60°C, or slightly higher, because it has the capacity to produce macromolecules that are intrinsically stable at this growth temperature.

C. thermoaceticum has a maximum growth temperature (T_{max}) at about 65°C and a minimum growth temperature (T_{min}) at about 45°C (Fontaine *et al.*, 1942). A case can be made that both the T_{max} and the T_{min} of *C. thermoaceticum* depend on properties of its proteins. Although an enzyme like formyltetrahydrofolate synthetase is relatively stable at 60°C, it denaturates rather rapidly at 65°C, the T_{max} of *C. thermoaceticum.* It is easily conceivable that if denaturation or inactivation of one or several physiologically important enzymes starts at 65°C, T_{max} of the bacterium may well be at that temperature.

That properties of enzymes may also determine the T_{min} of *C. thermoaceticum* is less obvious. There are many reports of enzymes that exhibit broken lines in Arrhenius plots. Although the interpretation as to the reason for these broken lines is quite a controversial subject, they appear in many cases to reflect conformational changes in the enzymes. Broken lines in Arrhenius plots have been obtained with formyltetrahydrofolate synthetase (Shoaf *et al.*, 1974), methylenetetrahydrofolate dehydrogenase (O'Brien *et al.*, 1973), acetate kinase (Schaupp and Ljungdahl, 1974), and formate dehydrogenase (Ljungdahl *et al.*, 1976). Table 1 is a list of the transition temperatures and the apparent activation energies for these enzymes as obtained from Arrhenius plots. As is evident, the activation energies below the transition temperatures are larger than above them. In other words, the enzymes appear to be less efficient catalysts below their transition temperatures than above them. It also follows that below the transition temperatures the activities of the enzymes rapidly decrease with decrease in temperature. Thus, at some point, by lowering the temperature, the efficiency of an enzyme or some enzymes in *C. thermoaceticum* may be lower than what is required for growth. Consequently, at that temperature, one would expect the T_{min}. The concept that T_{min} and T_{max} of bacteria may be controlled by properties of enzymes has previously been briefly discussed (Ljungdahl and Sherod, 1976).

TABLE 1
Enzymes in C. thermoaceticum *That Exhibit Broken Lines in Arrhenius Plots*

Enzyme	Transition temperature °C	*Apparent activation energy cal/mole*	
		Below trans. temp.	*Above trans. temp.*
Formyl-H_4folate synthetase[a]	43	10,400	8,900
Methylene-H_4folate dehydrogenase[b]	35	9,600	6,500
Formate dehydrogenase[c]	55	10,500	not linear
Acetate kinase[d]	40	10,500	3,110

[a] *Shoef* et al. *(1974)*.
[b] *O'Brien* et al. *(1973)*.
[c] *Ljungdahl* et al. *(1976)*.
[d] *Schaupp and Ljungdahl (1974)*.

THE QUARTERNARY STRUCTURE OF FORMYLTETRAHYDROFOLATE SYNTHETASE FROM *C. THERMOACETICUM*

That formyltetrahydrofolate synthetase from *C. thermoaceticum* undergoes a conformational change at the transition temperature observed in Arrhenius plots is supported by several findings (Shoaf *et al.*, 1974). Thus, plots of Km or log Km versus 1/T show line breaks at about 43°C. Fluorescence and absorbance of the enzyme respectively decreases and increases linearly with increasing temperature. In both cases, the lines change slopes between 40°C and 45°C. These changes are augmented by the presence of ammonium or potassium ions. The activity of the enzyme is greatly enhanced by these ions.

Recently, we have examined formyltetrahydrofolate synthetase from *C. thermoaceticum* using electron microscopy. The enzyme has a molecular weight of 240,000 and consists of four, apparently identical, subunits. Preparations of the enzyme, judged homogeneous by several criteria such as gel electrophoresis, gel filtration, and ultracentrifugation, were negatively stained with sodium phosphotungstate. As is shown in Fig. 1, the electron microscopy examination revealed two types of particles of which one marked with a large arrow will be referred to as the "H"-form, whereas the other marked with a small arrow will be called the "twin"-form. Figs. 2 and 3 show enlargements of the two forms. In Fig. 3, which shows the "H"-form, the four subunits are easily visualized. Our results also strongly indicate that the "twin"-forms are tetramers, and that each half of a "twin" consists of two subunits. This was concluded when we observed that the two structures under the same conditions are always present at the same ratio from different preparations; that the addition of ATP, ADP, or the inhibitor, $AMPPCH_2P$ (β-γ-methyleneadenosine triphosphate), converts the "H"-form to the "twin"-form (see Table 2); that this conversion is specifically mediated by adenosine di- and triphosphates; and that, on removal of these nucleotides by dialysis, the "twin"-form is converted back to the "H"-form.

It could be argued that the "twin"-form is a dimer which has arisen from the "H"-form. However, determinations of the molecular weight, the sedimentation constant, and the Stokes radius gave the same result in the absence, as well as in the presence, of ATP. This shows that in the presence of ATP, the tetrameric enzyme is not dissociated to a dimer.

The placement of the subunits in the "H"-form and the "twin"-form projections indicates that the binding between the subunits differs. Thus, the binding between two subunits

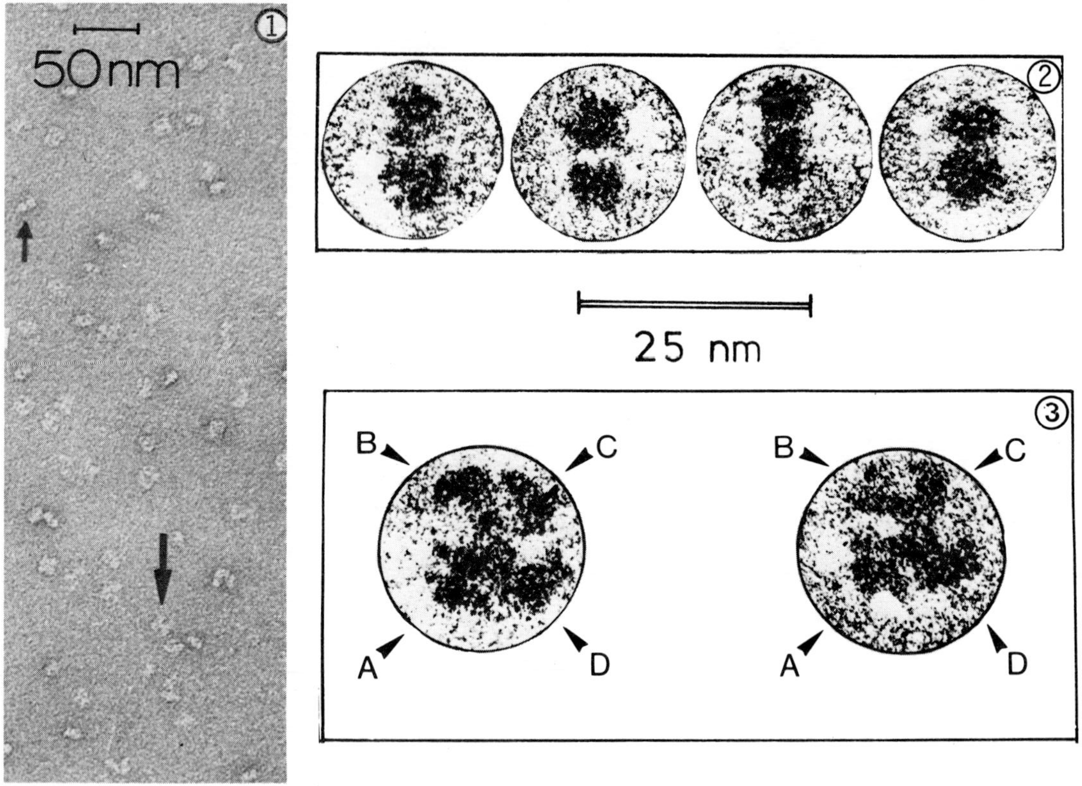

Electron Micrographs of negatively stained formyltetrahydrofolate synthetase from C. thermoaceticum.

FIGURE 1. Two conformational forms of the enzyme. The "H"-form is illustrated by the large arrow, whereas the "twin"-form is shown at the small arrow.

FIGURE 2. Enlargements of the "twin"-form.

FIGURE 3. Enlargements of the "H"-form.

TABLE 2
Effects of Substrates, Products and An Inhibitor on the Ratio of "H"-form to "Twin"-form Particles of Formyltetrahydrofolate Synthetase from C. thermoaceticum*

Additions	*Ratio "H":"twin"*
None	54:46
ATP, Mg	18:82
ADP, Mg	20:80
AMP, Mg	49:51
$AMPPCH_2P$, Mg	22:78
H_4folate	40:60
P_i, Mg	57:43
formate	44:56
ATP, Mg, H_4folate	22:78
ATP, Mg, formate	16:84
ATP, Mg, P_i	17:83
ATP, Mg, formate, H_4folate	42:58

**Enzyme, 30-50 μg in 1 ml of 0.1 M Tris-HCl, pH 7.6, containing 0.1 M KCl and with the addition as indicated of 5 mM nucleotides, 5 mM inorganic phosphate (P_i), 10 mM $MgCl_2$, 40 mM formate, and 2 mM ℓ-tetrahydrofolate (H_4folate) was negatively stained with sodium phosphotungstate. The numbers (at least 200 from each print) of "H"-form and "twin"-form particles were counted within a specified area from photographic prints of electron microscopic examinations.*

to form a dimer appears to be different from the binding between two dimers to form a tetramer. Harmony *et al.* (1975) and Harmony and Himes (1975) have shown that monomers (M) of formyltetrahydrofolate synthetase from *Clostridium cylindrosporum* are converted to active tetramers in the presence of monovalent cations (C^+). They showed that the stoichiometry of the reaction is $4M + 2C^+ \rightleftarrows M_4C_2^{2+}$ and suggested that a monovalent cation is sandwiched between two subunits. The projections seen in the electron microscope (Figs. 1-3) are clearly compatible with this suggestion. It is possible to imagine that a cation is situated between the two subunits constituting half of a "twin" or half of a "H" structure. The binding between the dimers may then occur without the involvement of a cation. We have examined formyltetrahydrofolate synthetase from *C. clyindrosporum* using electron microscopy. This enzyme has the same two projections as has the enzyme from *C. thermoaceticum*.

The results of the electron microscopy studies have been interpreted as follows: The "H"-form projections may not represent the real state of the protein molecules in solution. We suggest that the four "arms" of the "H"-form are flexible and that they are artificially flattened and attached to the carbon support film during the staining procedure. In solution, the "H"-form appears to be similar to the "twin"-form, as suggested by results from gel-filtration and ultracentrifugation experiments. The difference between the "H"- and the "twin"-forms would then be that the former have flexible arms, whereas, in the "twin"-form, the arms are in a stabilized position.

There appears to be an equilibrium between the "H"- and the "twin"-forms. This equilibrium is shifted to the stabilized "twin"-form by ATP or ADP. In the absence of ATP or at concentrations lower than 1 μM, the ratio of "H" to "twin" forms is 60/40 whereas, in the presence of 5 mM ATP or higher concentrations, this ratio is 20/80. It was possible to observe a correlation between the concentration of ATP and the ratio of "H"- to "twin"-form particles. In 0.1 mM ATP, 60% of the particles are in the "twin"-form, which is the midpoint between 40% "twin"-forms in the absence of ATP and 80% "twins" in the presence of 5 mM ATP. Interestingly, the Km for ATP of formyltetrahydrofolate synthetase from *C. thermoaceticum* is about 0.1 mM. This suggests a relation between the catalytic mechanism of the enzyme and the conversion between the "H"- and the "twin"-forms.

METHENYLTETRAHYDROFOLATE CYCLOHYDROLASE AND METHYLENETETRAHYDROFOLATE DEHYDROGENASE COMBINED FROM *C. THERMOACETICUM*

The purification and properties of methylenetetrahydrofolate dehydrogenase from *C. thermoaceticum* have been discussed in earlier papers (O'Brien *et al.*, 1973; and Ljungdahl *et al.*, 1976). Recently, in attempts to purify methenyltetrahydrofolate cyclohydrolase, we found that this activity co-purified with methylenetetrahydrofolate dehydrogenase.

By using ammonium sulfate fractionation and chromatography on DEAE-cellulose and Ultrogel AcA 44, we have obtained a homogeneous protein which exhibits both activities. The pure protein has a specific activity of 310 μmoles $min^{-1}mg^{-1}$ for cyclohydrolase and 200 μmoles $min^{-1}mg^{-1}$ for dehydrogenase. The molecular weight of the protein is about 55,000 and it consists of two subunits which appear to be identical. The properties of the enzyme are identical with those previously reported for methylenetetrahydrofolate dehydrogenase from *C. thermoaceticum* (O'Brien *et al.*, 1973) and we firmly believe that the previously and newly isolated proteins are identical. It is apparent that there is a protein present in *C. thermoaceticum* which possesses both methenyltetrahydrofolate cyclohydrolase and methylenetetrahydrofolate dehydrogenase activities. It is exciting and of interest that porcine liver (Cohen and MacKenzie, 1978), sheep liver (Paukert *et al.*, 1976), and yeast (Paukert *et al.*, 1977) contain homogeneous proteins which have three folate enzyme activities (formyltetrahydrofolate synthetase, methenyltetrahydrofolate cyclohydrolase, and methylenetetrahydrofolate dehydrogenase).

FERREDOXIN FROM *C. THERMOACETICUM*

As discussed in the introduction, *C. thermoaceticum* must have an electron transfer system. We have initiated studies of this system by isolating electron-carrying proteins. One of these, ferredoxin, is of particular interest since it functions in the transfer of electrons from pyruvate to NADP. NADPH is then used as the electron donor in the reduction of CO_2 to form formate. We have purified ferredoxin from *C. thermoaceticum* and determined some of its properties. Since we recently published these findings (Yang *et al.*, 1977), they will be only summarized here.

The *C. thermoaceticum* ferredoxin is unusual in that it contains only one $[Fe_4S_4]^*$ cluster, whereas ferredoxins from other clostridia contain two $[Fe_4S_4]^*$ clusters. Like other proteins from *C. thermoaceticum*, it has high thermostability and retains 50% of its activity after an exposure to 80°C for 60 min. Its molecular weight is about 7,300 and it contains a histidine residue which is uncommon for clostridial ferredoxins. Histidine has been found only in the thermostable ferredoxins from the thermophiles, *Clostridium tartarivorum* and *Clostridium thermosaccharolyticum* (Devanathan *et al.*, 1969). This suggests that histidine may contribute to the high thermostability of these proteins. Hase *et al.* (1976) have found a correlation between the number of glutamic acid residues and heat stability in ferredoxins. The *C. thermo-*

aceticum ferredoxin has a high number (nine) of such residues, which seems to be in agreement with the finding by Hase *et al.* (1976).

ISOLATION OF AN EXTREME THERMOPHILIC CLOSTRIDIUM FROM SOIL

Within the last year, we have looked for anaerobic thermophilic microorganisms in soil samples collected around Athens, Georgia, USA. The samples were taken from river and lake banks from composts of straw, bark, household, and garden refuse and horse manure. They were incubated at 60°C and at 74°C under argon. The numbered cellulytic, methanogenic, and sulfate-reducing bacteria were determined as well as the number of bacteria that grow in a complex medium. The results from a soil sample taken from a river bank are given in Table 3. Other samples gave similar results.

We have particularly been interested in the glycolytic bacteria that grow at 74°C and have now isolated nine strains, which appear very similar or identical. The isolation was performed using the Hungate technique. The isolates are rods and spore-formers, and the spores are terminal with swollen sporangia. They grow in plain phosphate buffer between pH 5.2-9.3 supplemented with 0.2% yeast extract and with one of the following carbohydrates: glucose, mannose, galactose, fructose, xylose, cellobiose, sucrose or starch. With 0.5% glucose as substrate in the presence of 0.2% yeast extract, the isolates grow between 42°C and 78°C. Maximum growth occurs about 68°C. The T_{max} value places them at the lower spectrum of the extreme thermophiles.

Preliminary results using one of the isolates indicates that it ferments glucose via the Embden-Meyerhof glycolytic pathway. The products are lactate, ethanol, acetate, CO_2, and H_2. The stoichiometry of the fermentation appears to be as follows:

$$\text{Glucose} \rightarrow 0.5 \text{ lactate} + \text{ethanol} + 0.5 \text{ acetate} + 1.5\ CO_2 + 2H_2 \qquad (5)$$

The isolate has the ability to produce hydrogen sulfide from trypton, sulfite, and thiosulfate, but not from sulfate. These properties are similar to those reported for *Clostridium thermohydrosulfuricum*, which previously was isolated from an Austrian sugar factory by Klaushofer and Parkkinen (1965) and studied by Hollaus and Sleytr (1972). Dr. Hollaus has kindly supplied us with strains of *C. thermohydrosulfuricum*. These strains appear to be very similar to our isolates. We

TABLE 3
Enumeration of Thermophilic Anaerobic Bacteria in 2.5 g of Soil from River Bank in Athens, Georgia, U.S.A.

Bacterial Group	*Sampling date*		
	Feb. 1977	*Sept. 1977*	*Feb. 1978*
Incubated at 60°C			
Cellulytic	1-10	300-600	
Methogenic	1-10	50-100	
Sulfate-reducing	10-100	10-50	
Growing on complex medium	5×10^5-1×10^6	5×10^5-1×10^6	
Incubated at 73/74°C			
Glycolytic	--	1-10	1-10

believe now that our isolates are strains of *C. thermohydrosulfuricum*.

Our attempts to isolate thermophilic anaerobic microorganisms have been triggered by the possibility of finding new bacteria, which may be used to ferment cellulose and other easily available raw materials to useful feedstock chemicals. It seems to us that the use of anaerobic thermophilic fermentative bacteria for such a purpose would be advantageous. The thermophiles are, as far as we know, non-pathogenic and not as easily contaminated as mesophilic bacteria.

ACKNOWLEDGEMENTS

The work discussed in this paper has been sponsored partly by U. S. Public Health Service Grant AM12913 from The National Institute of Arthritis and Metabolic Disease and by The Energy and Research Development Administration contract number EY-76-S-09-0888-M003. A stipend to Jürgen Wiegel from the Deutsche Forschungsgemeinschaft is gratefully acknowledged.

REFERENCES

Andreesen, J. R., Gottschalk, G. and Schlegel, H. G. (1970). Arch. Mikrobiol. *72*, 154.

Andreesen, J. R., Schaupp, A., Neurauter, C., Brown, A., and Ljungdahl, L. G. (1973). J. Bacteriol. *114*, 743.

Balch, W. E., Schoberth, S., Tanner, R. S., and Wolfe, R. S. (1977). Int. J. Syst. Bacteriol. *27*, 355.

Cohen, L. and MacKenzie, R. E. (1978). Biochim. Biophys. Acta *522*, 311.

Devanathan, T., Akagi, J. M., Hersh, R. T., and Himes, R. H. (1969). J. Biol. Chem. *244*, 2846.

Fontaine, F. E., Peterson, W. H., McCoy, E., Johnson, J. M. and Ritter, G. J. (1942). J. Bacteriol. *43*, 701.

Gottwald, M., Andreesen, J. R., LeGall, J. and Ljungdahl, L. G. (1975). J. Bacteriol. *122*, 325.

Harmony, J. A. K. and Himes, R. H. (1975). J. Biol. Chem. *250*, 8049.

Harmony, J. A. K., Himes, R. H. and Schowen, R. L. (1975). Biochemistry *14*, 5379.

Hase, T., Ohmiya, N., Matsubara, H., Mullinger, R. N., Rao, K. K. and Hall, D. O. (1976). Biochem. J. *159*, 55.

Hollaus, F. and Sleytr, U. (1972). Arch. Mikrobiol. *86*, 129.

Klaushofer, H. and Parkkinen, E. (1965). Z. Zuckerindustrie *15*, 445.

Ljungdahl, L. G. and Wood, H. G. (1963). Bacteriol. Proceedings Abstr. P58, 109.

Ljungdahl, L. G., Irion, E. and Wood, H. G. (1965). Biochemistry *4*, 2771.

Ljungdahl, L. G. and Wood, H. G. (1969). Ann. Rev. Microbiol. *23*, 515.

Ljungdahl, L. G., LeGall, J. and Lee, J.-P. (1973). Biochemistry *12*, 1802.

Ljungdahl, L. G. and Andreesen, J. R. (1976). In *Microbiol Production and Utilization of Gases* (H. G. Schlegel, G. Gottschalk and N. Pfennig, eds.), p. 163, E. Goltze, K. G., Göttingen, West Germany.

Ljungdahl, L. G. and Sherod, D. (1976). In *Extreme Environments: Mechanisms of Microbial Adaptation* (M. R. Heinrich, ed.), p. 147, Academic Press, N.Y.

Ljungdahl, L. G., Sherod, D., Moore, M. R. and Andreesen, J. R. (1976). In *Enzymes and Proteins from Thermophilic Microorganisms* (H. Zuber, ed.), p. 237, Birkäuser Verlag, Basel.

O'Brien, W. E. and Ljungdahl, L. G. (1972). J. Bacteriol. *109*, 626.

O'Brien, W. E., Brewer, J. M., and Ljungdahl, L. G. (1973). J. Biol. Chem. *248*, 403.

O'Brien, W. E., Brewer, J. M. and Ljungdahl, L. G. (1976). In *Enzymes and Proteins from Thermophilic Microorganisms* (H. Zuber, ed.), p. 249, Birkhäuser Verlag, Basel.

Parker, D. J., Wood, H. G., Ghambeer, R. K. and Ljungdahl, L. G. (1972). Biochemistry *11*, 3074.

Paukert, J. L., D'Ari-Straus, L. and Rabinowitz, J. C. (1976). J. Biol. Chem. *251*, 5104.

Paukert, J. L., Williams, G. R. and Rabinowitz, J. C. (1977). Biochem. Biophys. Res. Commun. *77*, 147.

Schaupp, A. and Ljungdahl, L. G. (1974). Arch. Microbiol. *100*, 121.

Schoberth, S. (1977). Arch. Microbiol. *114*, 143.

Schulman, M., Ghambeer, R. K., Ljungdahl, L. G. and Wood, H. G. (1973). J. Biol. Chem. *248*, 1973.

Schulman, M., Parker, D., Ljungdahl, L. G. and Wood, H. G. (1972). J. Bacteriol. *109*, 633.

Shoaf, W. T., Neece, S. H., and Ljungdahl, L. G. (1974). Biochim. Biophys. Acta *334*, 448.

Tanner, R. S., Wolfe, R. S. and Ljungdahl, L. G. (1978). J. Bacteriol. *134*, 668.

Thauer, R. K., Jungermann, K. and Decker, K. (1977). Bacteriol. Rev. *41*, 100.

Welty, F. K. (1976). Fed. Am. Soc. Expt. Biol. Procceedings *35*, 1749, Abstr. 2006.

Wood, H. G. (1952). J. Biol. Chem. *194*, 905

Yang, S.-S. and Ljungdahl, L. G. (1977). Am. Soc. Microbiol. Ann. Meeting Abstr. K 135.

Yang, S.-S., Ljungdahl, L. G. and LeGall, J. (1977). J. Bacteriol. *130*, 1084.

A
B
C 8
D 9
E 0
F 1
G 2
H 3
I 4
J 5